# 6th International Symposium on Bulk Nanomaterials (BNM-2019)



## From Fundamentals to Innovations

IOP Conference Series: Materials Science and Engineering
Volume 672

Ufa, Russia
25 - 27 September 2019

ISBN: 978-1-7138-0924-1
ISSN: 1757-8981

**Printed from e-media with permission by:**

Curran Associates, Inc.
57 Morehouse Lane
Red Hook, NY 12571

CURRAN ASSOCIATES INC.
proceedings.com

**Some format issues inherent in the e-media version may also appear in this print version.**



Institute of Physics
Dirac House, Temple Back
Bristol BS1 6BE UK

Phone: 44 1 17 929 7481
Fax: 44 1 17 920 0979

techtracking@iop.org

**Additional copies of this publication are available from:**

Curran Associates, Inc.
57 Morehouse Lane
Red Hook, NY 12571 USA
Phone: 845-758-0400
Fax: 845-758-2633
Email: curran@proceedings.com
Web: www.proceedings.com

# TABLE OF CONTENTS

IOP Conf. Series: Materials Science and Engineering **672** (2019) 011001 doi:10.1088/1757-899X/672/1/011001

# PREFACE

## Proceedings of the
## 6th International Symposium
## BULK NANOMATERIALS: from fundamentals to innovations.
## BNM-2019

The 6th International Symposium "BULK NANOMATERIALS: from fundamentals to innovations. BNM-2019" was held in Ufa, Russia on September 25-27, 2019.

The symposium was organized by Ufa State Aviation Technical University jointly with NanoMeT Ltd. (Ufa, Russia) and Technopark of Aviation Technologies (Ufa, Russia) under the organization and financial support of the Government of the Republic of Bashkortostan.

The BNM-2019 symposium, chaired by Prof. Ruslan Z. Valiev (Ufa, Russia), was focused on the demonstration of scientific achievements and developments in the design and study of novel bulk nanostructured materials (BNM) produced by severe plastic deformation (SPD), on solving a wide range of fundamental scientific problems of interdisciplinary character at the intersection of physical materials science, physics, solid-state chemistry, mechanics and materials processing technology, as well as the numerical computer simulation of the degradation phenomena in metals and alloys, such as oxidation, corrosion, fatigue, creep, radiation damage, etc. Special attention was given to discussing the fundamentals of the nanostructural design of new metallic, and composite materials with superior properties.

The outcome of the symposium is to inform the scientific community about the scientific achievements, the research results and their potential scientific and social-and-economic importance; strengthening the current scientific cooperation and establishing new collaborations, and informing potential investors about the latest scientific achievements in the area of BNM. It was possible due to the participation of the internationally recognized leading scientists, members of the academia and well-known researchers.

In the framework of the symposium, a special session devoted to the anniversary of Prof. Ruslan Z. Valiev was held. In this session, the results of his scientific activity of many years were summarized, and his pioneer achievements in the area of BNM in the last 25 years were presented.

The work of the symposium was organized according to the following sessions:
– New trends in nanoSPD treatment
– Grain-boundary engineering of UFG materials
– The evolution of the microstructure, phase transformations and texture analysis
– Modeling of nanomaterials
– Physical and mechanical properties of nanomaterials
– Biocompatible bulk nanostructured materials and coatings

During the sessions 31 key-note and invited reports were presented, as well as 32 oral reports and 69 poster presentations. In total, 184 people from 13 countries participated in the symposium, including young scientists and students.

The proceedings contain 68 papers presented at the symposium, which are arranged in the following order: first three papers are dedicated to Prof. Valiev and his achievements; then papers of the invited speakers of the symposium follow, then papers submitted by participants, presented as oral and poster reports in the alphabetic order, basing on the surname of the corresponding author. All papers have been peer-reviewed and passed several stages of editing. We express our sincere gratitude to all people who contributed to this process and did their best to enhance the scientific merit and quality of the Proceedings.

The organizing committee of the BNM-2019 symposium would like to express special gratitude to the guest editors of the special issue, namely to Prof. Terence Langdon (the United Kingdom), Prof. Michael Zehetbauer (Austria), Prof. Igor Alexandrov (Russia), Dr. Nariman Enikeev (Russia), Prof. Airat Nazarov (Russia), Mrs. Natalia Sayapina (Russia) for their contributions.

Special thanks also to Mrs. Natalia Sayapina for language corrections and Ms. Julia Bazhenova for preparing of final proofs of the publications and careful checking and meeting the formal requirements of the journal.

We are also grateful to the authors for their contributions, and to the speakers and participants for attending BNM-2019. We hope to welcome them along with new participants the next BNM-2021!

*Terence G. Langdon, Natalia A. Sayapina, Ruslan Z. Valiev and the other members of the Organizing Committee of BNM-2019*

# Peer review statement

All papers published in this volume of *IOP Conference Series: Materials Science and Engineering* have been peer reviewed through processes administered by the proceedings Editors. Reviews were conducted by expert referees to the professional and scientific standards expected of a proceedings journal published by IOP Publishing.

IOP Conf. Series: Materials Science and Engineering **672** (2019) 012001 doi:10.1088/1757-899X/672/1/012001

# Thirty years of collaboration and research from 1989 to 2019: a tribute to Ruslan Z. Valiev

**T G Langdon**

Materials Research Group, Department of Mechanical Engineering, University of Southampton, Southampton SO17 1BJ, U.K.

E-mail: langdon@soton.ac.uk

**Abstract**. A visit to the Institute of Problems of Superplasticity of Metals in Ufa in 1989 initiated a cooperation and collaboration with Ruslan Valiev that has continued for the last thirty years. This collaboration started with fundamental investigations of superplasticity, such as the demonstration that intragranular slip is necessary to accommodate grain boundary sliding in conventional superplastic flow, but it was then extended to cover the processing and properties of a wide range of metals produced using severe plastic deformation to achieve exceptional grain refinement. Highlights included the publication of a first comprehensive review of equal-channel angular pressing (now with more than 3600 citations on Google Scholar) and co-authoring a textbook (Bulk Nanostructured Materials: Fundamentals and Applications, published by TMS/Wiley in 2014).

## 1. Introduction

Creep refers to the permanent plastic flow that occurs in materials when they are subjected to a constant stress or a constant load over a long period of time and where the stress is insufficient to cause immediate fracture. In practice, creep is a diffusion-controlled process and this means it becomes important when testing at high temperatures, typically at temperatures above ~$0.5T_m$ where $T_m$ is the absolute melting temperature. Studies of creep may be traced back for more than 100 years [1] but the total strains experienced in tensile creep are generally quite small and ultimately the material breaks in catastrophic failure. Some early experiments reported quite large elongations in selected metals prior to fracture but these elongations were only of the order of a few hundreds of percent and it was only in 1934 that Pearson, working in England, reported the first demonstration of true superplasticity with a tensile elongation of ~1950% in a near-eutectic Bi-Sn alloy [2]. Unfortunately, this result received little attention in the west but it inspired extensive research in the Soviet Union and these results were later reviewed by Underwood [3] so that the process became known in other countries. It is interesting to note that the English word "superplasticity" is a direct translation from the Russian word *sverkhplastichnost*. This publicity in the west led to early experiments on vacuum forming and blow molding [4] and this progressed ultimately to the establishment of the superplastic forming industry. As of today, superplastic forming is now a major contributor in the processing of sheet metals into curved and complex shapes for use in many applications including in the aerospace, automotive and architectural sectors [5]. Early experiments showed that superplasticity required small grain sizes, typically less than ~10 μm [6], and therefore metals appropriate for superplastic forming were typically obtained by using thermo-mechanical processing to produce grain sizes in the range of ~2-5 μm.

IOP Conf. Series: Materials Science and Engineering **672** (2019) 012001 doi:10.1088/1757-899X/672/1/012001

## 2. Developments at the Institute of Problems of Superplasticity of Metals (IPSM) in Ufa

A critical publication appeared in the Russian literature in 1988 with the title "Low Temperature Superplasticity of Metallic Materials" [7]. This paper came from the Institute of Problems of Superplasticity of Metals (IPSM) of the Academy of Sciences of the Soviet Union in Ufa and I read the paper in translation in the library at the University of Southern California. I was amazed to discover that the authors were using an Al-4% Cu-0.5% Zr alloy with a grain size of 0.3 μm that had been produced by "shear under pressure". At that time my students at USC were superplastically testing the British commercial Supral 100 alloy (Al-2004, Al-6% Cu-0.4% Zr) but it was impossible for us to obtain a grain size smaller than ~3-5 μm. Accordingly, I decided it was important to find out more about these experimental procedures and therefore I made arrangements with Professor Oscar Kaibyshev, Director of IPSM, to visit his institute. In May 1989 my wife, Mady, and I set off from Los Angeles with intermediate stops in Copenhagen and Moscow. In Moscow we met Mikhail Myshlyaev who was a colleague from my research on creep, Zuzana Trojanová and Pavel Lukáč from Charles University, Prague, who were also on their way to IPSM and Ruslan Valiev who arrived to accompany us on our flight. Figure 1 shows us in our hotel in Moscow before the departure for Ufa. Arriving at IPSM we were given a remarkably warm welcome and I discovered I was only their second foreign scientist. Their first foreign visitor was Herbert Gleiter of Germany and they had also started the practice of drawing and then exhibiting a caricature of each foreign visitor.

**Figure 1.** Meeting at a hotel in Moscow on the way to Ufa in May 1989: standing (from left) Mikhail Myshlyaev, Ruslan Valiev, Zuzana Trojanová, Pavel Lukáč, Mady Langdon and the author.

An important outcome of this first visit to Ufa was the recognition that it would be extremely useful to invite Ruslan Valiev to spend some time working with me at USC so that we could establish a collaborative research activity. Accordingly, I visited IPSM again in 1990 and invited Ruslan to give a presentation at a symposium on "Hot Deformation of Aluminum Alloys" which I was co-organizing at a TMS conference in Detroit, Michigan, in October 1990 and afterwards to spend some time at USC. Ruslan accepted this invitation and made his first visit to California where he undertook research measuring the intragranular strain incurred in grains of the superplastic Pb-62% Sn eutectic alloy. The results from this research showed that grain boundary sliding is the dominant flow process in superplasticity but it is continuously accommodated by intragranular dislocation movement which occurs in an oscillatory manner and makes no net contribution to the total strain. These results were published in *Acta Metallurgica Materialia* in March 1993 as our first joint publication [8]. Figure 2 shows that we also took advantage of Ruslan's visit to spend some time exploring California.

**Figure 2.** Ruslan's first visit to California in 1990; on the road in central California with postdoc Jaroslav Fiala (left) and Ruslan (center). My red 1974 Volvo is visible at rear (I now have this car in U.K. after driving 800,000 km).

IOP Conf. Series: Materials Science and Engineering **672** (2019) 012001 doi:10.1088/1757-899X/672/1/012001

In November 1993 we published in *Journal of Materials Research* a paper on ultrafine-grained (UFG) materials processed by severe plastic deformation (SPD) and this was the first paper on these materials co-authored, at least in part, by researchers from outside of Russia [9]. This proved to be an important publication and it is now among the more than 100 papers that I have co-authored to date with Ruslan. A detailed report on my introduction to UFG materials and processing using SPD techniques was presented in a review published in 2007 [10].

**Figure 3.** Another visit to Ufa in 1994 with (from left) Oscar Kaibyshev (Founder and Director of IPSM), Tatyana Karimova (Head of Information Services), the author and Ruslan: drawings on the wall depict foreign visitors to IPSM.

Following these initial visits to Ufa and Los Angeles, there were fairly regular meetings between Ruslan and myself, either at our home institutions or at international conferences. Figure 3 was taken on one of my later visits to IPSM in 1994 and it shows (from left) Oscar Kaibyshev, the Founder and Director of IPSM, Tatyana Karimova, the Head of the Information Services at IPSM, myself and Ruslan. The numerous drawings on the wall behind us show the caricatures of the various foreign scientists that had visited IPSM up to 1994: the drawing of the first visitor, Herbert Gleiter, is shown as third from the right at the top and my caricature is shown as second from the right at the top. Another important development in 1994 was my election as a Foreign Academician of the Academy of Sciences of the Bashkortostan Republic, the republic where Ufa is the capital city, and figure 4 shows me standing at the entrance to the impressive academy headquarters building in Ufa.

**Figure 4.** The author (at center, bottom) standing outside the headquarters building of the Academy of Sciences of the Bashkortostan Republic in Ufa.

The increasing interest in NanoSPD materials led to the organization of a NATO Advanced Research Workshop which was held in Golitsino near Moscow in August 1999. A second conference on these materials was organized in Vienna, Austria, in December 2002 and this meeting was designated NanoSPD2. Subsequently, these conferences were held every three years with NanoSPD3 at Fukuoka, Japan, in 2005, NanoSPD4 at Goslar, Germany in 2008, NanoSPD5 at Nanjing, China, in 2011, NanoSPD6 at Metz, France, in 2014 and NanoSPD7 at Sydney, Australia, in 2017. The next meeting in this series, NanoSPD8, is scheduled to be held in Bangalore, India, in September 2020.

An important development at NanoSPD2 was the formation of an International NanoSPD Steering Committee which was designed to provide a focal point for the organization of meetings and to provide assistance for all activities within the broad field of SPD. This committee was initially constituted with Ruslan as Chairman and with five additional members: Yuri Estrin (Monash University, Australia), Zenji Horita (Kyushu University, Japan), Michael Zehetbauer (University of Vienna, Austria), Yuntian Zhu (North Carolina State University, USA) and me. In 2014 the committee was expanded by adding

IOP Conf. Series: Materials Science and Engineering **672** (2019) 012001 doi:10.1088/1757-899X/672/1/012001

**Figure 5.** A meeting of the original NanoSPD International Steering Committee: (from left) Zenji Horita, Michael Zehetbauer, Ruslan Valiev, Yuri Estrin, Yuntian Zhu and the author.

Roberto Figueiredo (Federal University of Minas Gerais, Brazil), Hyoung Seop Kim (Pohang University of Science and Technology, Korea), Terry Lowe (Colorado School of Mines, USA), Laszlo Toth (University of Lorraine-Metz, France) and Gerhard Wilde (University of Muenster, Germany). In 2018, Megumi Kawasaki (Oregon State University, USA) was also added to the committee. A photo of a meeting of the original committee is shown in figure 5.

## 3. The development of conferences in Ufa and the production of major review publications

By the early part of the 21$^{st}$ century it was clear that the processing and properties of nanostructured materials, produced through the application of severe plastic deformation, had become a major research topic within the field of Materials Science. The production and testing of UFG metals had now spread to laboratories around the world and many publications on this topic were receiving large numbers of citations in the scientific literature. Two further developments contributed to the overall interest in SPD processing.

First, a series of conferences was initiated in Ufa. These meetings attracted many foreign participants but especially they provided an opportunity for local Russian researchers to present their scientific results to an appreciative audience who were able to provide valuable comments and suggestions on the nature of the work. The first of these meetings was titled "Bulk Nanostructured Materials: from Fundamentals to Innovations – BNM-2007", it attracted participants from 23 different countries and it was followed by additional BNM conferences in later years. At the present time, the next conference will be BNM-2019 to be held in Ufa in September 2019. Figure 6 shows some of the participants at BNM-2007 when they attended a social activity at Ruslan's summer house near Ufa. These meetings also provided an

**Figure 6.** At Ruslan's summer house near Ufa in 2007: standing (from left) Jingtao Wang (from Nanjing), a friend, Mady Langdon, Marina (Ruslan's wife), the author, Igor Alexandrov and his wife, Cheng Xu (now at Chinese Academy of Sciences, Ningbo, China), Megumi Kawasaki (now at Oregon State University, USA) and Roberto Figueiredo (now at Federal University of Minas Gerais, Brazil).

IOP Conf. Series: Materials Science and Engineering **672** (2019) 012001 doi:10.1088/1757-899X/672/1/012001

opportunity to receive additional publicity for work in the area of UFG metals. An example is shown by the main page of the Bashkortostan paper shown in figure 7 and there were also opportunities for interviews with local and national news sources.

Fig. 7 Publicity in the Bashkortostan paper in 2007 for our research on UFG metals.

Second, it was important to publish major review articles documenting the principles of UFG metals and SPD processing. Accordingly, a review paper was published in 2006 in *Progress in Materials Science* describing the basic characteristics of equal-channel angular pressing (ECAP) [11]. To date, this review is listed on Web of Science as the fourth all-time most cited paper in this journal with more than 2800 citations and with more than 3600 citations on Google Scholar. It was also considered important to write an overview documenting the terminology and the procedures associated with SPD processing and UFG metals and this paper was written by the original six members of the NanoSPD International Steering Committee and published in *JOM* in 2006 [12]. This paper is now the all-time most cited paper to appear in this journal with a total of more than 1000 citations. In view of the success of this publication, the authors accepted an invitation to write a follow-up article exactly ten years later [13] and this paper, although only published in 2016, already has more than 100 citations and is the most cited *JOM* paper published in 2016. A textbook on the topic of Bulk Nanostructured Materials was co-authored with Ruslan and Alex Zhilyaev and published by Wiley/TMS in 2014 [14]. The cover of the book is shown in figure 8 and to date it has received over 270 citations.

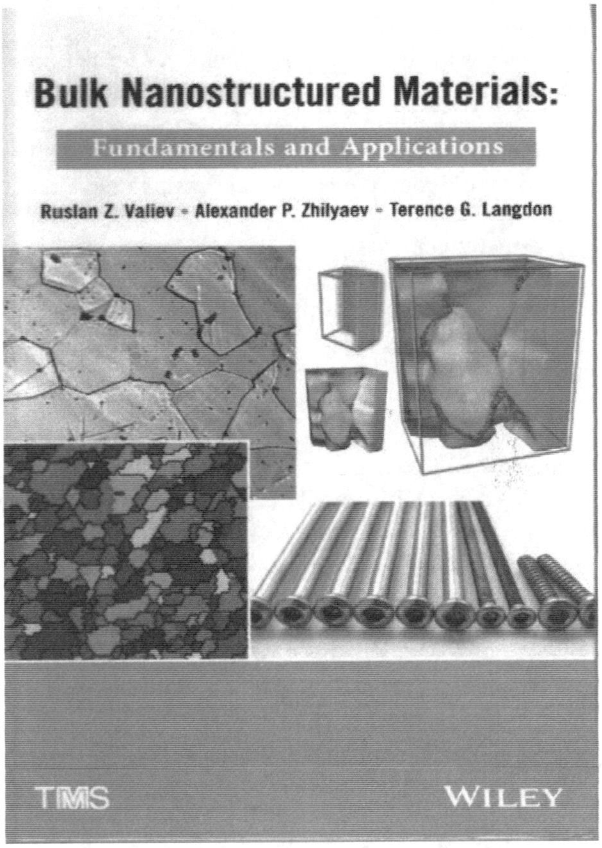


**Figure 8.** Cover of the book "Bulk Nanostructured Materials: Fundamentals and Applications," published by Wiley/TMS, Hoboken, NJ (2014).

A comprehensive review of citation statistics for papers published in the field of NanoSPD was presented in 2017 [15] and it shows that this scientific topic has many highly cited papers including the all-time most-cited papers appearing in both *Progress in Materials Science* [16] and *Scripta Materialia* [17]. These large numbers of citations confirm the world-wide interest in grain refinement through SPD processing.

## 4. Recognition for nanostructured materials through international awards

The outstanding research activities of Ruslan Valiev and the members of his team have been recognized by several awards. For example, he was the recipient of the Blaise Pascal Medal in Materials Science from the European Academy of Sciences at their annual meeting in Milan in 2011, he received the MSEA Journal Award from Elsevier at the TMS Annual Meeting in San Diego, California, in 2011 and he was co-recipient with Yuntian Zhu of the Sōmiya Award from the International Union of Materials Research Societies presented at Jeju Island, Korea, in 2015. Figure 9 shows Ruslan receiving the MSEA Journal Award from Enrique Lavernia, the Editor-in-Chief of *Materials Science and Engineering A*. In 2016, Ruslan, together with Rinat Islamgaliev and Igor Alexandrov of Ufa State Aviation Technical University and myself, were the recipients of the State Prize of the Republic of Bashkortostan in Science and Technology.

**Figure 9.** Ruslan (right) receiving the MSEA Journal Award from Enrique Lavernia (centre) at the TMS Annual Meeting in San Diego, California, in March 2011.

## 5. Summary and conclusions

My visit to the Institute of Problems of Superplasticity of Metals in the city of Ufa in Russia in 1989 led to a meeting with Ruslan Valiev and the initiation of a cooperation that has extended over the last thirty years and is continuing today. This cooperation has included multiple visits between our two institutes, co-authorship on more than 100 papers in the peer-reviewed scientific literature, collaborations in international conferences and the writing of extensive and widely-cited review articles and a major book.

## Acknowledgements

This work was supported by the European Research Council under ERC Grant Agreement No. 267464-SPDMETALS. Special thanks are due to my many colleagues and friends at IPSM and at Ufa State Aviation Technical University for their unreserved support over a period that now stretches through thirty years. Their support, and the inspiration they engendered by their research activities, did much to enhance my own research and provide new stimuli for my PhD students at USC and Southampton.

## References

[1] Andrade E N daC 1910 *Proc. Roy. Soc. London A* **81** 1
[2] Pearson C E 1934 *J. Inst. Metals* **54** 111
[3] Underwood E E 1962 *J. Metals* **14** 914
[4] Backofen W A, Turner L R and Avery D H 1964 *Trans. ASM* **57** 980
[5] Barnes A J 2007 *J. Mater. Eng. Perform.* **16** 440
[6] Langdon T G 1982 *Metall. Mater. Trans.* A **13** 689
[7] Valiev R Z, Kaibyshev O A, Kuznetsov R I, Musalimov R S and Tsenev N K 1988 *Dokl. Akad. Nauk SSSR* **301** 864
[8] Valiev R Z and Langdon T G 1993 *Acta Metall. Mater.* **41** 949
[9] Wang J, Horita Z, Furukawa M, Nemoto M, Tsenev N K, Valiev R Z, Ma Y and Langdon T G 1993 *J. Mater. Res.* **8** 2810
[10] Langdon T G 2007 *Int. J. Mater. Res.* **98** 4
[11] Valiev R Z and Langdon T G 2006 *Prog. Mater. Sci.* **51** 881
[12] Valiev R Z, Estrin Y, Horita Z, Langdon T G, Zehetbauer M J and Zhu Y T 2006 *JOM* **58** 33
[13] Valiev R Z, Estrin Y, Horita Z, Langdon T G, Zehetbauer M J and Zhu Y 2016 *JOM* **68** 1216
[14] Valiev R Z, Zhilyaev A P and Langdon T G 2014 *Bulk Nanostructured Materials: Fundamentals and Applications* (Hoboken: Wiley/TMS)
[15] Valiev R Z and Langdon T G 2017 *IOP Conf. Series: Mater. Sci. Eng.* **194** 012001
[16] Valiev R Z, Islamgaliev R K and Alexandrov I V 2000 *Prog. Mater. Sci.* **45** 103
[17] Iwahashi Y, Wang J, Horita Z, Nemoto M and Langdon T G 1996 *Scripta Mater.* **35** 143

# 20 years of INT KIT-USATU collaboration in the area of SPD

**H Hahn, Y Ivanisenko and A Kilmametov**

Karlsruhe Institute of Technology (KIT), Institute of Nanotechnology, PO Box 3640, 76021 Karlsruhe, Germany

E-mail: horst.hahn@kit.edu

**Abstract**. The field of Severe Plastic Deformation (SPD) has provided the scientific community with ultrafine grained and nanocrystalline materials. In contrast to other techniques for the synthesis of nanostructured materials, SPD processes allow the preparation of bulk quantities of metals and alloys. Technical applications have become a reality because commercial alloys can be transformed into novel nano- and microstructures, leading to superior mechanical properties, such as high strength and high ductility, resulting in many application areas. The field was pioneered at the Ufa State Aviation Technical University and has created worldwide interest. The report emphasizes some results of the close collaboration between Ufa State Aviation Technical University and the Institute of Nanotechnology at Karlsruhe Institute of Technology (KIT).

## 1. Introduction

The discovery of Severe Plastic Deformation (SPD) in its many different technical variations has opened a new field of materials research by allowing the preparation of many metastable metallic alloy systems in bulk form with ultrafine and nanometer grain sizes. The original idea to use severe plastic deformation to modify metallic materials and the establishment of the concept of non-equilibrium interfaces has been developed by Ruslan Valiev at Ufa State Aviation Technical University (USATU) [1], during the early period of nanocrystalline materials research [2]. The term SPD encompasses many well-established deformation processes, all leading to large plastic deformation and enormous modifications of the microstructure of the materials, resulting in large changes of the properties. Among these deformation processes are ECAP (Equal Channel Angular Pressing) [3], HPT (High Pressure Torsion) [4,5], ARB (Accumulated Roll Bonding) [6], and more. After years of mainly basic research the different processes have also been scaled up to fabricate larger amounts of materials, such as the continuous ECAP process (ECAP-Conform) [7], or more recently, the High Pressure Torsion Extrusion (HPTE) process [8]. The interesting basic science of severely deformed materials, the possibility to produce bulk samples with internal nanostructures with substantially larger volume compared to samples produced by other techniques such as Inert Gas Condensation (IGC) [9], and the opportunity to transfer such materials to industrial applications all led to a fast growth of this field in the past 20 years. Among the most exciting and pioneering scientific results that established a worldwide interest was the possibility to create materials with high strength and high ductility, not possible by conventional materials structures [10].

As a result of the complementary expertise at both institutions and the intensive exchange of scientists, the Institute of Nanotechnology at Karlsruhe Institute of technology (INT KIT) collaborates with the group of Ruslan Valiev at Ufa State Aviation Technical University for the past twenty years, a good time to summarize the results of this collaboration. In this respect the groups were successful in acquiring substantial research funding on both sides, leading to several joint publications. On the German side, the Deutsche Foschungsgemeinschaft (DFG) funded the project entitled "Plasticity in nanocrystalline metals and alloys" from 2006 to 2012 and the project "Basic Mechanisms of Mechanical Property Enhancement for Carbon Steels through Controlled Nanostructure Formation" from 2009 to 2012, leading to several joint publications [11-13]. On the other hand, the ideas in the area of SPD born at USATU were creatively developed at INT KIT, and the most prominent results are presented in this paper.

## 2. Instrumented High Pressure Torsion – a new tool to study deformation behavior of materials with limited ductility in a wide range of strains

As mentioned before, high pressure torsion (HPT) straining is a method of severe plastic deformation and it is conventionally applied as a technique for making ultrafine-grained materials, when a microstructure of coarse-grained metals and alloys is refined down to the grain size of 100–300 nm [4,5]. On the other hand, the application of HPT in a combination with torque measurement can be used as a method of mechanical testing. This is especially interesting for nanocrystalline materials with a grain size of 10–20 nm, because they show very limited ductility in tension, which restricts the experimentally accessible parameter space, whereas torsion testing under high pressure [14,15] allows practically unlimited shear strain without fracture. This technique was applied to a nanocrystalline Pd and Pd-10 at.% Au alloy prepared by IGC [16]. Figure 1 shows the graphs for IGC Pd with an initial grain size of 14 nm (as determined by TEM) deformed at three different strain rates. It is seen that the samples exhibit more rapid initial work hardening compared with coarse-grained Pd. Furthermore, the torque in the initial deformation stage (for e < 1, figure 1) is significantly rate dependent, with a clear trend for more torque at faster strain.

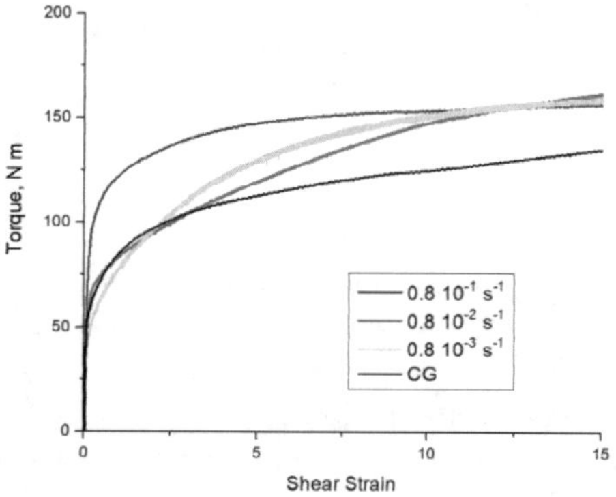


**Figure 1.** Shear Strain vs. Torque curves of coarse grained (CG) Pd deformed at $0.8\times10^{-2}$ $s^{-1}$ and NC Pd deformed at different shear strain rates.

The TEM observations of the microstructure after HPT revealed interesting changes resulted from severe straining. First of all, significant grain growth to 40 nm at the strain e = 15, more than twice the initial value. At the same time, the grains appear still equiaxed at this strain. The most remarkable microstructural features are structures such as those shown in figure 2a,b, when several neighbouring segments of grain boundaries are seen to align in parallel. These features were typically observed quite frequently in samples deformed to e = 15 at all studied strain rates, and were clearly absent in the as-IGC state. The formation of such arrays of coplanar grain boundaries confirms the predictions in Ref. [17], of a type of cooperative grain boundary sliding (GBS). Also significant strain rate sensitivity (figure 1) suggests that a thermally activated process like grain boundary sliding controls the deformation of nanocrystalline Pd. The subsequent studies of crystallographic texture development [18,19] confirmed that random texture persisted when the mean grain size was less than 20 nm. Such features as random texture and formation of mesoscopic shear planes extending over several grain boundaries represent a clear evidence of GBS at room temperature deformation in nanocrystalline Pd. And it could be revealed in material with limited ductility only thanks to application of instrumented HPT.

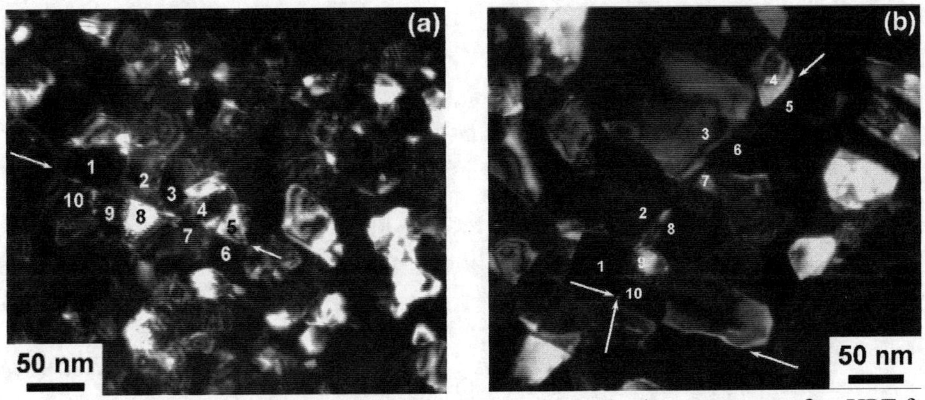


**Figure 2.** Dark field images using the 111 reflections of nc Pd microstructure after HPT for $\gamma$=15 at $\dot{\gamma}$ = 0.8 $10^{-2}$ $s^{-1}$ (a) and $\dot{\gamma}$ = 0.8 $10^{-1}$ $s^{-1}$ (b): (a) alignment of boundaries of 10 grains with formation of a mesoscopic shear plane; (b) another example of mesoscopic shear plane. Reprinted from [16] with permission of Elsevier.

## 3. HPT-driven phase transformations

HPT not only leads to grain refinement but also drives phase transformations that might be distinguished as diffusion-controlled and diffusionless (martensitic) ones.

### *3.1. HPT-induced diffusion-controlled phase transformations*

HPT processing allows an increase of the strain up to the extreme values without fracture. At steady state of deformation, the dynamic equilibrium between deformation-driven production of crystal defects and their relaxation (annihilation) is established. Such a strong influence on the material leads to accelerated mass transfer and, therefore, to various phase transformations [1,20-30] e.g. the formation or decomposition of a supersaturated solid solution, dissolution of phases, disordering of ordered phases, alloying in immiscible systems, low-temperature synthesis of new nanocrystalline alloys.

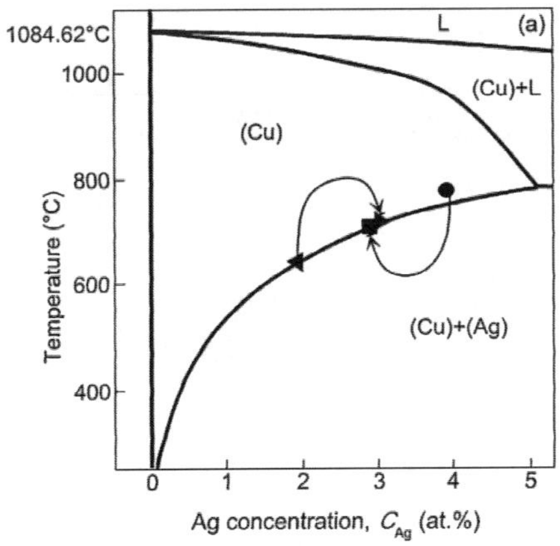


**Figure 3.** The Cu-rich part of the Cu-Ag phase diagram. Evolution of different compositions is shown for the same alloy before and after HPT.

Closer look at steady state processes revealed that the competition (and dynamic equilibrium) between the dissolution of precipitates and decomposition of supersaturated solid solution with precipitation of a second phase during HPT takes place in two-phase Cu-based alloys. As a result, a dynamic equilibrium between these two processes appears, and a certain steady-state concentration in a solid solution is reached (figure 3). This concentration is *equifinal* and equivalent to that if the sample annealed at the effective temperature $T_{eff}$. Therefore, the equivalent diffusion coefficient for the HPT-driven mass transfer [23,25] is comparable with the conventional diffusion coefficient at (elevated) $T_{eff}$. This coincidence additionally supports the view point about deep physical links between *equifinal* amount of defects appearing in steady-state during HPT at low temperature $T_{HPT}$ and defects *equilibrium* existing in the same material at elevated temperature $T_{eff}$. At the same time, this fact nicely correlates with the estimations of non-equilibrium vacancy concentrations during HPT which are as high as that values close to melting temperatures [31].

### *3.2. Diffusionless HPT-induced phase transformations*

Contrary to diffusive phase transformations, non-diffusive ones require atomic shuffling for rather smaller, much less than interatomic, distances. Ti and Ti-based alloys possess a high pressure ω-phase that can be observed as metastable one at ambient conditions after release of applied pressure or after HPT treatment as well. The effect of alloying of titanium with iron (up to 10 wt % Fe) on the ω-phase formation during HPT was studied using XRD analysis and HREM [26-30]. During shear deformation under uniaxial pressure, a pronounced orientation relationship between adjacent grains of α- and ω-phases is accompanied by a strong local preferred orientation, which is recognized as a "basal" crystallographic texture in α-phase. It appears favorable for the realization of the α-to-ω transformation under HPT conditions. A defect-rich high-pressure ω-phase forms after HPT and persists in the samples also after the pressure release. The amount of retained ω-phase after HPT depends on the iron concentration. It increases from 40% in pure titanium, reaches maximum of 95% at 4 wt % Fe and then decreases again to 10% at 10 wt. % Fe. It is because the addition of iron influences the lattice parameters in β- and ω-phases in a different manner. The minimal lattice mismatch between β- and ω-phases is reached at 4 wt. % Fe (figure 4). A good conformity between the lattices of the β- and ω-phases enhances the probability of the martensitic (diffusionless) β-to-ω transformation.

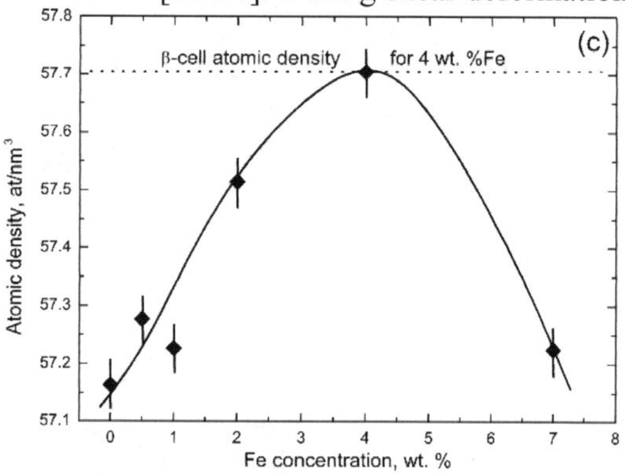


**Figure 4.** Dependence of cell density of ω-phase calculated from the lattice parameter data on iron concentration.

## 4. Scaling up of high pressure torsion

Until recently, a major disadvantage of the HPT method has been the small dimensions of the samples (typically disks 10-20 mm in diameter and about 1 mm in thickness), which hinders its commercial use for structural applications. Recently, the research team at INT has proposed to scale-up the HPT method for industrial processing of advanced materials with at least three order of magnitude larger yield using a new technical approach [32]. The novel machine allows continuous processing of long rods instead of the thin plates used in conventional HPT. The basic idea of the method, which was called high pressure torsion extrusion – HPTE, is to use a continuous transmission of the shear deformation zone along the metallic rod that is pushed through the die (figure 5). The strain accumulated in a specimen after one pass can be controlled by adjusting the specimen diameters ratio (D1/D0) and the velocities of the translational (v) and rotational (ω) motions of the tools according to the equation:

$$e = 2\ln\frac{D1}{D0} + 2\ln\frac{D1}{D2} + \frac{\omega \cdot R1}{\sqrt{3}\cdot v} \quad (1)$$

where R1=0.5D1 is the specimen radius.

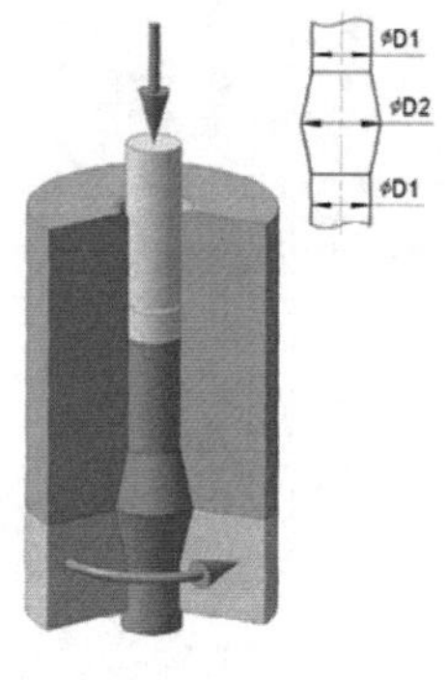


**Figure 5.** Schematic diagram of the HPTE process. After [8].

The proposed method has an important advantage over other large scale SPD techniques, i.e., equal channel angular pressing, accumulative roll bonding, twist extrusion, etc., as it enables processing of the billet in one pass without intermediate steps and thus allows for continuous fabrication of nanocrystalline rods. Furthermore, HPTE provides exciting new opportunities for the purposely design of nano- and microstructures and creating of hybrid materials [33].

Using the HPTE method, copper billets with the diameter of 12 mm and length of 35 mm were processed [8]. As a result, a gradient structure with coarse microstructure in the central part and uniform UFG microstructure with equiaxed grains and mean grain size of 350 nm in the rest of the sample was formed (figure 6). Figure 7 shows the results of microhardness measurements for the specimens processed using HPTE at different translational velocities V, i.e. different strain (equation (1)). It is clearly seen that the hardness distribution correlates well with the distribution of strain. In the central part of a specimen where the torsional deformation is negligible, the hardening occurs only due to the expansion-extrusion deformation. For the specimen processed at higher translational velocity, the microhardness gradually increases along the radius and reaches the value of about 116 HV at the sample´s periphery. In the case of specimen processed at lower velocity, the hardness saturates at about 120 HV already at the distance of 1 mm from the specimen axis. Therefore, using different processing parameters in HPTE one can easily adjust the volume fraction of the material with low or high hardness.

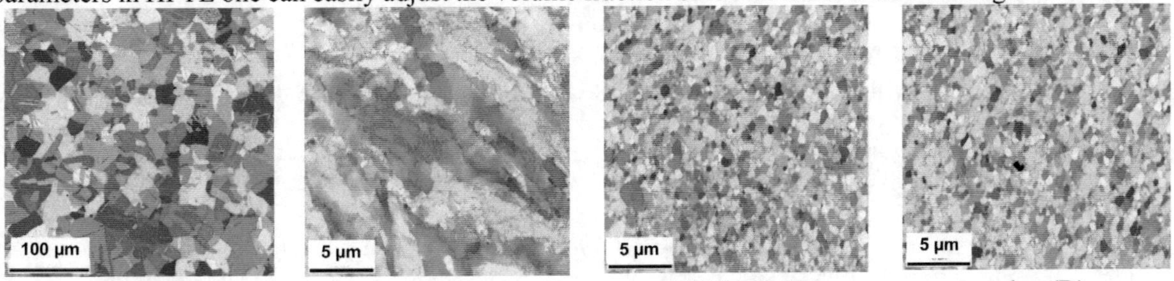


initial center middle (0.5R) edge (R)

**Figure 6.** Orientation maps (inverse pole figure maps) of a Cu specimen after annealing and processing by HPTE. Coloring corresponds to the projection of normal direction (ND) in the inverse pole figure.

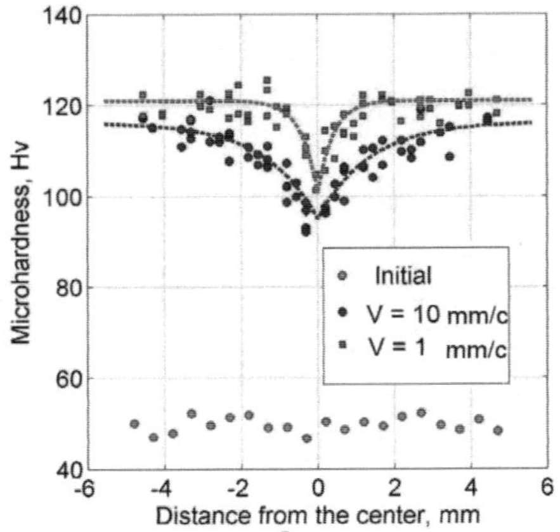


**Figure 7.** Variation of the microhardness of copper samples along the specimen radius in initial annealed state and after HPTE with rotational velocity ω = 1 rpm and different translational velocities.

## 5. Conclusions and outlook

While the field of Severe Plastic Deformation (SPD) is already more than 20 years old, it is still vibrant and many groups worldwide continue to develop new processing routes based on severe plastic deformation, to invent new materials and material combinations and address scale up aspects towards technological applications. The field includes strong basic science aspects on the structure formation, the phase stability under severe plastic deformation and detailed characterization of the resulting structures, as well as modelling contributions.

It should be noted that the collaboration between the INT KIT and Valiev‘s group at USATU resulted recently in another DFG-funded project entitled "Fundamental Parameters and Enhanced Properties of Amorphous Alloys Subjected to Severe Plastic Deformation (SPD)". In this project, the deformation behavior of metallic glasses during SPD will be examined with a particular emphasis on the shear bands formation in metallic glasses and their structural characterization.

**References**

[1] Valiev R Z, Korznikov A V and Mulyukov R R 1993 *Mater. Sci. Eng.* A **168** 141
[2] Gleiter H **1989** *Prog. Mater. Sci.* **33** 223
[3] Segal V M, Reznikov V I, Drobishevski F E and Kopilov V I 1981 *Izv. SSSR Met.* **1** 115
[4] Valiev R Z, Islamgaliev R K and Alexandrov I V 2000 *Prog. Mater. Sci.* **45** 103
[5] Zhilyaev A and Langdon T G 2008 *Prog. Mater. Sci.* **53** 893
[6] Saito Y, Utsunomiya H, Tsuji N and Sakai T 1999 *Acta Mater.* **47** 579
[7] Raab G J, Valiev R Z, Lowe T C and Zhu Y T 2004 *Mater. Sci. Eng.* A **382** 30
[8] Ivanisenko Yu, Kulagin R, Fedorov V, Mazilkin A, Scherer T, Baretzky B and Hahn H 2016 *Mater. Sci. Eng.* A **664** 247
[9] Birringer B, Gleiter H, Klein H-P and Marquardt P 1984 *Phys. Lett.* A **102** 365
[10] Valiev R 2004 *Nat. Mater.* **3** 511
[11] Kurmanaeva L, Ivanisenko Yu, Markmann J, Kübel C, Chuvilin A, Doyle S, Valiev R Z and Fecht H-J 2010 *Mater. Sci. Eng.* A **527** 1776
[12] Yang K, Ivanisenko Yu, Caron A, Chuvilin A, Kurmanaeva L, Scherer T, Valiev R Z and Fecht H-J 2010 *Acta Mater.* **58** 967
[13] Ning J L, Courtois-Manara E, Kurmanaeva L, Ganeev A V, Valiev R Z, Kübel C and Ivanisenko Yu 2013 *Mat. Sci. Eng.* A **581** 8
[14] Pippan R, Wetcher F, Hafok M, Vorhauer A and Sabirov I 2006 *Adv. Eng. Mater.* **8** 1046
[15] Vorhauer A and Pippan R 2008 *Metall. Mater. Trans.* A **39** 417
[16] Ivanisenko Yu, Kurmanaeva L, Weissmueller J, Yang K, Markmann J, Rösner H, Scherrer T and Fecht H-J 2009 *Acta Mater.* **57** 3391
[17] Hahn H, Mondal P and Padmanabhan K A 1997 *Nanostruct. Mater.* **9** 603
[18] Ivanisenko Yu, Skrotzki W, Chulist R, Lippmann T and Kurmanaeva L 2012 *Scripta Mater.* **66** 131
[19] Skrotzki W, Eschke A, Jóni B, Ungár T, Tóth L S, Ivanisenko Yu and Kurmanaeva L 2013 *Acta Mater.* **61** 7271
[20] Sauvage X, Chbihi A and Quelennec X 2010 *J. Phys.* **240** 012003
[21] Ivanisenko Y, MacLaren I, Sauvage X, Valiev R Z and Fecht H-J 2006 *Acta Mater.* **54** 1659
[22] Straumal B, Pontikis V, Kilmametov A, Mazilkin A, Dobatkin S and Baretzky B 2017 *Acta Mater.* **122** 60
[23] Straumal B, Kilmametov A, Ivanisenko Y, Kurmanaeva L, Baretzky B, Kucheev Y, Zieba P, Korneva A and Molodov D 2014 *Mater. Lett.* **118** 111
[24] Straumal B, Kilmametov A, Ivanisenko Y, Mazilkin A, Kogtenkova O, Kurmanaeva L, Korneva A, Zieba P and Baretzky B 2015 *Int. J. Mat. Res.* **106** 657
[25] Kilmametov A, Khristoforov A, Wilde G and Valiev R 2007 *Z. Krist.* **41** 339
[26] Ivanisenko Y, Kilmametov A, Rosner H and Valiev R 2008 *Int. J. Mat. Res.* **99** 36
[27] Kilmametov A, Ivanisenko Y, Straumal B, Mazilkin A, Gornakova A, Kriegel M, Fabrichnaya O, Rafaja D and Hahn H 2017 *Scripta Mater.* **136** 46
[28] Kilmametov A, Ivanisenko Y, Straumal B, Gornakova A, Mazilkin A and Hahn Y 2018 *Metals* **8** 1
[29] Kilmametov A, Ivanisenko Y, Straumal B, Mazilkin A, Gornakova A, Kriegel M, Fabrichnaya O, Rafaja D and Hahn H 2018 *Acta Mater.* **144** 337
[30] Kilmametov A, Kulagin R, Mazilkin A, Seils S, Boll T, Heilmaier M and H. Hahn 2019 *Scripta Mater.* **158** 29
[31] Kilmametov A, Vaughan G, Yavari A, LeMoulec A, Botta W and Valiev R 2009 *Mater. Sci. Eng.* A **503** 10
[32] Fedorov V Y, Ivanisenko Yu, Baretzky B and Hahn H: Vorrichtung und Verfahren zur Umformung von Bauteilen aus Metallwerkstoffen. German patent DE 10 2013 213 072.4, European patent EP 2 821 156 B1
[33] Beygelzimer Y, Estrin Y and Kulagin R 2015 *Adv. Eng. Mater.* **17** 1853

# Progress in NanoSPD research through scientometric analysis

**Z Safargalina**[1,2] **and I V Alexandrov**[3,4]

[1] Institute of Physics of Advanced Materials, Ufa State Aviation Technical University, 12 K. Marx str., Ufa 450008 Russia
[2] Laboratory for Mechanics of Bulk Nanomaterials, Saint Petersburg State University, 198504, Universitetskiy prospekt 28, Peterhof, St. Petersburg, Russia
[3] Department of Physics, Ufa State Aviation Technical University, 12 K. Marx St., Ufa 450008, Russia
[4] Sino-Russian Joint Laboratory of Functional Nanostructured Materials, Changzhou University, Changzhou 213164, P.R. China

E-mail: Zarsaf82@gmail.com

**Abstract**. In this paper, we use the scientometric approach based on the Web of Science Core Collection to analyze the data on publication activity in the area of nanostructured materials processed by severe plastic deformation (SPD) techniques on the following three main topics: ultrafine-grained materials, severe plastic deformation and bulk nanostructured materials. The results marked early 1990s as the starting period for first publications in question as well as their fast development. Pioneer works by Prof. Valiev and his laboratory at that time and their role in the growth of nanoSPD-related research were outlined.

## 1. Introduction

NanoSPD research, that is the study of nanostructured materials produced by severe plastic deformation techniques, has been attracting growing interest in the scientific community all over the world and this phenomenon only tends to progress over the recent decade due to fundamental findings and considerable innovation potential. This is easily demonstrated by the vast number of international nanospd-related events held worldwide. In recent years the development of nanoSPD topics has become one the most attractive research directions in materials science, as materials nanostructuring can provide new and unusual properties for a wide range of metals and alloys [1]. With new research comes growing publication activity, in addition there is interest towards the very origin and history of first discoveries in the field and the analysis of future trends; the latter is possible by means of a separate field of study that deals with measuring and analyzing scientific literature, i.e. scientometrics. Its major research issues include the measurement of the impact of research papers and academic journals, the understanding of scientific citations, and the use of such measurements in policy and management contexts [2]. Modern scientometrics is mostly based on the work of Derek J. de Solla Price and Eugene Garfield. The latter created the Science Citation Index, which in turn led to the Web of Science (which is an online subscription-based scientific citation indexing service originally produced by the Institute for Scientific Information (ISI), later maintained by Clarivate Analytics) [3].

Citations in science serve to link similar research items and help to locate publications having most significant impact in a particular field. It also helps to assess current trends and emerging fields of

IOP Conf. Series: Materials Science and Engineering **672** (2019) 012003 doi:10.1088/1757-899X/672/1/012003

research. In this case, it is of particular interest to apply such approach to the analysis of nanoSPD-related topics so as to follow the development of this field and see its future tendencies.

## 2. Materials and methods

For the purpose of this research we were using Web of Science Core Collection, including the Emerging Sources Citation Index. To analyze the research output in the topics as a measure of scholarly interest to nanoSPD, we have made topical searches for three main nanoSPD-related topics: "Ultrafine-grained materials", "Severe plastic deformation" and "Bulk nanostructured materials", which effectively returns records containing those words combinations in the paper titles, abstracts, or keywords.

To analyze the most important research outcomes in this topic, we were using the measure of total citation counts according to Web of Science Core Collection. Modern bibliometricians try to refrain from using total citation counts and instead prefer to compare counts for papers of similar age, document type, and in comparable fields. Although, as Van Noorden, Maher and Nuzzo pointed out in [4] the old-fashioned hall of fame based on total citation counts still has value.

## 3. Results and discussions

In the framework of the present work the following three nanoSPD-related topics [5,6] served as the focus of analysis:

1. Ultrafine-grained materials, i.e. polycrystalline materials having grain size less than 1 micron and mostly high-angle grain boundaries;
2. Severe plastic deformation, i.e. heavy straining under high imposed pressure, which is a metal forming procedure for producing ultrafine-grained and nanostructured materials;
3. Bulk nanostructured materials, that is UFG materials having various nanostructural features – nanoparticles, nanoclusters, segregations at grain boundaries, etc.

The topics were applied to conduct the quantitative analysis of publications published between 1900 and 2019 indexed within Web of Science Core Collection. The obtained results are presented in the graphs clearly demonstrating that active publications in these areas started in early 1990s.

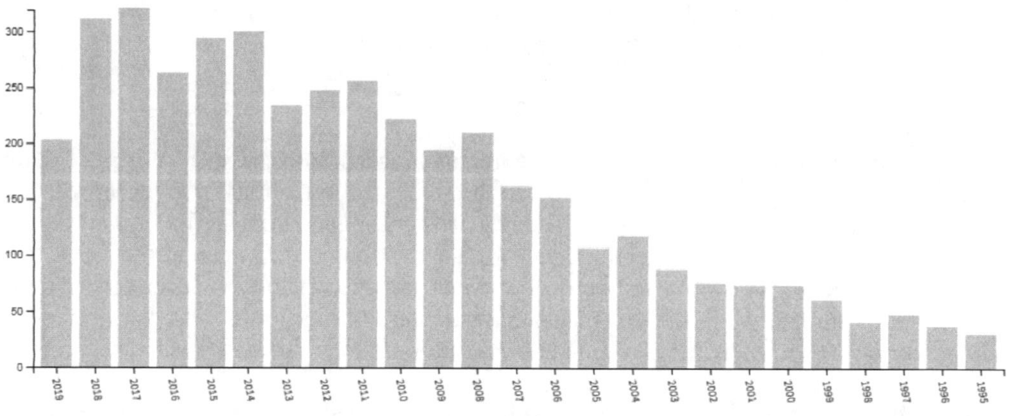


**Figure 1.** Showing 4,155 records for topic "ultrafine-grained materials". The record count in each column is the total number of articles published in the selected year. The count includes Early Access articles that are fully peer-reviewed, citable, and published but have not been assigned a volume/issue/page number.

As is seen, although first publications for the topic "ultrafine-grained materials" had appeared in 1977 and 1980 (one publication per year) [7], "real" publication activity started only in the early 1990s (with 10-12 papers per year) with steady growth until about 2000 when this number experienced a sudden increase, and then another leap can be observed in 2008.

IOP Conf. Series: Materials Science and Engineering **672** (2019) 012003 doi:10.1088/1757-899X/672/1/012003

However, the content analysis for the early works published before 1990 (for examples, see [7]) demonstrated that despite the presence of a specific term “ultrafine grains” in the title or abstract or keywords, the paper text was in fact dealing with grain sizes exceeding 1 micron, i.e. considerably larger than included in the present-day definition of ultrafine-grained materials. Yet at the same time the publications of early 1990s include the pioneer research by R.Z. Valiev with co-authors [8-10] as well as a number of international papers [11, 12] authored by T.G. Langdon and Z. Horita and having high citation numbers.

The analytic analysis of the topic “severe plastic deformation” (figure 2) traces the start of publications back to 1964 with 1-2 papers appearing per year until 1991 and 2000 (the number boosted to 14 and 113 accordingly). Over the years, publication activity on this topic has experienced steady and quick growth (+about 100 publications a year), now reaching and maintaining a very high level of productivity such as over 900 publications per year (in 2017 and 2018).

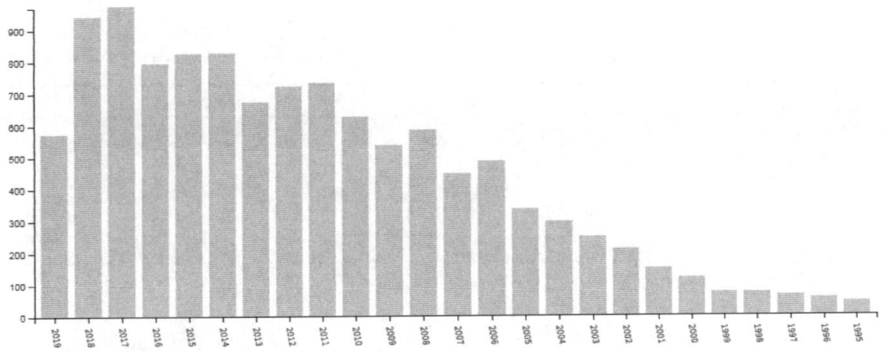


| Field: Publication Years | Record Count | % of 11,321 |
|---|---|---|
| 2019 | 567 | 5.008 % |
| 2018 | 935 | 8.259 % |
| 2017 | 970 | 8.568 % |
| 2016 | 788 | 6.961 % |
| 2015 | 819 | 7.234 % |
| 2014 | 820 | 7.243 % |
| 2013 | 668 | 5.901 % |
| 2012 | 717 | 6.333 % |
| 2011 | 728 | 6.431 % |
| 2010 | 621 | 5.485 % |
| 2009 | 531 | 4.690 % |
| 2008 | 579 | 5.114 % |
| 2007 | 440 | 3.887 % |
| 2006 | 479 | 4.231 % |
| 2005 | 331 | 2.924 % |
| 2004 | 291 | 2.570 % |
| 2003 | 243 | 2.146 % |
| 2002 | 204 | 1.802 % |
| 2001 | 143 | 1.263 % |

| Field: Publication Years | Record Count | % of 11,321 |
|---|---|---|
| 2000 | 113 | 0.998 % |
| 1999 | 66 | 0.583 % |
| 1998 | 67 | 0.592 % |
| 1997 | 56 | 0.495 % |
| 1996 | 47 | 0.415 % |
| 1995 | 36 | 0.318 % |
| 1994 | 17 | 0.150 % |
| 1993 | 16 | 0.141 % |
| 1992 | 8 | 0.071 % |
| 1991 | 14 | 0.124 % |
| 1990 | 2 | 0.018 % |
| 1986 | 1 | 0.009 % |
| 1983 | 1 | 0.009 % |
| 1982 | 1 | 0.009 % |
| 1969 | 1 | 0.009 % |

**Figure 2.** Showing 11,321 records for topic “Severe plastic deformation”.

The analysis of total citations for "severe plastic deformation" provides the following top 5:

| Publication | Times cited |
|---|---|
| Valiev, Islamgaliev, Alexandrov, 2000 [13] | 4,640 |
| Valiev, Langdon, 2006 [14] | 2,675 |
| Meyers, Mishra, Benson, 2006 [15] | 2,596 |
| Zhilyaev, Langdon, 2008 [16] | 1,664 |
| Kumar, Swygenhoven, Suresh 2003 [17] | 1,055 |

Obviously, the topic of "bulk nanostructured materials" is rather "young" in contrast to other subjects with first publications appearing in the early 1990s with steady growth until 2009 when the number suddenly increased to 269 and keeps at about the same level up until today (figure 3). The analysis of paper citations provides the following top 5:

| Publication | Times cited |
|---|---|
| Valiev, Islamgaliev, Alexandrov, 2000 [13] | 4,640 |
| Poudel, Hao, Ma et al, 2008 [18] | 3,096 |
| Hochbaum, Chen, Delgado, et al [19] | 2,860 |
| Valiev, Langdon, 2006 [14] | 2,675 |
| Meyers, Mishra, Benson, 2006 [15] | 2,596 |

Thus, the scientometric analysis of publications for topics "ultrafine-grained materials" and "severe plastic deformation" showed that their origin and increase in the total number per year over the study period can be contributed to the articles of early 1990s by the Ufa group of authors from the laboratory of Prof. R.Z. Valiev and to the series of international publications launched by Profs. T.G. Langdon (UK) and Z. Horita (Japan). Further, "bulk nanostructured materials" appeared as a new and exciting topic and has been steadily becoming an essential and integral part of active studies in the field of nanotechnologies and nanomaterials (http://nanospd.asso.univ-lorraine.fr/). Here it is important to note that the most popular devices for imposing large and severe plastic strains were introduced in scientific literature even earlier; that is by Prof. P. Bridgman in 1935 for high pressure torsion [20] and by Prof. V.M. Segal and colleagues in 1981 for equal channel angular pressing/extrusion (ECAP/ECAE) [21]. It was only in the end of 1980s-beginning of 1990s that the implementation of SPD approaches for producing UFG materials and increasing their properties took place, which served as the start of nanoSPD research [22]. Here, a particular publication [12] should be noted that introduced pioneer results of first studies – thus playing a crucial role in forming the entire nanoSPD subject - and outlined new tasks and challenges for further research; the number of its citation presently heads towards 5000, which is among record performance in modern materials science.

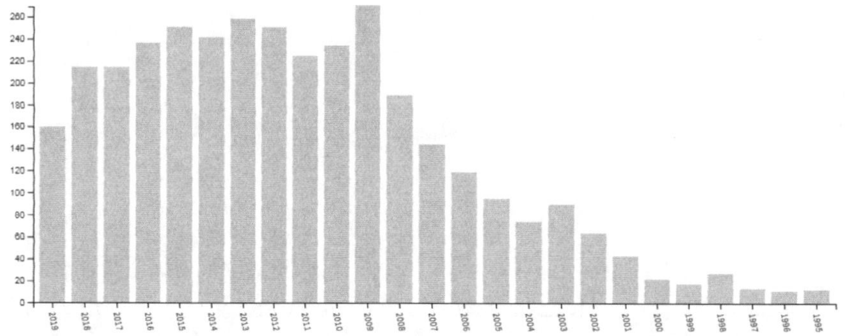

| Publication Years | Record Count | % of 3,435 |
|---|---|---|
| 2019 | 157 | 4.571 % |
| 2018 | 213 | 6.201 % |
| 2017 | 213 | 6.201 % |
| 2016 | 235 | 6.841 % |
| 2015 | 249 | 7.249 % |
| 2014 | 240 | 6.987 % |
| 2013 | 257 | 7.482 % |
| 2012 | 249 | 7.249 % |
| 2011 | 223 | 6.492 % |
| 2010 | 232 | 6.754 % |
| 2009 | 269 | 7.831 % |
| 2008 | 187 | 5.444 % |
| 2007 | 142 | 4.134 % |

| Publication years | Record count | % of 3,435 |
|---|---|---|
| 2006 | 117 | 3.406 % |
| 2005 | 93 | 2.707 % |
| 2004 | 72 | 2.096 % |
| 2003 | 88 | 2.562 % |
| 2002 | 62 | 1.805 % |
| 2001 | 41 | 1.194 % |
| 2000 | 20 | 0.582 % |
| 1999 | 16 | 0.466 % |
| 1998 | 25 | 0.728 % |
| 1997 | 12 | 0.349 % |
| 1996 | 9 | 0.262 % |
| 1995 | 10 | 0.291 % |
| 1994 | 2 | 0.058 % |
| 1993 | 2 | 0.058 % |

**Figure 3.** Showing 3,435 records for topic “bulk nanostructured materials”.

## 4. Conclusions

The analysis of nanoSPD research, that is the development of bulk nanostructured materials produced by severe plastic deformation, conducted in the present work shows that scientometric analysis may be an efficient tool to study the history behind the development of a scientific trend as well as to anticipate its further growth and priority tasks. In particular, the analysis reveals the key role of pioneer research by Prof. Ruslan Valiev and his colleagues in the early 1990s in the formation of this subject.

## Acknowledgement

The present work is dedicated to Prof. Ruslan Valiev's honorary session at the 6th International Symposium "BULK NANOMATERIALS: from fundamentals to innovations. BNM-2019" (Ufa, Russia). The author also acknowledges considerable contribution from Clarivate Analytics' expert, Pavel Kasyanov, for information and inspiring discussion about citation analysis and Prof. Terry G. Langdon for many interesting talks on the subject.

The work on the paper was possible due to partial financial support from Saint Petersburg State University in the framework of Call 3 project (id 26130576).

## References

[1] Valiev R Z, Estrin Y, Horita Z, Langdon T G, Zehetbauer M J and Zhu Y T 2016 *Mater. Res. Lett.* **4** 1
[2] Leydesdorff L and Milojevic S 2015 Scientometrics *International Encyclopedia of Social and Behavioral Sciences* Section 8.5: Science and Technology Studies, Subsection 85030, ed M Lynch, vol 21 (Oxford: Elsevier) pp 322–327
[3] https://en.wikipedia.org/wiki/Web_of_Science
[4] Van Noorden R, Maher B and Nuzzo R 2014 *Nature* **514** 550
[5] Valiev R Z, Estrin Y, Horita Z, Langdon T G, Zehetbauer M J and Zhu Y T 2006 *JOM* **58** 33
[6] Valiev R Z, Zhilyaev A P and Langdon T G 2014 *Bulk Nanostructured Materials: Fundamentals and Applications* (Hoboken: Wiley/TMS)
[7] Kaibyshev O A and Kazachkov I V 1980 *Dokl. Akadem. Nauk SSSR* **252** 351

[8] Valiev R Z, Kaibyshev O A, Kuznetsov R I, Musalimov R S and Tsenev N K 1988 *Dokl. Akad. Nauk. SSSR* **30** 864
[9] Valiev R Z, Krasilnikov N A and Tsenev N K 1991 *Mater. Sci. Eng.* A **137** 35
[10] Valiev R Z, Korznikov A V and Mulyukov R R 1993 *Mater. Sci. Eng.* A **168** 141
[11] Wang J T, Horita Z, Furukawa M, Nemoto M, Tsenev N K, Valiev R Z, Ma Y and Langdon T G 1993 *J. Mater. Res.* **8** 2810
[12] Wang J, Iwahashi Y, Horita Z, Furukawa M, Nemoto M, Valiev R Z and Langdon T G 1996 *Acta Mater.* **44** 2973
[13] Valiev R Z, Islamgaliev R K and Alexandrov I V 2000 *Prog. Mater. Sci.* **45** 103
[14] Valiev R Z and Langdon T G 2006 *Prog. Mater. Sci.* **51** 881
[15] Meyers M A, Mishra A and Benson D J 2006 *Prog. Mater. Sci.* **51** 427
[16] Zhilyaev A P and Langdon T G 2008 *Prog. Mater. Sci.* **53** 893
[17] Kumar K S, Van Swygenhoven H and Suresh S 2003 *Acta Mater.* **51** 5743
[18] Poudel B *et al.* 2008 *Science* **320** 634
[19] Hochbaum A I, Chen R K, Delgado R D, Liang W J, Garnett E C, Najarian M, Majumdar A and Yang P D *Nature* 2008 **451** 163
[20] Bridgman P W *Phys. Rev.* 1935 **48** 825
[21] Segal V M, Reznikov V I, Drobyshevskiy A E and Kopylov V I 1981 *Russ. Metall.* **1** 99
[22] Edalati K and Horita Z 2016 *Mater. Sci. Eng.* A **652** 325

# Strength and electroconductivity of bulk nanostructured copper alloys

**I V Alexandrov**[1,3] **and W Wei**[2,3]

[1] Ufa State Aviation Technical University, Ufa, 450008, Russia
[2] School of Materials Science and Engineering, Changzhou University, Changzhou 213164, P. R. China
[3] Sino-Russian Joint Laboratory of Functional Nanostructured Materials, Jiangsu Key Laboratory of Materials Surface Science and Technology, Changzhou University, Changzhou 213164, P. R. China

E-mails: igorvalexandrov@yandex.ru, benjamin@yandex.ru

**Abstract**. The results of a study of the strength and conductivity of electrotechnical bulk nanostructured copper alloys obtained by methods of severe plastic deformation is presented. The contribution of the microstructure parameters to the formation of high strength while maintaining sufficient electrical conductivity is discussed.

## 1. Introduction

Modern progress is impossible without the use of electrical materials. To transfer electric energy from the manufacturer to the consumer, to create electric motors and generators, devices and circuits for micro- and nanoelectronics, etc., metal materials are widely used, including copper alloys. The choice in favor of copper alloys as conductors of electric current is due to their inherent high electrical conductivity, strength and ductility. At the same time, the requirements for electrical materials are constantly growing due to changes in their operating conditions. As an example, it is requisite for particularly high-strength copper alloys in high-speed rail transport, heat exchangers, welding electrodes, etc. (figure 1).

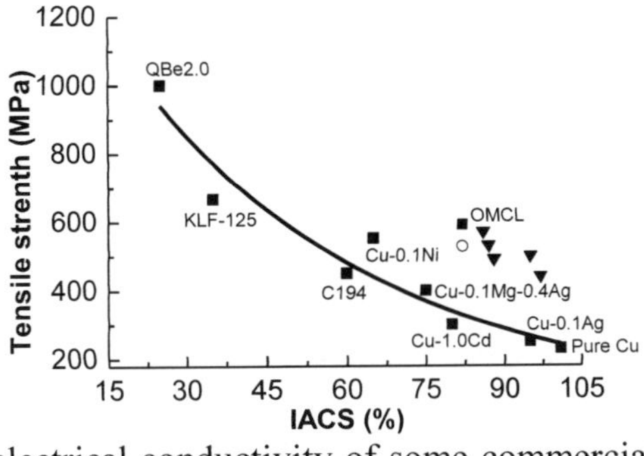


**Figure 1.** Tensile strength vs. electrical conductivity of some commercial Cu-based alloys. KLF-125 corresponds to the Cu-3.2Ni-0.7Si-1.25Sn-0.3Zn alloy, C194 to the Cu-2.3Fe-0.1Zn-0.03P alloy, and OMCL to the Cu-0.3Cr-0.1Zr-0.005Mg-0.02Si alloy [8].

Promising from this point of view are dispersion hardened alloys of the Cu-Cr-Zr system with a Cr content from 0.5 to 1.5 wt.% and Zr from 0.02 to 0.20 wt.% [1,2]. As a result of applying traditional methods of thermo-mechanical treatment, including hardening with high temperature and subsequent aging, for these alloys it is possible to achieve combinations of strength and electrical conductivity of 450 MPa and 80% IACS or 590 MPa and 75% IACS, respectively [3-6]. Moreover, it is characteristic that the increase in strength is accompanied by a decrease in electrical conductivity.

The use of severe plastic deformation (SPD) methods allows the formation of dispersion-hardened copper alloys with high strength bulk nanostructured states with grain sizes of tens to hundreds of nanometers and nanodispersed particles. In this case, the increase in the electrical conductivity of these states is promoted either by aging following SPD [7–9], or by increased SPD temperatures that activate dynamic aging [10–12] or continuous dynamic recrystallization [13], or by high speeds and low SPD temperatures, leading to the formation of deformation nanotwins, in addition to grains of the nanoscale range, [14]. As a result of the above effects on the microstructure of the alloys of the Cu-Cr-Zr system, bulk nanostructured states can be formed, which are characterized by the simultaneous manifestation of high strength and electrical conductivity.

This article presents the results of an analysis of the influence of microstructure parameters on the strength and electrical conductivity of electrotechnical bulk nanostructured copper alloys obtained by SPD methods.

## 2. SPD and subsequent aging

It is well known that deformation processing by traditional methods leads to an increase in the strength of copper alloys. However, the electrical conductivity decreases significantly. In this state, copper alloys are of no interest from the point of view of industrial applications. The high strength is caused by the small grain size, subgrain structure inside them, deformation twins, a high dislocation density in the grain body, elastic microdistortions of the crystal lattice, etc. These changes in the microstructure during plastic deformation can also affect the electrical conductivity, significantly reducing it. At the same time, subsequent aging, which is a traditional method of heat treatment, can significantly affect both strength and electrical conductivity, significantly increasing the latter. The reason for this is the precipitation of dispersed particles of the second phases from the supersaturated solid solution. In this case, the content of atoms of the alloying elements in the solid solution will decrease, as a result, the elastic stresses in the crystal lattice of the matrix will also decrease, and the scattering of conduction electrons by the atoms of the alloying elements dissolved in the matrix will become less significant.

The degree of influence on the strength and electrical conductivity of the formed microstructure is determined to one degree or another by the nature of the wrought alloys (chemical and phase composition), the nature of the deformation effect (stress-strain state diagram, degree, speed and temperature of deformation), temperature and time of subsequent aging. Thus, it is possible to influence the indicated properties of copper alloys in order to achieve their optimal combination, inter alia, by choosing the optimal combination of parameters of plastic deformation and subsequent aging.

The microstructure of bulk nanostructured copper alloys obtained by SPD methods is significantly different from the microstructure of metals and alloys deformed by traditional methods. Bulk nanostructured states are characterized by a very small grain size with large-angle grain boundaries, a low density of dislocations in the grain body, however, a very high density of extrinsic grain-boundary dislocations, long-range elastic stress fields created by these dislocations, a crystal lattice with a changed parameter, a high concentration of twins, and increased solubility of alloying elements, specific crystallographic texture. Aging of bulk nanostructured alloys obtained by SPD methods, due to the characteristics of the formed ultrafine-grained (UFG) structure, can have certain features.

In [7], the Cu-0.5wt.% Cr alloy billets were subjected to SPD by equal channel angular pressing (ECAP). The strength of the state formed as a result of 4 ECAP passes turned out to be equal to 460 MPa and exceeded the strength of the initial solid-solution quenched state equal to ~180 MPa by almost 2.6 times. The ECAP strain rate was 0.2 $mm^{-1}$. Additional cold rolling of this state was carried out to a compression ratio of 90% at room temperature. At the same time, pancake shaped microstructure changed to fragmentary one with less sharp boundaries. The formed microstructure can be attributed to UFG structures. The dislocation density as a result of cold rolling increased. Dislocations fragmented the lamella boundaries formed as a result of ECAP. The strength of the condition after 4 passes of ECAP as a result of cold rolling increased to 570 MPa. It was concluded that the high strength of the workpieces subjected to ECAP followed by cold rolling is mainly due to the release of a small grain size, the presence of alloying atoms in a solid solution, a high density of dislocations, and high-angle grain boundaries.

The electrical conductivity of the initial state, equal to 37% with increasing number of ECAP passes and during subsequent rolling, remained almost unchanged, remaining at a low level, and amounted to only 35%. Since the number of ECAP passes and subsequent rolling did not affect the level of electrical conductivity, it was concluded that the dislocation density, which increased as a result of ECAP and subsequent cold rolling, has a smaller effect than alloying atoms in solid solution. An analysis of the obtained experimental results indicated that the indicated regularities in the change in strength and electrical conductivity are associated with depletion of the solid solution by Cr atoms, enlargement of precipitate particles, and annihilation of dislocations of opposite signs. Using this approach, it was possible to achieve a combination of tensile strength of 554 MPa with an electrical conductivity of 84% IACS in samples subjected to ECAP and cold rolling.

In [15], analytical modeling was used to assess the contribution of structural parameters to the strength and electrical conductivity of this alloy. Peierls-Nabarro stress, dislocation hardening, solid solution hardening, and hardening by particles of the secondary phase were considered as possible sources of contributions to the strength characteristics. At the same time, the electron – phonon interaction, the interaction of free electrons with impurity atoms and particles of the secondary phase, and also with dislocations were considered as parameters of the microstructure responsible for the electrical conductivity.

The calculation results showed that the contribution to the strength characteristics of the Cu-0.5wt.%Cr alloy subjected to 4 ECAP passes can be caused, first of all, by dislocation hardening. Its share was estimated at 82%. An assessment of the Peierls-Nabarro stress contribution indicated its contribution of 18%. The contribution of solid solution hardening turned out to be negligible and equal to less than 1%. The contribution of the particles of the secondary phases was estimated to be equal to zero due to their absence in the alloy.

At the same time, in this state, the main contribution to the electrical resistance is due, primarily, to the chromium atoms dissolved in the copper matrix. This contribution exceeded 63%. The second largest was the contribution of thermal vibrations in the crystal lattice, amounting to more than 36%. The contribution of dislocations turned out to be negligible.

Thus, it is possible to affect the strength, first of all, by increasing the density of dislocations, which practically do not affect the electrical conductivity of the alloy. From this it follows that by increasing the density of dislocations due to an increase in the degree of deformation before the start of the processes of recovery and recrystallization or lowering the temperature of deformation, an increase in strength can be achieved. In this case, the conductivity will not fall. Indeed, an increase in the degree of SPD, for example, due to an increase in the number of passes with ECAP, leads to an increase in strength characteristics. In this case, the electrical conductivity does not change. True, at the same time, its value is small and equal to only about 37% of IACS.

As a result of aging following SPD, small particles of chromium were precipitated from the copper matrix in the microstructure of the Cu-0.5wt.%Cr alloy [7]. An increase in the aging temperature up to 450°C with a holding time of 1 hour led to an increase in strength with its subsequent drop at higher temperatures. An increase in aging time in excess of 1 hour led to a drop in microhardness. However, an increase in temperature and duration of aging is accompanied by an increase in electrical conductivity. Thus, experimental studies indicate that up to a certain temperature and exposure time due to the decomposition of the solid solution and the precipitation of fine particles, both strength and electrical conductivity increase.

The resulting increase in strength is due to various multidirectional competitive processes. During aging, the contribution of solid solution hardening decreases, which was negligible anyway. Dislocations annihilate and their density decreases as a result. Accordingly, the role of dislocation hardening decreases. At the same time, the concentration of particles of the secondary phases increases, which resist the slip of dislocations, and accordingly increase their contribution to the strength characteristics. On the other hand, electrical conductivity increases with aging. This is due to the fact that impurity atoms dissolved in the matrix lattice begin to participate in the formation of particles of secondary phases. Thus, the main factor of low electrical conductivity associated with the presence of alloying elements in the matrix of atoms loses its significance. Moreover, the next largest factor, which is the thermal vibrations of atoms in the crystal lattice of the matrix, remains unchanged.

Analytical modeling confirms this conclusion. Estimates made for this alloy in the state after 4 ECAP passes, rolling by 90% and aging at 450 °C with a holding time of 1 hour indicate the following. The main contribution to the strength of this state is made by dispersed chromium particles precipitated from the solid solution. This contribution is about 54%, i.e. it grew from almost zero value, characteristic for the state formed as a result of 4 ECAP passes. At the same time, the contribution of dislocation hardening fell from 82% to ~30%. The Peierls stress contribution has remained unchanged. On the other hand, after aging, the main contribution to the electrical resistance is made by the thermal vibrations of the atoms of the matrix. Its value was about 83%. However, it is not possible to influence the magnitude of this factor. The second largest is the contribution of chromium atoms in the matrix lattice. It amounted to about 12%, i.e. fell three times compared with the state after 4 passes ECAP. The contribution of the released chromium particles was less than 5%. Thus, a change in the balance between the number of atoms of impurity atoms in the matrix and in the composition of the released particles can optimize the electrical conductivity. The contribution of the dislocations turned out to be negligible.

Thus, dispersed particles of secondary phases are the main parameter of the microstructure that can be purposefully controlled. Their concentration and size, as well as the distribution in the body or grain boundaries, should determine the combination of strength properties and electrical conductivity.

It should be noted that the use of SPD in combination with subsequent aging was used by other researchers as well for this and other alloys of the Cu-Cr system, including with the addition of atoms of various kinds of alloying elements. For example, in [16], the properties of the Cu-0.80%Cr-0.080% Zr alloy (wt.) were optimized. As a result of 8 ECAP passes along the $B_C$ route, followed by aging at a temperature of 425 °C for 240 minutes, the microstructure of the initial annealed and quenched coarse-grained state became UFG and predominantly equiaxed, but slightly elongated grains (subgrains) were also observed in it. The ultimate tensile strength of the annealed, hardened and aged state, on the one hand, and the hardened state after ECAP and aging, on the other hand, are 396 and 670 MPa, respectively, and the electrical conductivity reached 75% and 74% IACS, respectively.

Thus, the potential of achieving high bulk electrical conductivity strength corresponding to the initial quenched and aged state in bulk nanostructured materials obtained by SPD and subjected to subsequent aging has been demonstrated.

## 3. Dynamic aging and dynamic recrystallization in SPD

As shown by experimental studies [10-12], the formation of bulk nanostructured states, characterized by the simultaneous manifestation of high strength and electrical conductivity, can occur as a result of dynamic aging during SPD. The samples of the Cu-0.5wt.%Cr alloy with small additions of Ag, Fe, P, and Si atoms were studied. SPD by means of high pressure torsion (HPT) was implemented at an elevated temperature of 300 °C. As a result of this treatment, an UFG structure with nanodispersed particles was formed. Thus, instead of a two-stage process of deformation with subsequent aging, a one-stage process is proposed that combines both of these processes. The fact of dynamic aging is indicated by the fact that annealing after HPT does not lead to significant changes in microhardness and electrical conductivity. In this case, the samples demonstrate a high ultimate tensile strength of 790-840 MPa and a high electrical conductivity of 81%-85% IACS. Based on the studies, the authors came to the conclusion that the optimal combination of high strength and electrical conductivity corresponds to a microstructure with an average grain size of about 200 nm and a particle size of 10 nm. Such a microstructure is also characterized by increased thermal stability up to 500 °C. Thus, the particles of the secondary phases released during the dynamic aging process, in this case, determine the optimal combination of strength and ductility.

The authors of [13] also used elevated SPD temperatures to control the strength and electrical conductivity of the Cu-0.8%Cr-0.06%Zr (wt.) Alloy. In this case, SPD was implemented by ECAP at various elevated temperatures in the range 473-673 K. The formation of the UFG structure was considered the result of continuous dynamic recrystallization. The increase in strength during SPD from 215 MPa to 480 MPa and 535 MPa at ECAP temperatures of 673 K and 473 K, respectively, was justified by grain refinement and an increase in the dislocation density. Moreover, the contribution of the latter component was assessed as predominant. At the same time, electrical conductivity was 73% IACS and 66% IACS, respectively. Its slight decrease from the 79% IACS characteristic of the initial state was associated with the formation of a structure with a high dislocation density in the grain body during ECAP. However, the authors do not note the role of atoms of alloying elements in solid solution and particles of the secondary phases in the formation of an attractive complex of strength and electrical properties, as is the case in other articles devoted to this subject.

## 4. Dynamic plastic deformation

High speeds and low temperatures of plastic deformation significantly affect the microstructure of metallic materials. In this case, a significant refinement of the structural elements takes place, additional deformation modes are activated, the dislocation density increases, and deformation twinning processes are activated. At the same time, dynamic recovery processes are difficult. All this helps to increase the strength characteristics. Moreover, the evolution of the electrical conductivity of materials subjected to such influences is of interest.

In [14], as a result of dynamic plastic deformation at a temperature of liquid nitrogen (LNT-DPD) with velocities of $10^2$–$10^3$ $s^{-1}$ in the Cu-1wt.%Cr-0.1%Zr alloy, a nanostructured state was formed, which is characterized by record strength values equal to 700 MPa, and electrical conductivity equal to 78.5% IACS. The formed nanostructured state was characterized by an average nanograin size of about 47±19 nm and an average nanotwin size of about 25±14 nm, with a volume fraction of about 20%. In this case, considered as the annealed state, the strength was slightly less than 400 MPa. Annealing was performed to precipitate Cr particles from the solid solution and increase the electrical conductivity up to 84.2% IACS. For comparison, along with dynamic deformation, the ingots in the initial state were subjected to quasistatic deformation by compression at a rate of $10^{-3}$ $s^{-1}$ at room temperature (RT-QSC). The authors concluded that the deformation nanotwins characteristic of the nanostructure obtained by the LNT-DPD method are responsible for high strength. Moreover, they have a significantly smaller effect on the drop in electrical conductivity than the boundaries of nanograins. It is noted that the contribution of dispersion hardening in alloys of the Cu-Cr-Zr system subjected to SPD can be reduced due to the strongly deformed microstructure. As a result, the high strength of the alloy subjected to LNT-DPD is mainly due to a mixed type microstructure consisting of nanograins, nanotwins and Cr particles. Moreover, the contribution of these parameters to the increase in strength is difficult to single out.

The authors associated the slight decrease in electrical conductivity with an increase in the degree of deformation during LNT-DPD with the propagation of defects in the crystal structure, including dislocations and boundaries. It is believed that twins have significantly lower electrical resistance than boundaries with different misorientation angles.

An analysis of the obtained experimental results using analytical modeling [17] showed the following. In the initial annealed state, low strength is determined by the low dislocation density, and high electrical conductivity by a small contribution to the resistance of the particles of the secondary phases. This contribution is comparable with the contribution of dislocations. The increase in strength due to RT-QSQ is mainly due to dislocation hardening. The decrease in electrical conductivity is associated with the scattering of conduction electrons at grain boundaries and vacancies. The contribution of the latter is approximately 13 times less than that of the former. In the state after LNT-DPD, an even greater increase in strength is associated with the formation of twin boundaries. At the same time, a slight decrease in electrical conductivity is due to electron scattering, primarily at grain boundaries. The contribution of electron scattering on deformation vacancies is approximately 4 times smaller, and at the twin boundaries it is still almost 18 times smaller. Thus, the simulation results indicate a significant role of twin boundaries in the formation of a high-strength state as a result of LNT-DPD. At the same time, twins have little effect on electrical conductivity. In this case, the role of electron scattering at numerous grain boundaries of the nanostructured state is significant.

It is clear that a high density of grain boundaries in the nanostructured state cannot be avoided. However, it is possible to form bulk nanostructured states with a high density of nanotwins either by increasing the speed and/or lowering the temperature of the SPD, or by choosing a more alloyed alloy with a lower stacking fault energy, more prone to twin deformation. Furthermore, a new process of ECAP and deep cryogenic treatment (DCT) was proposed to achieve a combination of tensile strength ~423 MPa and electrical conductivity ~97.9% IACS simultaneously [18].

## 5. Conclusions

The above results of experimental studies and analytical modelling indicate the possibility of simultaneously achieving high strength and electrical conductivity due to the targeted effect on bulk microstructure parameters of bulk nanostructured copper alloys for electrical purposes.

**Acknowledgments**
The work was carried out with the financial support of the Ministry of Education and Science of Russia within the framework of the project № 16.1969.2017/4.6. W. Wei is grateful to be supported by the National Natural Science Foundation of China under Grant No. 51561001 and the Science and Technology Project of Changzhou, P. R. China under grant No.CZ20180016 and CE20170028.

## References

[1] Hang C-Q 2001 *China Railway Science* **22** 1
[2] Loginov V Yu and Mysik RK 2014 *Journal of Siberian Federal University. Techn.* **7** 3
[3] Edwards D J, Singh B N and Tahtinen S 2007 *J. Nucl. Mater.* **367** 904
[4] Watanabe C, Monzen R and Tazaki K 2008 *J. Mater. Sci.* **43** 813
[5] Zhou H T, Zhong J W, Zhou X, Zhao Z K and Li Q B 2008 *Mater. Sci. Eng.* A **498** 225
[6] Gosh G, Miyake J and Fine M E 1997 *JOM* **49(3)** 5
[7] Wei K X, W Wei, Wang F, Du Q B, Alexandrov I V and Hu J 2011 *Mater. Sci. Eng.* A **528** 1478
[8] Wei W, Wei K X, Du Q B, Musin F F, Hu J and Alexandrov I V 2015 *Handbook of Mechanical Nanostructuring*, ed M Aliofkhazraei (Weinheim: Wiley-VCH) p 115
[9] Vinogradov A, Suzuki Y, Ishida T, Kitagawa K and Kopylov V I 2004 *Mater. Trans.* **45** 2187
[10] Islamgaliev R K, Nesterov K M, Bourgon J, Champion Y and Valiev R Z 2014 *IOP Conf. Series: Mater. Sci. Eng.* **63** 012118
[11] Islamgaliev R K, Nesterov K M, Champion Y and Valiev R Z 2014 *J Appl Phys* **115** p 194301
[12] Murashkin M Yu, Sabirov I, Savage X and Valiev R Z 2015 *J. Mater. Sci.* **51(1)** 33
[13] Mishnev R, Shakhova, I, Belyakov A and Kaibyshev R 2015 *Mater. Sci. Eng.* A **629** 29
[14] Sun L X, Tao N R and Lu K 2015 *Scripta Mater.* **99** 73
[15] Chembarisova R G, Aleksandrov I V and Yamileva A M 2019 *Tech. Phys.* **64** 162
[16] Purcek G, Yanar H, Demirtas M, Alemdag Y, Shangina D V and Dobatkin S V 2015 *Mater. Sci. Eng.* A **649** 114
[17] Chembarisova R, Alexandrov I, Wei W 2018 *Defect and Diffusion Forum* **385** 256
[18] Wei K X, Chu Z Q, Wei W, Du Q B, Alexandrov I V, Hu J 2019 *Adv. Eng. Mater.* **21** 1801372

IOP Conf. Series: Materials Science and Engineering **672** (2019) 012005 doi:10.1088/1757-899X/672/1/012005

# The influence of the mode of interaction of compact samples made of nonpassivated iron nanopowders with air on their microstructure

**M I Alymov, B S Seplyarskii, S G Vadchenko, R A Kochetkov, N M Rubtsov, V A Zelensky and A S Schukin**

Merzhanov Institute of Structural Macrokinetics and Materials Science, Russian Academy of Sciences, Chernogolovka, Russia

E-mail: alymov@ism.ac.ru

**Abstract.** In this work, various modes of interaction (spontaneous ignition or a combustion mode) of compact samples made from nonpassivated iron nanopowders with air are found out. The experiments showed that the modes of interaction of the samples with air after their extraction from a weighing bottle depended on the media (argon or air), in which the weighing bottle was placed before extraction of the samples, as well as on the duration of the exposure of the weighing bottle in the air. It is shown that the process of interaction of the sample with air has a surface character; the inner layer of the sample contains unreacted nanopowder. The microstructure of the transverse fracture of the samples after spontaneous ignition and combustion represents a porous aggregation of agglomerates. Integral energy-dispersive analysis showed a higher content of oxygen in the near-surface layer of the sample about 300÷500 microns thick. The analysis of the microstructure of the fracture of a separate agglomerate showed the presence of a layer with an elevated oxygen content on the surface of the agglomerates. Particles in the agglomerates retain the dimensions of passivated iron nanopowders.

## 1. Introduction

Nanopowders of metals are pyrophoric, i.e. these are capable of self-ignite on contact with air due to high chemical activity and a large specific surface [1-2]. To make safe further processing of nanopowders to products, they are passivated. Passivation means the synthesis of a thin protective film on the surface of nanoparticles, which prevents self-ignition of metal nanopowders. The processes of passivation of iron and nickel nanopowders were investigated in [3-7].

However, it is obvious that there can also be situations in which the passivation of the nanopowder is technically impossible or undesirable, though technical operations with the nanopowder need to be carried out. Therefore, the relevant task is the development of new methods of obtaining compact products from nanopowders allowing to provide the required fire and explosion safety level in the processing of both nanopowders and their products.

Literary data on the regularities of spontaneous ignition and self-heating of compacted samples obtained from nonpassivated nanopowders are quite limited. In the works [8, 9], a research of ignition of nanosystems with the sizes of particles of reagents in the range of 40-80 nm showed that the temperature and energy of ignition can be much lower than in mixes of micropowders (1-100 microns).

IOP Conf. Series: Materials Science and Engineering **672** (2019) 012005 doi:10.1088/1757-899X/672/1/012005

In [10], the features of flame propagation within tablets made of mixes of Al/CuO nanopowders (the so-called nanothermites) depending on the density were investigated at laser initiation of combustion. Similar measurements [11] were carried out with the samples of $Al/MoO_3$ nanopowders; the obtained results were qualitatively similar to the results on Al/CuO nanothermites.

In the work, the interaction of compact samples made of nonpassivated iron nanopowders with the air was experimentally investigated. The main attention was paid to the influence of the interaction mode on the microstructure of the samples after spontaneous ignition and combustion.

## 2. Experimental

In the work, iron nanopowders obtained by the chemicometallurgical method were used for experimental studies of ignition and passivation of compact samples. The main stages of the synthesis of a metal nanopowder are the metal hydroxide sedimentation, drying and reduction [3-5]. These powders are pyrophoric.

All operations on obtaining compact samples, namely opening the vessels with nanopowders, weighing and pressing them, were carried out in a sealed box (hereinafter - the box) filled with inert gas (argon), equipped with a lock to extract and replace powders and samples. The oxygen concentration in the box did not exceed 0.1 vol.%. Special experiments showed that at such an oxygen concentration, nanopowders remained pyrophoric. Compact cylindrical samples made of iron nanopowder with a diameter of 5 mm, a length of 7 ÷ 12 mm and a density of 2.5 ÷ 4 g/cm$^3$ were obtained and investigated. After pressing, each sample was placed in a separate weighing bottle.

The weighing bottle with the samples was extracted out of the box via the lock. Cylindrical samples were extracted out of the weighing bottle just before the experiment and set vertically for 3-5 seconds on a boron nitride support. The scheme of the experiment is shown in figure 1. A Flir 60 infrared camera was used for monitoring the time dependence of the temperature within a sample surface and for determination of the maximum temperature at a given time point as well. A SONY HDR-CX330 video camera was used for measuring the combustion velocity (propagation velocity of the oxidation reaction across the surface of the samples). The dynamics of the temperature change on the surface of the sample was also controlled by two tungsten-rhenium thermocouples 40 microns thick, which were tightly touching the lateral surface of the sample.

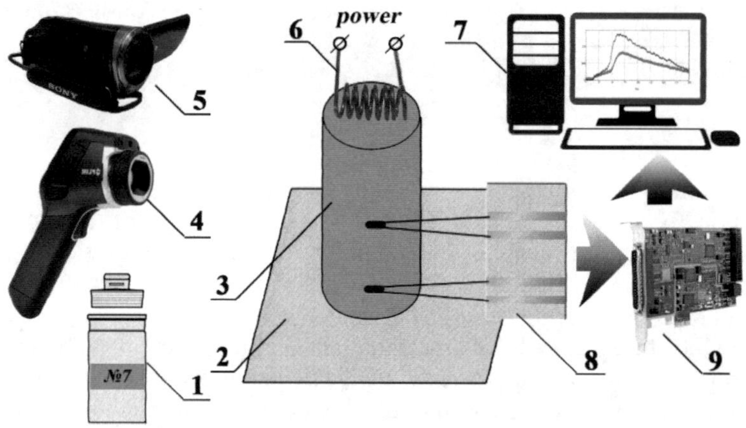


**Figure 1.** The scheme of the experiment on measurements of the temperature characteristics of samples in air: 1 – weighing bottle with a compact sample after extraction from the box, 2 – boron nitride support, 3 – sample under investigation, 4 – Flir E60 infrared camera, 5 –digital video camera, 6 – tungsten spiral, 7 – computer for processing A/D converter data, 8 – the contact platform with fixed thermocouples, 9 – A/D converter.

IOP Conf. Series: Materials Science and Engineering **672** (2019) 012005 doi:10.1088/1757-899X/672/1/012005

The microstructure of the samples after self-ignition and combustion was investigated by the scanning electron microscopy (SEM) method with a Carl Zeiss Ultra Plus microscope. The phase structure of the samples was studied using a DRON-3M diffractometer. The diffractograms were recorded in a mode of step scanning in the range of angles $2\theta = 30^o \div 100^o$ with a step $0.2^o$. The obtained diffractograms were analyzed using the PDF-2 database.

## 3. Preliminary results

To determine the specific surface and average sizes of the iron nanopowder, the nanopowder was previously passivated. Depending on the passivation time, the specific surface S of nanopowders determined by the BET method was $9 \div 9.6$ $m^2/g$. The average diameter d of iron particles was calculated according to the following formula $d = 6 / (\rho S)$, where $\rho$ is the density of iron 8.96 $g/cm^3$. At $S = 9.0$ $m^2/g$, the diameter of the particles of the iron powder was equal to 85 nm.

## 4. Results and discussion

In figure 2, the results of video filming of the process of self-heating in air (the external initiation is missing) of a cylindrical sample pressed from a nonpassivated iron nanopowder are presented. After extraction out of the box, the closed weighing bottle with the samples was in the argon atmosphere; then the samples were extracted to air as described above. Such a mode of reaction with air is observed for all samples, if the weighing bottle after extraction out of the box was in argon atmosphere.

**Figure 2.** Frames of video filming of a sample during self-heating.

Self-heating of the sample has a non-uniform character, though it begins at the same time within the entire surface of the sample. Infrared video filming showed that the maximum temperature is located near the top end face of the sample practically at all stages of oxidation. The main reason of heterogeneity of a warming up is non-uniformity of the sample density along its height; the top part of the sample always had a minimum density.

In another series of experiments, samples of nonpassivated iron nanopowder were pressed in argon atmosphere, however, after extraction from the box the closed weighing bottle with the samples was placed in air. The experiments showed that the modes of interaction of the samples with the air after they were extracted from the weighing bottle depended on the duration (t) of the presence of the weighing bottle in air. The first sample (t is less than 4 min) showed itself similar to the samples from the first series of experiments. Thus, self-heating of the samples occured, and they changed their color. The next sample (t is 10 min) was warmed up to only 55 °C and did not change color. The samples, which were in the weighing bottle in air for more than 20 min, showed no temperature rise after extraction and did not change color. To test the hypothesis that passivation with the maintenance of chemical activity occurred while the weighing bottle was in air, the following experiments were performed. The samples, which showed no temperature rise after extraction from the weighing bottle, were ignited from the top end face with a tungsten spiral. This procedure led to the propagation of a combustion wave over the sample. The combustion velocity for a sample density 3.07 $g/cm^3$ makes $\approx$ 0.025 cm /s.

In the following experiments, the investigation of the microstructure of the transverse fracture of a cylindrical sample after self-heating and cooling was performed. The microstructure of the transverse fracture of a cylindrical sample represents an aggregation of agglomerates of iron nanoparticles. The agglomerates have the form of "crocks" (figure 3), i.e. parts of spheres of irregular form.

The integral energy-dispersive analysis (EDA) from the surface of the fracture showed a higher oxygen content in the near-surface layer (about 300÷500 microns thick) of the sample. When examining the experimental EDA data of, it is necessary to take into account that the transverse fracture of the sample, but not the polished section, was studied. Therefore, the results of EDA have a qualitative character rather than quantitative one; these only provide insights into the magnitude of the oxygen content in various areas of the sample and the distribution of oxygen inside and on the surface of a separate agglomerate. The average content of oxygen in the near-surface area of the sample is 45÷50 atomic percent, and in the central part of the sample it is about 20 atomic percent.

In figure 3 d-c, the microstructure of a fracture of a separate agglomerate in the near-surface area of the sample is presented. In the figure, one can see the existence of a pore-free layer on the surface of the agglomerates ("crust") about 300 nanometers thick. Apparently, the formation of a dense oxide film on the surface of the agglomerates prevents the diffusion of oxygen into the agglomerate and further oxidation of the iron nanopowder as well. The internal part of the agglomerate represents a porous aggregation of iron particles about 100 nanometers in size.

In figure 4 the microstructure of the internal part of the sample is presented.

According to a local EDA, the oxygen content on the surface of the agglomerates in the central part of the sample is 15÷20 atomic percent, inside the agglomerates it is about 5 atomic percent. On the surface of the agglomerates of the sample there is no dense "crust", which is characteristic of the near-surface part of the sample.. Only a small increase in density to a depth about a single particle is observed. The internal part of the agglomerates does not differ from the similar agglomerates in the near-surface part of the sample.

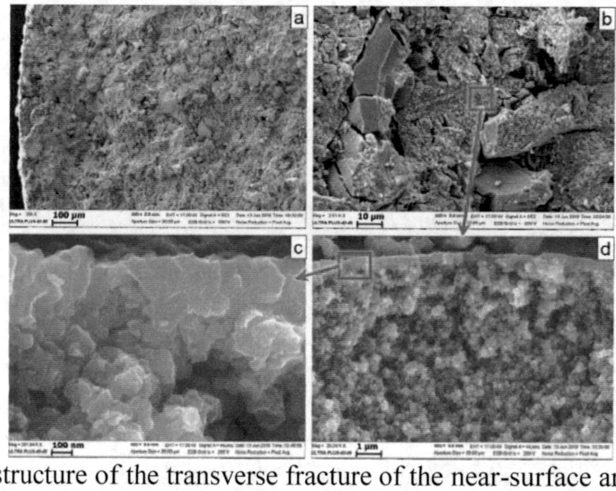


**Figure 3.** The microstructure of the transverse fracture of the near-surface area of the sample after self-heating and cooling recorded with a different magnification.

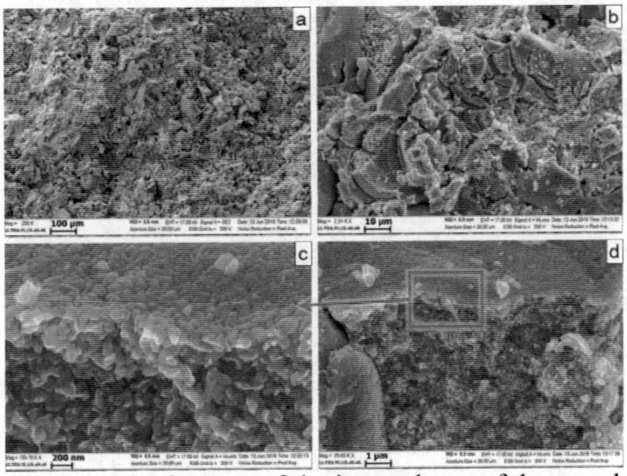

**Figure 4.** The microstructure of the fracture of the internal part of the sample recorded with a different magnification.

Apparently, the pores between the agglomerates serve as the main transport way, by which the air penetrates into the sample, oxidizing incidentally the surface of the agglomerates. The amount of oxygen, which penetrates into the internal part of the sample, is less than into its near-surface part, therefore, the oxidation and passivation of the powder occurs more slowly; these processes do not lead to the formation of dense oxide layers. At the same time, diffusion of oxygen into agglomerates and oxidation of iron nanoparticles occur.

The microstructure of the transverse fracture of the sample, which was ignited from the top end face with a tungsten spiral, is shown in figure 5. Self-heating of this sample after extraction from the weighing bottle was not observed. The near-surface area about 300÷400 microns thick with a higher average content of oxygen (to 45 atomic percent) in comparison with the central part of the sample (about 25 atomic percent of $O_2$) can bee seen in the structure of the transverse fracture of the sample.

Within the near-surface area, the agglomerates are covered with the oxide "crust" about 300÷400 nanometers thick (figure 5 d, c). The oxygen content in the "crust" makes about 60 atomic percent, in the agglomerate it makes 15÷30 atomic percent.

**Figure 5.** The microstructure of the transverse fracture of the near-surface area of the sample after combustion and cooling recorded with a different magnification.

EDA of the microstructure of the internal part of the sample showed that the oxygen content makes 25 atomic percent. The thickness of the sample of the oxidized layer ("crust") does not exceed the size of the single particle, of which the agglomerates consist, on the surface of the agglomerates of the internal part of the sample.

RFA investigation of the samples after self-heating and ignition showed that in both cases iron oxide $Fe_3O_4$ is observed in addition to the iron phase. In the sample in which self-heating was observed, the ratio of the oxide phase is much lower than in the sample after ignition; it correlates with the EDA data obtained using SEM.

**Acknowledgements**

This work was supported by the Russian Science Foundation (grant number 16-13-00013П).

**References**

[1] Bouillard J, Vignes A, Dufaud O, Perrin L and Thomas D 2010 *J. of Hazardous Materials* **181** 873
[2] Pivkina A, Ulyanova P, Frolov Y, Zavyalov S and Schoonman J 2004 *Propellants, Explosives, Pyrotechnics* **29** 39
[3] Alymov M I, Rubtsov N M, Seplyarskii B S, Zelensky V A and Ankudinov A B 2016 *Mendeleev Commun.* **26** 452
[4] Alymov M I, Rubtsov N M, Seplyarskii B S, Zelensky V A and Ankudinov A B 2017 *Inorganic Materials* **53** 911
[5] Alymov M I, Rubtsov N M, Seplyarskii B S, Zelensky V A and Ankudinov A B 2016 *Mendeleev Commun.* **26** 482
[6] Alymov M I, Rubtsov N M, Seplyarskii B S, Zelensky V A, Ankudinov A B, Kovalev I D, Kochetkov R A and Schukin A S 2017 *Dokl. Chem.* **477** 261
[7] Alymov M I, Rubtsov N M, Seplyarskii B S, Kochetkov R A, Zelensky V A and Ankudinov A B 2017 *Mendeleev Commun.* **27** 631
[8] Dong S, Cheng H, Yang H, Hou P and Zou G 2002 *J. Phys: Condensed Matter.* **14** 11023
[9] Hunt E M and Pantoya M L 2005 *J. Applied Physics* **98** 034909
[10] Saceleanu F, Idir M, Chaumeix N and Wen JZ 2018 *Front. Chem.* **6** 465
[11] Gromov A A and Teipel U 2014 *Metal Nanopowders: Production, Characterization, and Energetic Applications* (Weinheim: Wiley-VCH)

IOP Conf. Series: Materials Science and Engineering **672** (2019) 012006 doi:10.1088/1757-899X/672/1/012006

# Microstructure evolution in a CuZr alloy and CP Ti processed by a novel technique of free bending in rotating rollers

**M Janeček[1], T Krajňák[1], J Veselý[1], P Minárik[1], D Preisler[1], J Stráský[1], A G Raab[2], G I Raab[2] and R N Asfandiyarov[2]**

[1] Charles University, 5 Ovocný trh, Prague, 116 36 Czech Republic
[2] Ufa State Aviation Technical University, 12 K. Marx st., Ufa 450008 Russia

E-mail: milos.janecek@mff.cuni.cz

**Abstract.** A novel technique referred to as free bending in rotating rollers was employed to refine the initial structure and to produce gradient fine-grained microstructures in the Cu0.5Zr alloy and commercially pure titanium (CP Ti). A newly designed die allowing a continuous and repetitive movement of a workpiece through a set of rollers in a channel bent by 90 degrees was employed to produce a series of billets after different numbers of passes resulting in different strains imposed on the billet. Microstructure evolution with increasing number of passes was characterized by precise measurements of microhardness (HV) distribution along the cross section of individual billets. The heterogeneous distribution of HV observed after a single pass was transformed to almost a uniform one in the whole cross section of the billets after 8 passes. EBSD analysis and TEM observations revealed a significant microstructure refinement in both materials after the final stage of processing. A homogeneous microstructure with an average grain size of about 300 nm was observed in CP Ti, whereas the microstructure of the Cu0.5Zr alloy was not fully refined and exhibited a bimodal grain size distribution. The experimental results were compared with the theoretical modelling of strain distribution by FEM.

## 1. Introduction

In the last decades, the grain size strengthening was widely used for increasing the mechanical performance of metallic materials. New methods of material processing and preparation were developed in order to obtain the refined microstructures in the ultrafine or even nanometer range [1,2]. Severe plastic deformation (SPD) techniques, which imposed a large plastic deformation of hundreds of percentages was found to be successful in this regard [3,4]. SPD-processed materials with FCC and BCC crystal structures usually exhibit improved strength at the expense of deterioration of ductility [5,6]. On the other hand, SPD methods were successfully employed to increase both the strength and ductility of magnesium alloys with a HCP crystal structure due to the grain refinement and the texture formation, which enables the easier activation of non-basal slip systems [7,8].

Preparation of a gradient structure was found to be another way of increasing both strength and ductility. Materials with a gradient structure, e.g. the grain size in the nanoscale level on the surface and a coarse-grained structure in the center of the billet, were found to be attractive for many practical applications. The combination of resistant surface layers with a plastic core provides enhanced strength and wear resistance simultaneously with increasing ductility and bending fatigue resistance [9,10].

IOP Conf. Series: Materials Science and Engineering **672** (2019) 012006 doi:10.1088/1757-899X/672/1/012006

The main aim of the present work is two-fold: i) to prepare a gradient structure by free bending in rotating rollers [11] in the Cu0.5Zr alloy and CP Ti, and ii) to investigate the possibility to produce a homogeneous, refined microstructure in the whole cross section of the billet by applying a higher number of passes.

## 2. Experimental material and procedures

The Cu-0.5 wt% Zr alloy and commercially pure Ti (grade 4) were supplied in the condition produced by hot rolling and subsequent mechanical treatment. Billets of the cross section of 10x10 mm were pressed through a die schematically shown in figure 1. This die contains a set of rollers forming a channel in which the sample is bent by 90 degrees during a single pass. A series of billets after 1 and 8 passes using rotation of the billet by 90 degrees after each pass (equivalent to route Bc used in the ECAP technique) was prepared. Cu0.5Zr and CP Ti specimens were processed at a temperature of 300 °C. Microstructure characterization of the initial state and samples after 8 passes was performed by transmission electron microscopy and EBSD on specimens taken from the central part of the cross section. TEM microscopy was performed on a Jeol 2000FX transmission electron microscope operating at 200 kV. EBSD maps were collected by a SEM FEI Quanta 200 electron microscope equipped with an EDAX EBSD camera. Microstructure evolution with increasing number of passes was investigated in detail by microhardness measurements along the plane perpendicular to the processing direction – the billet cross section. Vickers microhardness mapping under an applied load of 500 g and a dwell time of 10 s was measured using an automatic microhardness tester Qness Q10a. A regular network of indents consisting of an orthogonal square net with the lateral distances of 300 μm (1 and 8 passes) and 500 μm or 1000 μm (initial states of CP Ti and Cu0.5Zr) where used. Prior to all measurements, the surface of the Cu0.5Zr alloy was mechanically polished successively using 1200, 2400 and 4000-grit SiC papers followed by polishing on diamond suspensions with a particle size of 3 μm and 1 μm. The final treatment consisted of polishing by 0.04 μm colloidal silica suspension. CP Ti samples were first ground using 1200, 2400, 4000-grit SiC papers and then polished on the colloidal alumina suspension with decreasing particle size down to 0.02 μm for over 16 hours on a Vibromet machine. Mechanically thinned TEM foils to a thickness about 120 μm were prepared by twin jet electropolishing in a solution of 6% $HClO_4$ and 33% butyl alcohol in methanol at a temperature of -20 °C.

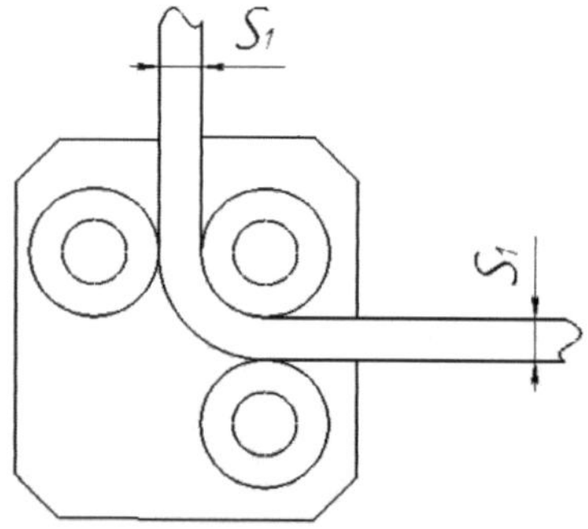

**Figure 1.** Principle of free bending in rotating rollers.

## 3. Experimental results and discussion

### *3.1 Microhardness*

Figure 2 shows the Vickers microhardness maps of the initial samples and those processed by free bending in rotating rollers. The variations of microhardness across the cross section of the billets are clearly displayed by the color code at these maps. The variation of the HV values after a single pass is similar for both materials. The microhardness increases preferably at the top part of the billet (deformed in compression). Further, a gradual decrease of HV from the edge to the central part is apparent. On the other hand, the bottom part of the billet (deformed in tension) exhibits a comparable increase of HV only in a very thin layer (thickness less than 1 mm) and outside this layer the HV rapidly decreases towards the central part. This dominant strengthening in the peripheral parts of the cross section is consistent with the results of numerical simulations performed for Cu-0.18 wt% Zr alloy, where the maximal values of strain were observed at the peripheral areas of the cross section [12]. After a total of 8 passes both materials exhibit almost a uniform distribution of HV in the whole cross section of the billet. The microhardness of CP Ti increases from 240 (initial state) to about 290 (8 passes). Such an increase in HV is comparable with the CP Ti processed via the ECAP technique [13]. On the other hand, a much higher increase in HV of about 200% was observed in the Cu0.5Zr alloy.

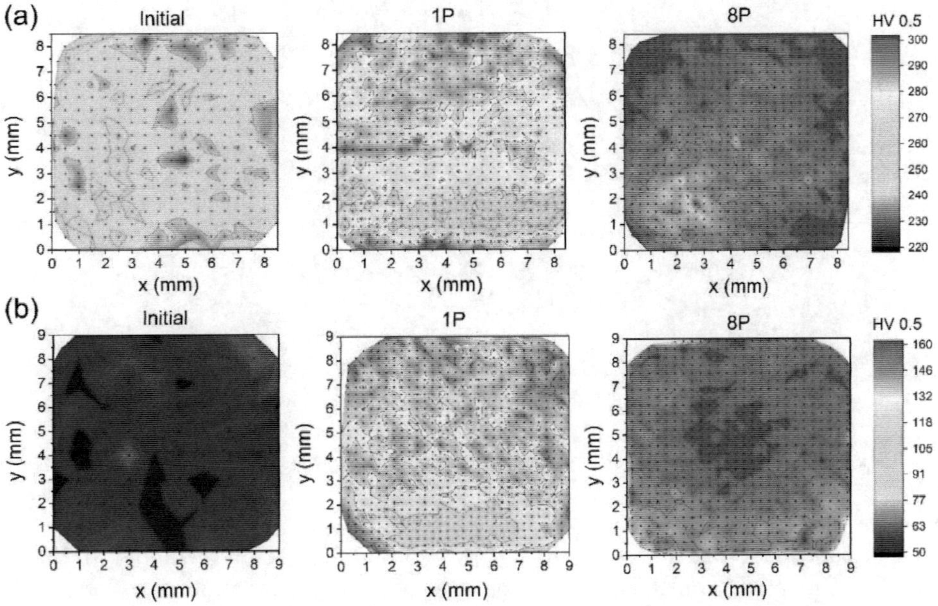


**Figure 2.** Microhardness maps measured on the cross section of (a) CP Ti and (b) Cu0.5Zr alloy. Black dots inside the maps represent the individual measured points.

*3.2 Microstructure*
Microstructure observation revealed a significant grain refinement in both materials processed by 8 passes via free bending in rotating rollers, cf. figure 3 and figure 4. As it is apparent from figure 3b, a fully refined, homogenous microstructure was observed in the central part of CP Ti. The microstructure consists of equiaxed grains separated by HAGBs with a size below 400 nm. A typical grain size obtained in CP Ti processed by ECAP is usually about 300 nm [14]. Moreover, numerous heavily deformed grains, containing a high dislocation density, are clearly visible in the micrograph.

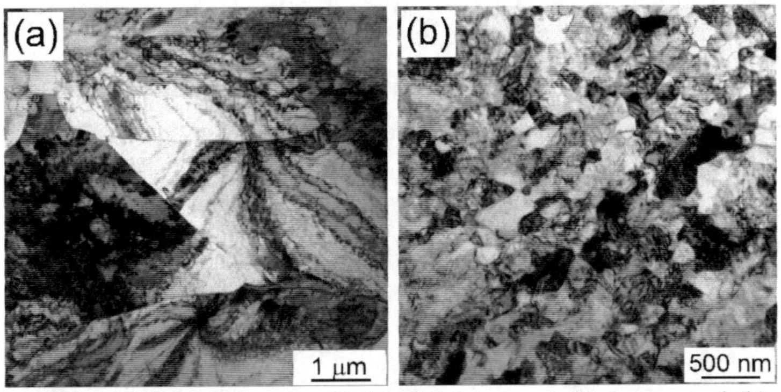


**Figure 3.** Bright-field TEM micrographs of CP Ti (a) in the initial state and (b) after 8 passes.

On the other hand, the microstructure of the Cu0.5Zr alloy after 8 passes exhibits a bimodal grain size distribution, cf. figure 4b. The microstructure contains a high fraction of low-angle grain boundaries (missorientation lower than 15 degrees) highlighted by the white color, indicating the dislocations arranged into dislocations cells or walls. Such a microstructural feature is common for FCC materials processed via methods of SPD [15]. The average grain size in the Cu0.5Zr alloy (taking into account grains separated by HAGBs) decreased from about 150 μm in the initial state to 4 μm after 8 passes of free bending in rotating rollers.

**Figure 4.** Inverse pole figure maps of the Cu0.5Zr alloy (a) in the initial state and (b) after 8 passes.

## 4. Conclusions

The present paper investigated the microhardness and microstructure evolution in the Cu-0.5 wt% Zr alloy and commercially pure Ti processed by free bending in rotating rollers. The following conclusions can be drawn from this experimental investigation:

- The gradual decrease of HV from the upper edge to the central part of the billet after a single pass of free bending in rotating rollers indicates the presence of a gradient type microstructure in both materials.
- The formation of a gradient fine-grained microstructure during one pass is consistent with the theoretical modelling of strain distribution by FEM.
- The application of 8 passes leads to an almost homogeneous distribution of microhardness values in the whole cross sections of the samples.
- A UFG microstructure with grains having an average size of 200-500 nm was achieved in CP Ti processed by 8 passes.

## Acknowledgements

Financial support by the Ministry of Education, Youth and Sports of the Czech Republic (project no. LTARF18010) is gratefully acknowledged. The Russian co-authors acknowledge funding from the Ministry of Science and Higher Education of the Russian Federation under grant agreement No. 14.586.21.0059 (UIN: RFMEFI58618X0059).

## References

[1] Zhilyaev A P and Langdon T G 2008 *Prog. Mater. Sci.* **53** 893
[2] Valiev R Z, Estrin Y, Horita Z, Langdon T G, Zehetbauer M J and Zhu Y T 2006 *JOM* **58(4)** 33
[3] Valiev R Z, Islamgaliev R K and Alexandrov I V 2000 *Prog. Mater. Sci.* **45** 103
[4] Čížek J, Janeček M, Krajňák T, Stráská J, Hruška P, Gubicza J and Kim HS 2016 *Acta Mater.* **105** 258
[5] Shih M H, Yu C Y, Kao P W and Chang C P 2001 *Scripta Mater.* **45** 793
[6] Máthis K, Krajňák T, Kužel R and Gubicza J 2011 *J. Alloys Comp.* **509** 3522
[7] Krajňák T, Máthis K, Janeček M and Gubicza J 2014 *IOP Conf. Ser.: Mater. Sci. Eng.* **63** 012058
[8] Krajňák T, Minárik P, Stráská J, Gubicza J, Máthis K and Janeček M 2019 *J. Mater. Sci.* **54** 3469
[9] Raab A G, Chukin M V, Aleshin G N and Raab GI 2014 *IOP Conf. Ser.: Mater. Sci. Eng.* **63**

012008
[10] Raab G I, Gunderov D V, Shafigullin L N, Podrezov Yu M, Danylenko M I, Tsenev N K, Bakhtizin R N, Aleshin G N and Raab A G 2015 *Mater. Phys. Mech.* **24** 242
[11] Utyashev F Z and Raab G I 2013 *Deformation Methods for Fabrication and Processing of Ultrafine-Grained and Nanostructured Materials* (Gilem, Ufa) [in Russian]
[12] Raab A G, Aksenov D A, Asfandiyarov R N, Kodirov I S and Raab G I 2018 *IOP Conf. Ser.: Mater. Sci. Eng.* **447** 012088
[13] Qarnia M J, Sivaswamy G, Rosochowski A, Boczkal S 2017 *Materials & Design* **122** 385
[14] Stolyarov V V, Zhu Y T, Alexandrov I V, Lowe T C and Valiev R Z 2003 *Mater. Sci. Eng.* A **343** 43
[15] Janeček M, Čížek J, Dopita M, Král R and Srba O 2008 *Mater. Sci. Forum* **584-586** 440

# The effect of heat treatment on the structure and mechanical properties of additively manufactured stainless steels

**G F Korznikova**[1], **V V Smirnov**[2], **D V Gunderov**[2, 3] **and R R Mulyukov**[1]

[1] Institute for Metals Superplasticity Problems, Russian Academy of Sciences, 39 Khalturin st., Ufa 450001, Russia
[2] Ufa State Aviation Technical University, 12 K. Marx str., Ufa 450008, Russia
[3] Institute of Molecule and Crystal Physics - Subdivision of the Ufa Federal Research Centre of the Russian Academy of Sciences, prospekt Oktyabrya 151, Ufa, 450075, Russia

E-mail: korznikova@anrb.ru

**Abstract**. Selective laser sintering was applied for stainless steel samples additive manufacturing. It was found that in the as fabricated state, the samples consist of martensite and residual austenite, contained significant dislocation density and residual stresses. Additively manufactured samples had an ultrafine-grained structure with predominantly high angle grain boundaries and were free of crystallographic texture and residual porosity. The ultimate tensile strength of the as-built steel was 960 MPa. Annealing caused a decrease in residual austenite, but did not significantly affect the size and morphology of the grains. The ultimate strength after annealing increased markedly while maintaining plasticity.

## 1. Introduction

Selective laser sintering (SLS) at the moment is one of the most promising methods of additive production. This is associated with its ability to facile making of complex three-dimensional geometries with quite moderate manufacturing effort without interjacent operations of thermomechanical treatment. It is well known that set of the final product properties including its mechanical characteristics and physicochemical properties is defined by the microstuctural features formed under SLS, which, in turn can be affected by varying such parameters as initial powder size and quality, fusion modes and some others conditions implied during SLS process [1,2].

A major drawback of the SLS process is the presence of impurities and porosity accompanied by the structure heterogeneity leading to degradation of the final product properties. Studies aimed to improve both the structural characteristics and mechanical properties of steel samples obtained by SLS is a relevant issue due to the fact that steel is the key material in different industrial applications [1-3].

The allotropy of Fe-C alloys is an important issue which can promote new insightful ways of governing the structure parameters during SLS process in the presence of significant temperature gradients in the fusion zone. For instance, the retained austenite present in martensite-aging steels after quenching at high cooling rates can significantly affect on the overall complex of properties of SLS produced EOS PH1 steel.

This paper presents the results of study of microstructural and mechanical characteristics of the SLS produced samples of EOS PH1 martensite aging steel in as produced and annealed states.

IOP Conf. Series: Materials Science and Engineering **672** (2019) 012007 doi:10.1088/1757-899X/672/1/012007

One should note that the investigation of SLS produced steels counterparts is of particular importance for both fundamental material science and for practical applications due to the fact that the obtained results contribute to the development of new approaches aimed on improving the properties of existing products and can provide new insights for design of new manufacturing concepts.

## 2. Experimental

SLS processing of the powder of EOS stainless steel PH1 was done by EOSINT commercial equipment system. The processed samples were in the form of cylinders with a 10 mm diameter of and a 30 mm height. The chemical composition of the powder corresponds to the DIN 1.4540 compositions (table 1).

**Table 1.** Chemical composition of EOS Stainless Steel PH1 (wt.%).

| Fe | Cr | Ni | Cu | Mn | Si | Mo | N |
|---|---|---|---|---|---|---|---|
| balance | 14-15.5 | 3.5-5.5 | 2.5-4.5 | ≤1 | ≤1 | ≤1 | 0.15-0.45 |

Microstructural analysis of the samples processed by SLS was performed using a scanning electron microscope (SEM) Tescan Mira 3LMH using the back-scattered electrons (BSE) and secondary electrons (SE) detectors. XRD phase analysis of the studied samples was conducted on a DRON-4-07 diffractometer using Cu-Kα radiation. Tescan VEGA 3 SBH SEM with an energy dispersive X-ray spectroscopy (EDS) attachment was used for the chemical composition analysis. Electron backscattered diffraction (EBSD) investigations were realized using the Tescan Mira 3 LMH SEM. Small grains comprising three or fewer pixels were automatically not considered in the analysis in order to improve the reliability of the EBSD data. 2° lower limit boundary misorientation cut off was used for exclusion of orientation noise induced spurious boundaries. Differentiation of low-angle and high-angle boundaries was based on A 15° criterion. The grain size was estimated by determining each grain area and its circle-equivalent diameter calculation, the method is also known as grain reconstruction [4].

Mechanical properties were investigated by means of tensile tests on the specimens with a cross-section of 1.9×0.4 $mm^2$ and a length of 4 mm. The tests were conducted at ambient temperature and a nominal strain rate of $10^{-3}$ $s^{-1}$ on the Instron 5982 testing machine.

The microhardness of the additively manufactured samples in the as-processed state, as well as after heat treatment, was measured on the cross section of cylindrical samples along the diameter with a step of 1 mm on a Vickers microhardness tester (MKV-h21) for 15 s at an applied load of 100 gf.

## 3. Results and discussion

XRD analysis revealed the initial powder was composed of 90% martensite phase (*bcc*) and 10% austenite phase (*fcc*) which is probably due to high rate of quenching during the preparation of powder. In the additively manufactured sample, XRD analysis (figure 1) revealed the absence of crystallographic texture, and phase composition, similar to initial structure consisting of 90% martensite and 10% austenite phase, increased lattice parameters values and high internal stresses which is the result of high cooling rates that are usually used during the SLS process.

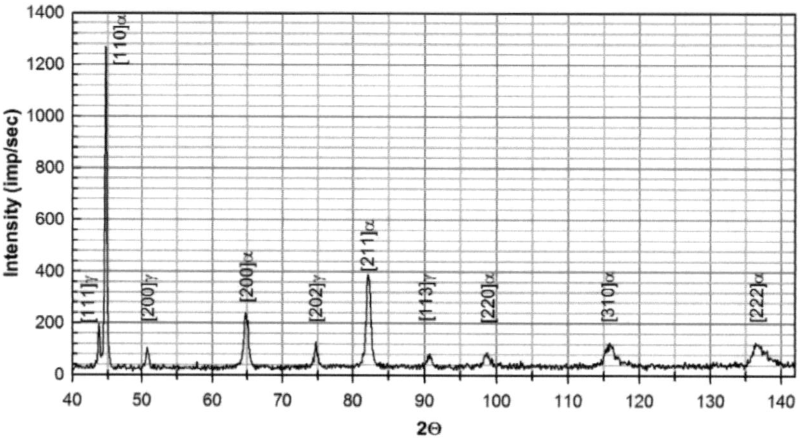


**Figure 1.** X-ray diffraction pattern of Stainless Steel PH1 sample in the as fabricated state.

Figure 2a presents the results of microstructural analysis of the PH1 Stainless Steel sample processed by SLS in the longitudinal sections. The bulk monolithic structure without any visible microcracks and pores can be observed. Both grain size and the chemical composition of the samples revealed the high degree of homogeneity along the analyzed area. One could observe a large fraction of elongated grains, but determination of preferable elongation direction seems not to be possible in this case.

a

b

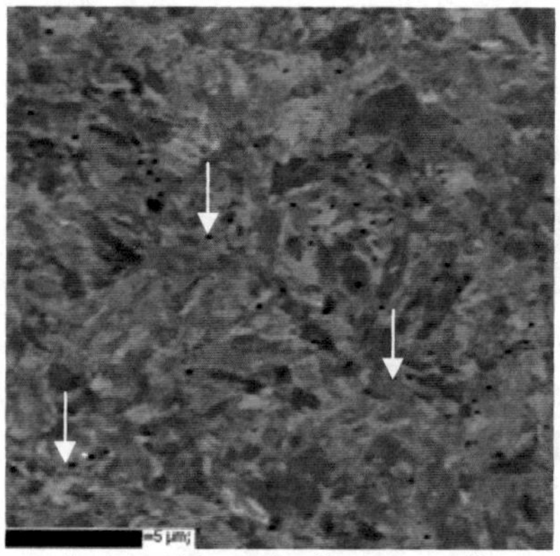

**Figure 2.** BSE image of SLS sample in the as fabricated state (a) and after annealing at 480 °C for 3 hours.

Since EOS Stainless Steel PH1 is characterized by having outstanding mechanical properties, specifically in the precipitation hardened state, some of the samples after SLS were heat treated at 480 °C for 1, 2, 3 and 4 hours. The white arrows in figure 2b point small dark intermetallic precipitates that appeared after hardening heat treatment.

The X-ray diffraction patterns exhibit 5% of the *fcc* austenite phase after annealing. Besides, the level of internal stresses and values of lattice parameters decreased significantly with increasing duration of annealing, while coherent scattering region increased.

EBSD mapping (figure 3) revealed mainly *bcc* and few *fcc* grains in the longitudinal and transverse sample sections. The average size of both martensite and austenite grains was estimated to be about 0.4 μm. Low and high angle boundaries (LAB) and (HAB) are shown by white and black lines respectively. The fraction of *bcc* phase HABs was about 65%, therewith 67% of them represent the boundaries between the variants that corresponds to the orientation relationship of Kurdyumov-Sachs.

This mechanism of martensitic transformation is a result of the phase instability due to high local stress fields present in the crystal lattice.

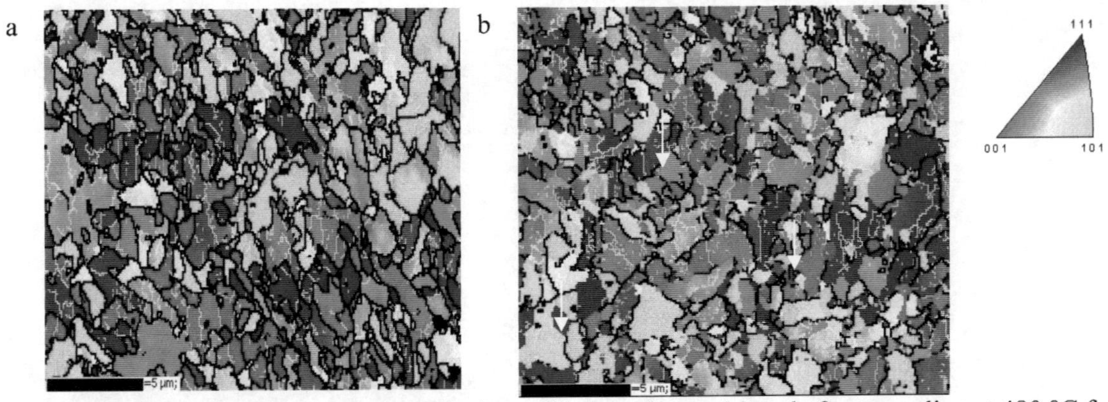


**Figure 3.** EBSD maps of SLS samples in the as fabricated state (a) and after annealing at 480 °C for 3 hours (b).

EBSD analysis allowed us to calculate the average grain size and the fraction of HABs and LABs in the samples under study (figure 4). It is necessary to note that the grain size was calculated taking into account LAB. Annealing caused some increase in grain size and the amount of HABs, while the percentage of LABs decreased (figure 4b).

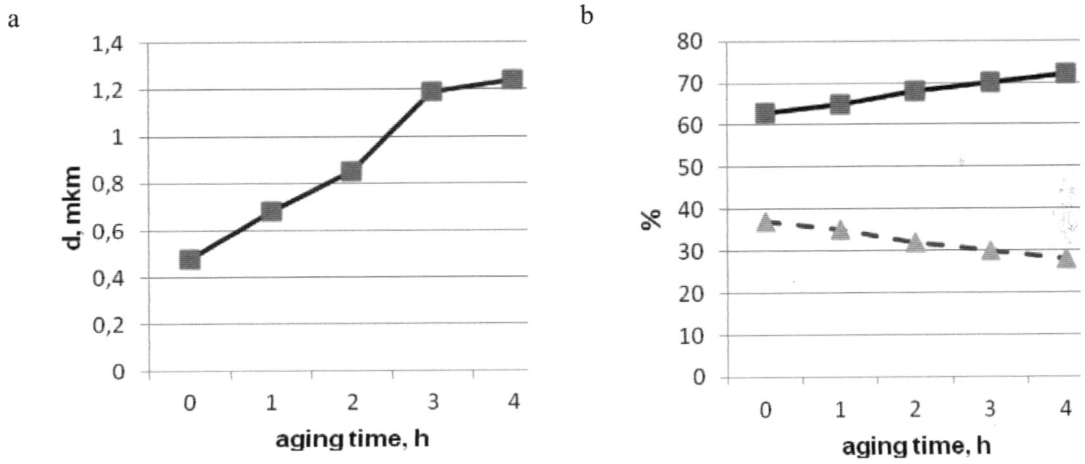


**Figure 4.** Dependence of the grain size *d* (a) and fractions % of HAB (——) and LAB (- - - -) (b) on the duration of annealing at 480 °C.

Martensitic aging steels are highly alloyed steels with a low carbon content. They are known for possessing superior strength and toughness without losing ductility. A low carbon content helps to prevent thermal cracking upon cooling [5]. The main strengthening of this type of steels occurs during aging heat treatments, causing precipitation of intermetallic compounds. According the data sheet of EOS Stainless Steel PH1, an aging temperature of 480 °C was chosen.

Figure 5 indicates the evolution of microhardness (Hv) as a result of aging. It is obvious that $H_V$ in the as processed sample is about 300 $H_V$ with some elevated values on the edges of the cross section that may be related to the cooling rate on the side surface of the sample during SLS.

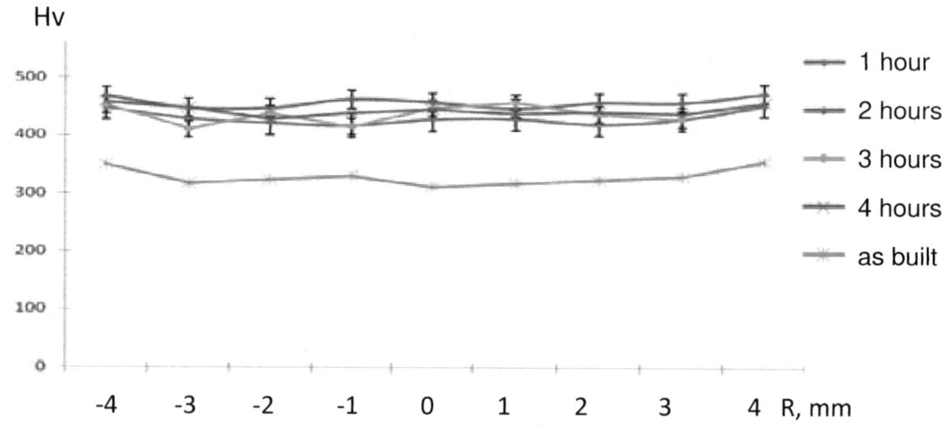


**Figure 5.** Distribution of microhardness along the diameter over the cross section in the as-received state and after precipitation annealing at 480 °C for 1, 2, 3 and 4 hours.

Annealing at 480 °C led to a substantial (almost 30%) increase in microhardness. At the same time, the duration of annealing practically did not affect the value of microhardness - all the microhardness distribution curves after annealing are grouped at the same level about 450 HV.

The results of tensile tests at room temperature are summarized in table 2.

**Table 2.** Tensile tests results of EOS Stainless Steel PH1.

| Structural state | YS, MPa | UTS, MPa | Elong., % |
|---|---|---|---|
| As built | 925 | 960 | 21 |
| Aging time 1 hour | 1195 | 1240 | 16 |
| Aging time 2 hours | 1160 | 1190 | 18 |
| Aging time 3 hours | 1190 | 1240 | 15 |
| Aging time 4 hours | 1080 | 1210 | 18 |

It is obvious that the yield strength ($\sigma_y$) as well as tensile strength ($\sigma_t$) increased substantially after aging, while the elongation value ($\delta$) decreased. The strain hardening induced by heat treatment can be attributed to the precipitation of intermetallic compounds in the form of small particles, well distinguishable in the BSE images (figure 2b). It can also be observed that EOS Stainless Steel PH1 show a rapid aging response. During the first hour the strength and the hardness increase to maximum values.

The obtained values of tensile strength for SLS samples are close to those for samples of the same composition produced by rolling and subjected to the same thermal treatments [6].

## 4. Summary

The reported results provide a strong evidence that the additive manufacturing treatment of PH1 Stainless Steel by means of SLS-processing results in formation of an ultrafine grained material with predominantly high angle grain boundaries and free of residual porosity and crystallographic texture. High temperature gradients present during SLS process are responsible for formation of non-equilibrium microstructures in the as-fabricated state, e.g. retained austenite in martensite, nonequilibrium values of lattice parameters and significant residual stresses. Subsequent aging at 480°C raises rapidly strength and microhardness to maximum values during the first hour of heat treatment. The mechanical properties of SLS samples are not worse than those of the samples obtained by conventional rolling

**Acknowledgments**
The present work was performed according to the state assignment of IMSP RAS. TEM and SEM studies were performed using the facilities of the shared services centre Structural and Physical-Mechanical Studies of Materials at the IMSP RAS. Authors are grateful to Dr. S. Mironov for assistance in EBSD studies.

**References**

[1] Herzog D, Seyda V, Wycisk E and Emmelmann C 2016 *Acta Mater.* **117** 371
[2] Guo P, Zou B, Huang Ch and Gao H 2017 *J. Mater Process. Technol.* **240** 12
[3] Salzbrenner B C, Rodelas J M, Madison J D, Jared B H, Swiler L P, Shen Yu-L and Boyce B L 2017 *J. Mater Process. Technol.* **241** 1
[4] Humphreys F J 1999 *J. Microscopy* **195** 170
[5] Kempen K, Thijs L, Yasa E, Van Humbeeck J, and Yasa E 2011 *Phys. Procedia* **12** 255
[6] Zhou T, Faleskog J, Babu R P, Odqvist J, Yua H and Hedström P 2019 *Mater. Sci. Eng.* A **745** 420

IOP Conf. Series: Materials Science and Engineering **672** (2019) 012008 doi:10.1088/1757-899X/672/1/012008

# Effects of nano and submicrocrystalline states under a wide range of loads

**O B Naimark**

Institute of Continuous Media Mechanics UB RAS, 1 Acad. Korolev str., 64013 Perm, Russia

E-mail: naimark@icmm.ru

**Abstract**. The scale in nanomechanics ($10^{-10}$-$10^{-9}$ m) has a fundamental physical meaning and can be considered as a control parameter when considering qualitative changes in the properties of polycrystalline materials when the grain structure changes in a wide range of load intensities. The existence of this parameter can lead to effects beyond the scale, which are traditionally considered in mechanics, and are accompanied by the influence of quantum effects on the deformation properties and failure of materials. New ideas about submicrocrystalline and nano-scale structure of materials achieved by the use of high-resolution methods (synchrotron tomography, soft X-rays, nanoindentation) have led to qualitative changes in approaches to the study and description of the properties of materials, including biomaterials. This may provide new information on the processes of deformation and fracture at the micro- and nano-scale level and lead to new approaches in the design and evaluation of the reliability.

## 1. Intoduction

New ideas about nanoscale structure of materials achieved using high-resolution methods (e.g., synchrotron tomography, soft x-ray radiation, nanoindentation) have led to qualitative changes in approaches to study and describe the properties of materials, including biomaterials. Being an intensively developing area with promising applications, research has stimulated a number of transdisciplinary directions in modeling, such as mechano-thermochemistry, mechanobiology, which suggests the development of new principles of modeling. As applied to materials science, this can provide new information on the processes of deformation and fracture of materials at the micro- and nanoscale levels and lead to new approaches in the design and assessment of the reliability of construction in aviation, space, power energy techniques. Characteristic scale was introduced in [1] for crystalline solid

$$\xi = \left(\frac{h}{\rho (G/\rho)^{1/2}}\right)^{1/4} = \left(\frac{h}{C\rho}\right)^{1/4} \sim (10^{-9} - 10^{-10})m$$

which includes the Planck constant $h$ and the classical parameters of the medium: density $\rho$, elastic modulus $G$, and longitudinal elastic waves velocity $C$. The ratio of $\xi$ and the interatomic distance $r_0$ for various crystalline materials gives $\alpha = r_0/\xi \sim 2.4 - 5.0$. The scale $\xi$ can be considered as the length determined by both quantum and classical (mechanical) effects.

The failure problem demonstrates a number of statements for which understanding of the structure evolution at the nanostructural level is crucial, since it naturally relates effects on macroscopic, meso-, nano- and quantum scales. In this case, one of the key questions is what quantum effects can extend to other large-scale levels. An example is the correspondence of quantum dots in optoelectronics to their sizes, the critical length of which is close to 10 nm. The illustration of these links is the correlations between the laws of fractoluminescence and the stages of failure [3], which is an illustration of the critical behavior of systems at the quantum and microscopic levels with the corresponding scaling laws.

Descriptions of the relationship between atomic structures and mechanical behavior (plasticity, failure) were developed in the works of R. Peierls, F. Nabarro, V.V. Novozhilov, D. Dugdale, G.I. Barenblatt and made it possible to substantiate approaches that take into account the formation of cohesive zones and effects non-locality, as well as to explain the multiscale effects of plastic deformation and fracture under very high cycle fatigue and intense loads.

## 2. Critical submicrocrystalline scales in fracture and plastic deformation

Critical lengths correspond to qualitative changes in the behavior of materials, such as violation of the Hall-Petch law upon transition to the nanocrystalline state [2], stages of damage-failure transition in very high cycle fatigue [3], initiation of localized (adiabatic) shear bands under dynamic loading [4], the structure of cohesive zones, the formation of failure waves under shock-wave loading of brittle materials [5]. The formation of a nano- and submicrocrystalline structure under deformation has characteristic signs of a critical phenomenon and is associated in [6] with non-equilibrium (structural-scaling) transitions in ensembles of mesodefects, which result in multiscale fragmentation of the crystalline phase. The development of the fragmentation process and grain size distributions are determined by the intensity of the deformation process (plastic strain rate), which is localized on corresponding spatial scales. Bimodal grain size distribution at moderate strain rates, for example ECAP processing, is characteristic of the scenario of multiple plastic strain localization regions. Dynamic deformation, accompanied by the adiabatic shear banding, includes regions with crystalline fragments close to the nanostructured scale. The formation of these regions indicates the involvement of multiscale mechanisms of plastic relaxation. Features of plastic fragmentation as a critical phenomenon manifest themselves in the submicrocrystalline range of grain sizes in the form of anomalies in the mechanical energy absorption, that is typically of the first kind phase transitions [7].

The existence of a characteristic scale $\xi$ can be associated with a "damage quantum", that is a scale at which relaxation mechanisms are "exhausted". The signs of achieving a nanostructured scale in fragmentation processes correspond to a qualitative change in the distribution statistics of fragments and fractoluminescence signals (from bimodal to unimodal) during dynamic fracture of brittle materials. The statistical laws of grain fragmentation, corresponding to unimodal and bimodal distributions, also manifest themselves as morphological features of failure surfaces in Very High Cycle Fatigue (VHCF) [3], figure 1.The existence of a critical scale explains the self-similar laws of nucleation and growth of cracks under conditions of VHCF, are related to the ratio of characteristic scales (damage localization $L_c$, "cohesion zones" determining the growth kinetics of small and macroscopic cracks) $L_{pz}$, and scale $\xi$: $\lambda_c = L_c/\xi >> 1$ and $\lambda_{pz} = L_{pz}/\xi >> 1$, which suggests the existence of self-similar intermediate asymptotics, respectively for the stage crack initiation ("fish-eye") and crack propagation, figure 1.

IOP Conf. Series: Materials Science and Engineering **672** (2019) 012008 doi:10.1088/1757-899X/672/1/012008

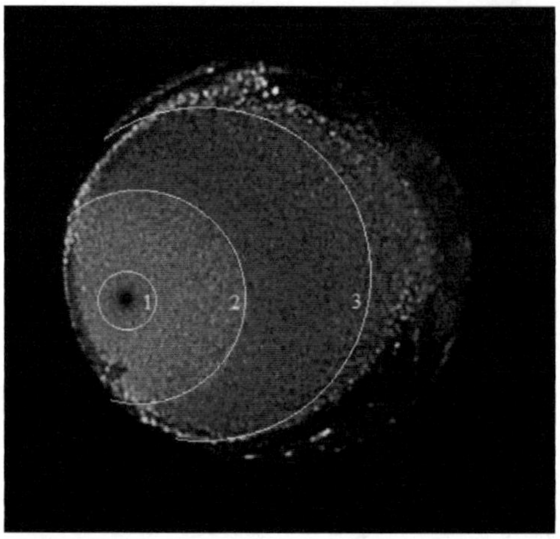


**Figure 1.** Fracture surface of the Ti6Al4V titanium alloy sample under VHCF ($10^9$ cycles) with characteristic fracture zones: 1 — "fish-eye" region of fracture initiation, 2 — propagation of "small cracks", 3 — crack propagation according to the Paris kinetics [5].

### 3. Self-similar fragmentation laws under dynamic and shock-wave loading

The statistical features of the fragmentation dynamics were studied in in-situ experiment on dynamic loading of fused quartz rod with recording fractoluminescence associated with fracture zone formation. By analyzing spatiotemporal distributions (fragment sizes, time intervals of fracture luminescence signals), we found the power-law distribution that is typical of non-equilibrium systems revealing the self-organized criticality (figure 2). The power-law universality of the fragmentation statistical laws can be interpreted as the "degeneration" of spatial and temporal scales up to the macroscopic sample length. The study of load intensity and material structure on multiple site fracture in ceramics was based on analyzing fragmentation statistics data. Rod samples were dynamically compressed (the split Hopkinson-Kolsky pressure bar test [8]), and tubular samples were loaded by shock pulse [9].

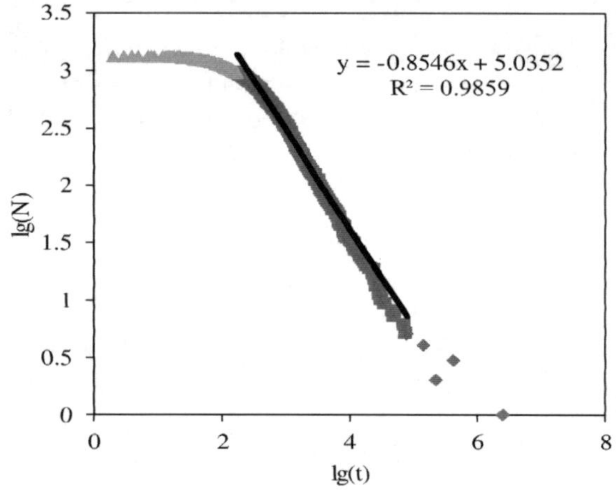


**Figure 2**. Cumulative distribution function of fractoluminescence time intervals (fused quartz sample).

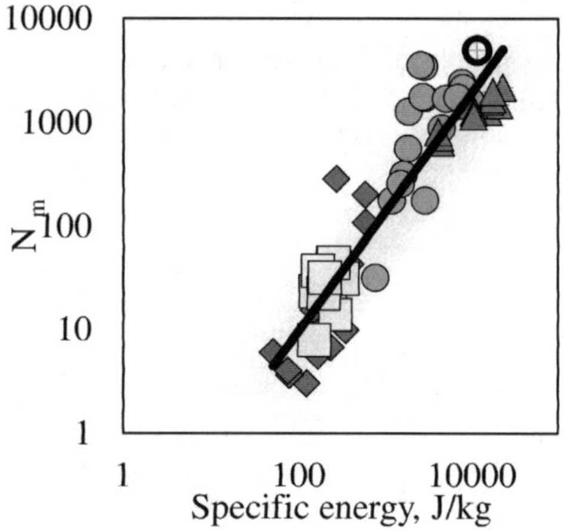


**Figure 3.** Specific quantity of fragments vs energy density: ♦ — quartz; — granite; ● — $ZrO_2$; ▲ — $Al_2O_3$;O— SiC [9].

Figure 3 shows the dependence of the specific quantity of fragments on the impact load intensity (strain energy density). The processing of experimental data for five materials indicates that an increase in the number of fragments per unit mass Nm with the loading rate increase is described by the power law.

## 4. Failure waves as the limiting case of fragmentation

The limiting case of fragmentation is the initiation of "failure waves" in shocked fused quartz that has the nature of "delayed failure" and was explained as the "resonance" failure mode [5]. The "failure wave" was experimentally studied in a Taylor test [6] with high-speed framing of shadow images in a fused quartz rod after its collision with a rigid target at the velocity 534 m/s (figure 4).

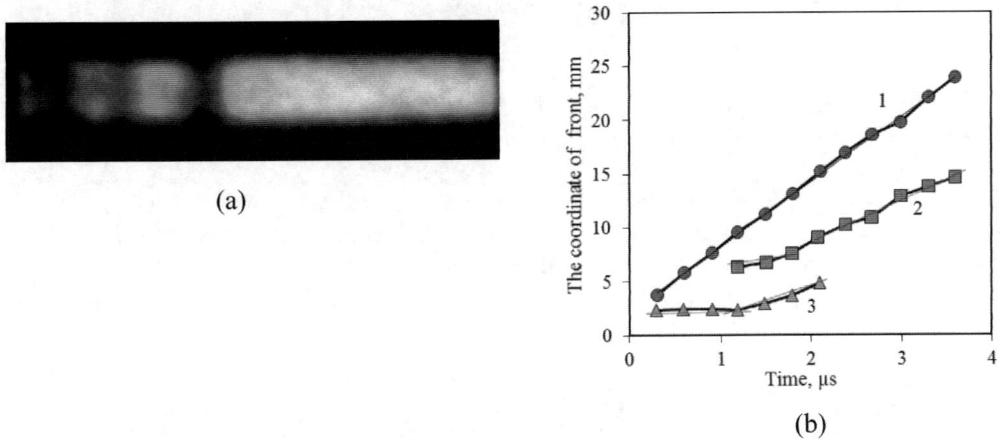


**Figure 4.** Failure wave in a fused quartz rod: (a) - high speed framing in the Taylor test; (b) *X-t* diagram *1* — compression wave, *2* — failure wave, *3* — collision surface.

The first shadow region corresponds to the compression wave travelling with the velocity $V \approx 6.6$ km/s. The second wave follows with constant lag with the velocity $V_{fw} \approx 4$ km/s, that is a wave front separating the solid and the fully comminuted material. It was shown in [10] that a failure wave propagates as "delayed" failure governed by "resonance" excitation of the collective blow-up modes of defects. Distributions close to unimodal should be observed during the formation of failure waves, as the limiting case of fragmentation with the formation of fragment sizes close to $\xi$ .

## 5. Structural investigation of localized plastic shear mechanisms and grain refining

The localized plastic shear effects at high strain rates and grain refining were studied for Al6061 and AlMg6 aluminum alloys, and steel during collision of a long rigid projectile with a plane target, in situ recording of temperature field dynamics (CEDIP Silver 450M infrared camera) and free surface velocity (the Doppler interferometry (VISAR)) on the back side of the target [4]. Multiscale structural relaxation mechanisms governed by the collective behavior of defect ensembles, plastic flow localization, and failure was investigated by structural analysis of the surface profile by a NewView-5010 interferometer-profilometer with subsequent calculation of the scale invariant (the Hurst exponent) and spatial scales of a region of correlated microshear behavior. The calculation of the scale invariants (structural scaling exponents) and corresponding scales allows us to find a relationship between shear localization stages with the signs of "criticality" due to correlated behavior of defects and the sizes of strain localization and adiabatic shear failure zones.

The recovered Al6061 targets after dynamic loading were examined by optical microscopy in the diameter cross-section after mechanical polishing and electrochemical etching (figure 5). Micro-etching for different resolutions revealed a sharp increase in the density of slip bands, which reaches a

IOP Conf. Series: Materials Science and Engineering **672** (2019) 012008 doi:10.1088/1757-899X/672/1/012008

critical density in the vicinity of the fracture surface. The sequence of coarse slip bands consisting of a system of 15–20 fine bands demonstrates spatial similarity, which is confirmed by the calculation of scale invariants. The fracture surface formation is accompanied by the formation of multiscale porosity and its clustering.

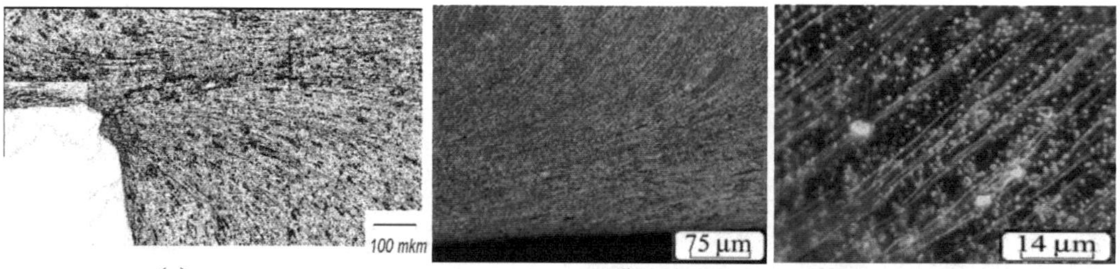


(a) (b) (c)

**Figure 5.** Aluminum alloy microstructure near the area of "plug" formation (optical microscopy, ×500) (a); subsurface layer microstructure: coarse (b) and fine slip bands (c).

Microstructural analysis by transmission electron microscopy has shown that the projectile penetration into the target causes the elongation of subgrains into bands and their fragmentation in a thin material layer (figure 6). One can clearly see dislocation pile-ups inside crystallites; the crystallite boundaries are smeared. The initial coarse-grained structure is refined into an ultrafine-grained one with the grain size ~ 300 nm from the initial 1.52 mcm.

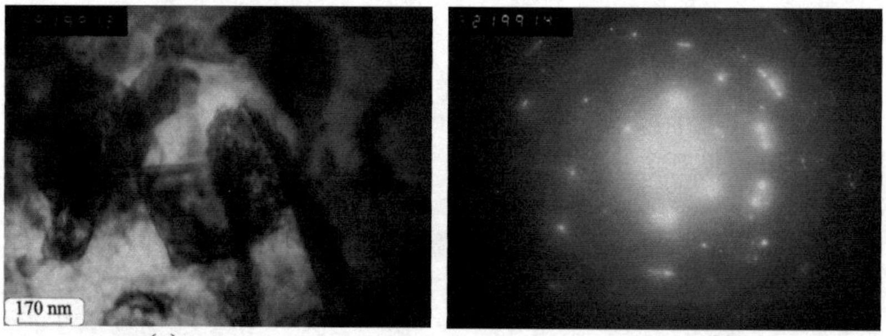


(a) (b)

**Figure 6.** Structure of the deformed layer near the impact surface: electron microscopic (light field) image (a), microdiffraction pattern (b).

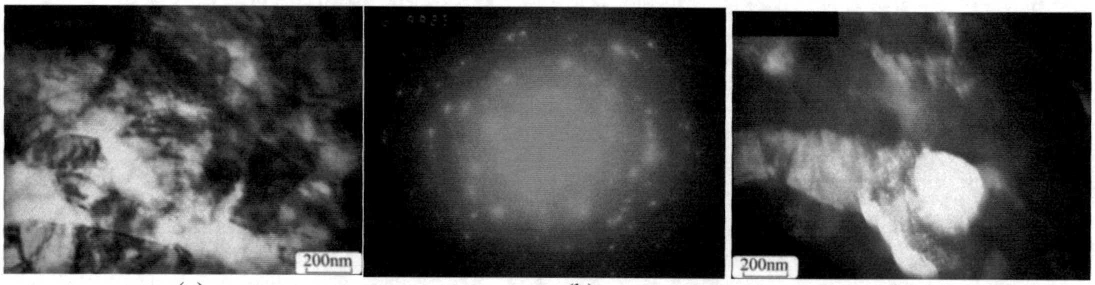


(a) (b) (c)

**Figure 7.** Electron microscopic image of the structure of highly deformed layer: light field (a), electron diffraction pattern (b), dark field (c).

The micro-diffraction patterns of such a structure have a great number of reflection points (figure 7), bearing witness to the appearance of high-angle grain misorientations. It can thus be concluded that, along with low-angle boundaries, high-angle boundaries are formed within initial subgrains due to rotational modes that might be associated with the ordering of dislocation structures.

## 6. Results and discussion

The basic aspects of fragmentation refer to statistical and energetic regularities of non-equilibrium "critical systems". Important features of the "critical behavior" are manifested in statistical fragmentation laws depending on the state, initial structure of the material, and load intensity when "narrow" (exponential) and "wide" (power-law) statistical distributions as well as their combinations are observed. The developed concepts of multiscale defect evolution as a critical phenomenon, such as structural scaling transitions, revealed a relationship between gauge (scale-invariant) properties of a "solid with defects" as a nonequilibrium system, self-similar solutions characterizing multiscale collective defect modes. The appearance of the defect modes defines the structural relaxation mechanisms (localized plasticity) and damage localization mechanisms with the "blow-up" mode kinetics that causes crack nucleation. The statistical fragment distribution in the material regions traversed by "failure waves" must reflect the maximum fragmentation level achieved within the acoustic range of energy density and approach monodisperse fragmentation with minimum particle size.

## Acknowledgements

Research was supported by the Program of Presidium of RAS, project n. 18-11-1-7.

## References

[1] Barenblatt G I, Montairo P 2010 *Phys. Mesomechanics* **13** 41
[2] Naimark O B 1997 *Phys. Met. Metallogr.* **84** 327
[3] Betekhtin V I, Kadomtsev A G, Narykova M A, Bannikov M V, Abaimov S G, Akhatov I S, Palin-Luc T, Naimark O B 2017 *Phys. Mesomech.* **20** 78
[4] Froustey C, Naimark O B, Panteleev I A, Bilalov D A, Petrova A N, Lyapunova E A 2017 *Phys. Mesomechanics* **20** 31
[5] Naimark O B 2016 *Int. J. Fracture* **202** 271
[6] Naimark O B 2004 Defect Induced Transitions as Mechanisms of Plasticity and Failure in Multifield Continua *Advances in Multifield Theories of Continua with Substructure* ed. G Capriz and P Mariano (Boston: Birkhäuser) pp 75–114
[7] Naimark O B 2010 *Phys. Mesomech.* **13** 306
[8] Davydova M M, Uvarov S V and Naimark O B 2016 *Phys. Mesomech.* **19** 86
[9] Bannikova I A, Uvarov S V and Naimark O B 2014 *AIP Conf. Proc.* **1623** 59
[10] Naimark O, Uvarov S, Radford D, Proud W, Field J, Church P, Cullis I and Andrews T 2003 The failure front in silica glasses *Behavior of Dense Media under High Dynamic Pressures* vol 2, ed A Delpuech (Cambridge) pp 65–74

IOP Conf. Series: Materials Science and Engineering **672** (2019) 012009 doi:10.1088/1757-899X/672/1/012009

# Mechanical properties and thermal stability of an ultrafine-grained near β titanium alloy

**E V Naydenkin, I P Mishin, I V Ratochka and O N Lykova**

Institute of Strength Physics and Materials Science SB RAS, pr. Akademicheskii 2/4, Tomsk 634055, Russia

E-mail: nev@ispms.tsc.ru

**Abstract**. Comparative studies of the structure, mechanical properties and thermal stability of an ultrafine-grained near β titanium alloy of the transition class Ti-5Al-5Mo-5V-1Cr-1Fe obtained by the methods of multiple pressing and radial-shear rolling with subsequent aging have been carried out. It is shown that as a result of multiple pressing, the formation of a homogeneous micro-duplex ultrafine-grained structure with an average grain (phase) size of ~0.2 μm is observed. Radial shear rolling followed by aging leads to the formation of an inhomogeneous structure with precipitates of the α phase particles and a subgrain size of the β phase of about 0.5 μm inside which thin-acicular precipitates of the metastable α" phase are observed. Despite the smaller imposed plastic deformation and the larger size of grain-subgrain structure, alloy samples after radial-shear rolling and aging demonstrate higher strength properties at room temperature compared to the alloy obtained by multiple pressing (the ultimate strength of the alloy after these treatments is 1695 and 1600 MPa, respectively.) At the same time, the alloy obtained by the method of multiple pressing shows higher ductility and thermal stability of the mechanical properties due to the micro-duplex ultrafine-grained structure. The drop in strength as a result of annealing is due to the development of recrystallization in the case of the alloy after pressing and the dissolution of the strengthening α" phase in the case of the alloy after radial-shear rolling and aging.

## 1. Introduction

In the past two decades, bulk ultrafine-grained (submicro- and nanostructured) metallic materials obtained by methods of severe plastic deformation (SPD) [1-3] have been actively developed and investigated. By the SPD effect, a grain-subgrain structure with an element size of less than 1 micron is formed in metals and alloys. Obtaining this structure can significantly change the mechanical and physical properties of such materials [1-4]. Due to this, they can find wide application in various industrial conditions. At the same time, ultrafine-grained (UFG) metals and alloys, obtained by the SPD methods have a high density of deformation defects, a large length of highly non-equilibrium interfaces and increased diffusion coefficients [1,2,5,6]. As a result, such materials show, as a rule, low thermal stability that requires additional research.

To obtain an UFG structure in metallic materials, several methods of SPD treatment have been developed, such as: high-pressure torsion, equal-channel angular pressing, multiple (or *abc*) pressing and others [1–4]. These methods provide the accumulation in the material of high degrees of plastic strain (usually $e > 4$-6) for refining the structure. In recent years, along with the aforementioned, new methods of deformation influence appear, which allow processing to be carried out in conditions close to industrial ones. Such methods include radial-shear rolling, through which it is possible to carry out deformation processing of cylindrical rods [7,8]. As was shown in [7,8], the combination of radial-shear rolling followed by aging allows the structure to be significantly reduced and the mechanical characteristics of the near β titanium alloy to be significantly improved. At the same time, it should be noted that comparative studies of the structure and mechanical characteristics of samples obtained by different SPD methods were not previously carried out using the example of near β titanium alloys. In this regard, in the present work, a comparative study of the structure, mechanical properties, and thermal stability of ultrafine-grained near β titanium alloy Ti-5Al-5Mo-5V-1Cr-1Fe obtained by the methods of multiple pressing and radial-shear rolling with subsequent aging is carried out.

## 2. Material and methods

A commercial near β titanium alloy (Russian alloy VT22) with a chemical composition: Ti-4.74 wt.% Al-5.57 wt.% Mo-5.04 wt.% V-0.81 wt.% Cr-0.98 wt.% Fe was used for the investigations. An ultrafine-grained structure in the alloy was obtained by multiple (*abc*) pressing with a change in the deformation axis [9] on an IP-2000 press in the temperature range 800-450 °C in billets with the dimensions of 25×25×40 $mm^3$. The strain for one pressing was about 0.5 and the total accumulated strain *e* was about 8. Radial-shear rolling was carried out on a rolling mill "14-40" in the temperature range 850-750 °C. The initial and final diameter of the bars before and after the rolling was 40 and 17 mm, respectively. Thus, the extraction ratio (strain) of the bars by radial-shear rolling *e* was 1.7. Aging of the alloy after radial-shear rolling was carried out in a SNOL-16251/11 air furnace at a temperature of 480 °C for 5 hours. Annealing of the samples of the alloy with an ultrafine-grained structure was carried out in the same furnace in the temperature range 500-800 °C for one hour.

Tensile tests of samples in the form of a double blade with a gage size of 5×1.7×0.8 $mm^3$ were carried out on a PV-3012M machine equipped with a strain gauge load measurement system s with an initial strain rate $6.9 \cdot 10^{-3}$ $s^{-1}$ at room temperature. Samples for tensile tests were cut by the electric-spark method. Before testing, a layer about 100 μm thick was removed from the surface of the samples by mechanical grinding and subsequent electrolytic polishing with an electrolyte with the composition: 20% HClO4+80% CH3CO2H. Transmission electron microscopy studies of thin foils were performed using a Jeol JEM-2100 microscope at the Shared Use Center "NANOTEH" of ISPMS SB RAS. Foils for electron microscopy were prepared by the standard method on the "Micron-103" jet polishing unit using an electrolyte of the following composition: 20% HClO4+80% CH3CO2H. The study of fractographic raptures after tensile tests was performed by scanning electron microscopy using a Quanta 200 3D microscope. The phase composition of the investigated titanium alloy after different SPD treatments was studied on a Shimadzu XRD-6000 diffractometer using $CuK_\alpha$ radiation, equipped with a monochromator.

## 3. Results and discussions

### *3.1. Microstructural investigations*

As shown on figure 1, in the result of severe plastic deformation by the method of multiple pressing a homogeneous ultrafine-grained structure with an average grain size of 0.2 μm was formed in the near *β* alloy. A complex deformation contrast inside the grains (figure 1a) indicates high internal stresses. Microdiffraction patterns obtained using a selector diaphragm with an area of ~ 1.6 $\mu m^2$ show a large number of individual reflexes located around the circumference, which indicates predominantly high angle misorientations of the neighboring grains (figure 1a).

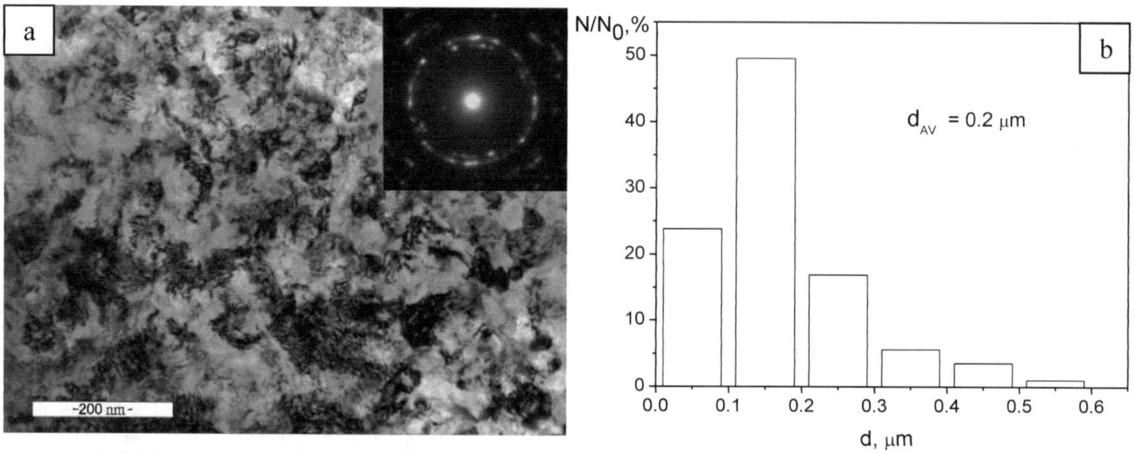


**Figure 1.** Microstructure of the near *β* titanium alloy after multiple pressing: a – bright field image with microdiffraction pattern; b – histogram of grain size distribution.

Radial-shear rolling under the above mentioned regime followed by aging at a temperature of 480 °C for 5 hours leads to the formation of a heterogeneous hierarchical microstructure in the Ti-5Al-5V-5Mo-1Cr-1Fe alloy (figure 2). In the volume of the initial large β-phase grains (~150 μm) separated as a result of deformation into sub-grains about 0.5 microns in size precipitates of α-phase particles with sizes of 0.2-0.7 μm formed during the process of radial-shear rolling at temperatures below the temperature of complete polymorphic transformation (transus) are observed [10]. Inside the β subgrains a fine-acicular structure is formed (figure 3b), which is a consequence of the formation of the metastable α" phase during the decomposition of the solid β solution at annealing (aging). Figure 2b shows an example of α"-phase precipitation in β subgrains.

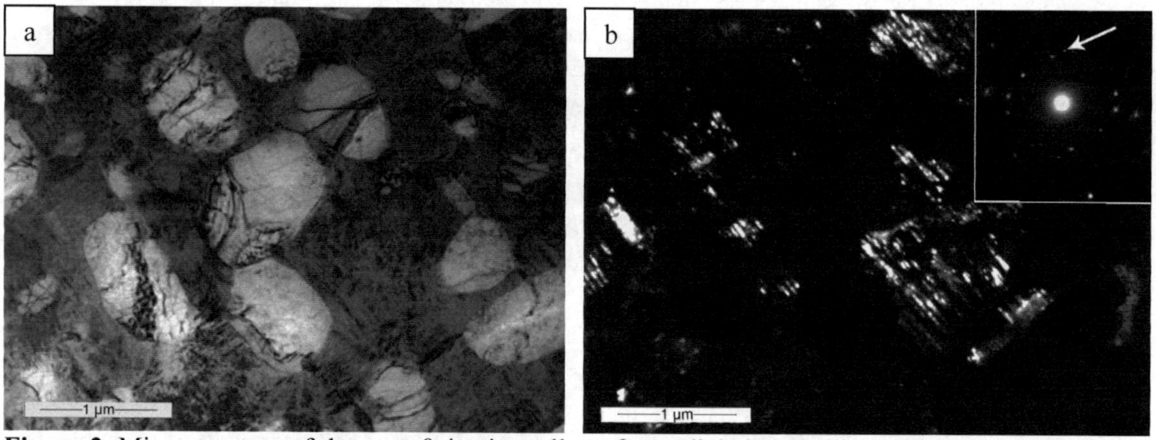


**Figure 2.** Microstructure of the near *β* titanium alloy after redial-shear rolling with subsequent aging: a – bright field image; b – dark field image in reflex $(111)_{\alpha''}$ with microdiffraction pattern.

An X-ray studying showed that after both treatments, the UFG alloy is a mixture of two phases *α* / *β*. The ratio of the phases in the alloy produced by multiple pressing and radial-shear rolling with subsequent aging is 62/38 and 69/31, respectively. Such a difference in the phase composition of the alloy samples obtained by the two SPD methods indicated above may be due to the difference in their mechano-thermal processing.

Thus, as a result of treatment of the Ti-5Al-5Mo-5V-1Cr-1Fe alloy using the SPD methods indicated above, a UFG structure is formed in it, which differs significantly in homogeneity and size of the structural elements in the considered states.

## *3.2. Mechanical behaviour and fracture*

Despite the larger size of the elements of the grain-subgrain structure detected by the TEM method, alloy samples after radial-shear rolling and subsequent aging demonstrate higher strength properties at room temperature compared with the alloy obtained by multiple pressing (figure 3). The tensile strength in the first case reaches 1695 MPa, while in the second - 1600 MPa.

These properties significantly exceed the corresponding properties obtained in the UFG Ti-6Al-4V alloy produced by the method of multiple pressing [11] and ECAP [12]. At the same time, the character of the tensile curves of the alloy after these treatments varies considerably. Therefore, the alloy after multiple pressing demonstrates the deformation behavior typical for UFG materials obtained by SPD methods with a short hardening stage (<1%) and a long (~9%) softening stage (figure 3) [2, 3]. In the case of samples of an alloy obtained by radial-shear rolling with subsequent aging, the tensile curve demonstrates a longer stage of work hardening (up to 2%) with a slightly lower coefficient and a shorter stage of softening (figure 3). At the same time, the ductility of the alloy after multiple pressing is more than twice higher than the corresponding value in the alloy after rolling and subsequent aging (figure 3).

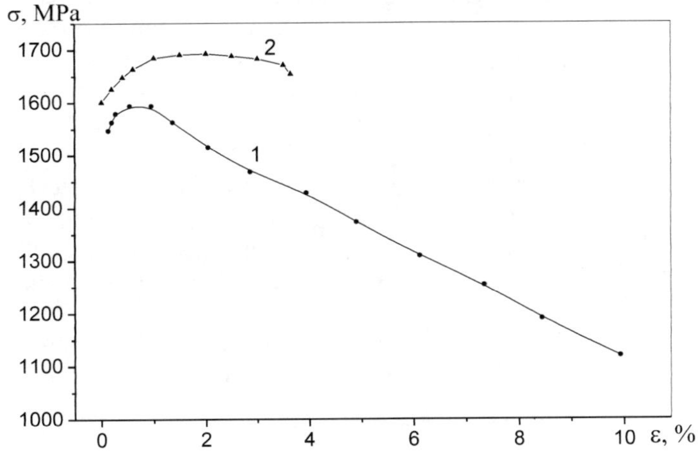


**Figure 3.** Tensile curves of UFG Ti alloy produced by *abc* pressing (1) and radial-shear rolling with subsequent aging (2).

The study of the fracture surfaces of alloy samples after tensile tests showed that in both cases there is a quasi-viscous rapture with a smaller dimple size in the case of the alloy obtained by the method of multiple pressing, which corresponds to the size ratio of the structural elements (figure 4). The character of the rapture patterns suggests that in the case of the alloy obtained by the method of multiple pressing there is a more viscous fracture than in the case of the alloy after rolling and aging (figure 4), which agrees well with the flow curves of these materials (figure 3).

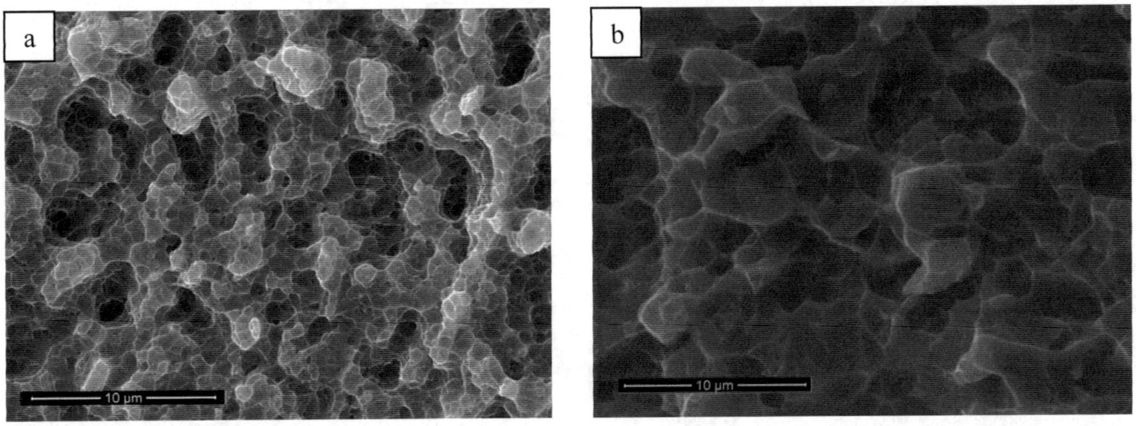


**Figure 4.** Fractographic raptures of the near $\beta$ titanium alloy produced by multiple pressing (a) and radial-shear rolling with subsequent aging (b) after tensile tests.

### *3.3. The effect of anneals*

Studying the effect of hourly annealing in the temperature range of 500-800 °C on the mechanical properties of the UFG alloy obtained by various SPD methods showed that the alloy obtained by the method of multiple pressing exhibits higher thermal stability compared to the alloy obtained by the method of radial-shear rolling with subsequent aging (figure 5). So, after annealing at a temperature of 500 °C, the ultimate strength of the alloy obtained by the method of multiple pressing is practically unchanged, whereas in the alloy after rolling and aging it decreases noticeably. With a further increase in the annealing temperature, a significant drop in the strength properties of the alloy in both structural states is observed (figure 5). At that, if in the titanium alloy obtained by radial-shear rolling with subsequent aging, an increase in ductility with decreasing strength is observed, then in the alloy after multiple pressing, the ductility after anneals changes in that or in the other direction (figure 5).

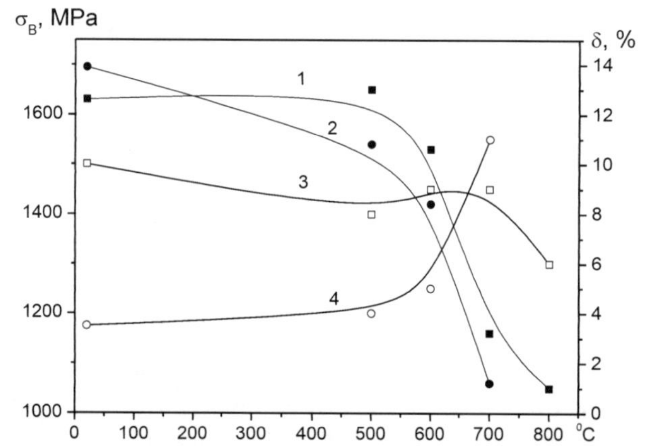


**Figure 5.** Dependence of ultimate strength (1,2) and ductility (3,4) on annealing temperature for pressed (1,3) and rolled (2,4) alloy.

Structural studies of the alloy showed (figure 6) that the reasons for the decrease in its strength properties as a result of annealing in different structural states are different. So in the case of the alloy obtained by the method of multiple pressing after annealing at a temperature of 700 °C, the development of the recrystallization process is observed (figure 6a). At the same time, a globular structure with an average grain size of about 0.5 μm is retained. Annealing at the same temperature of the alloy obtained by radial-shear rolling with subsequent aging does not lead to a noticeable growth of the elements of the grain-subgrain structure (figure 6b). At the same time, the thin-acicular precipitations of the metastable α"-phase, which observed in the alloy after aging, are not detected after the annealing, which apparently, leads to a significant drop in its strength properties under tension at room temperature.

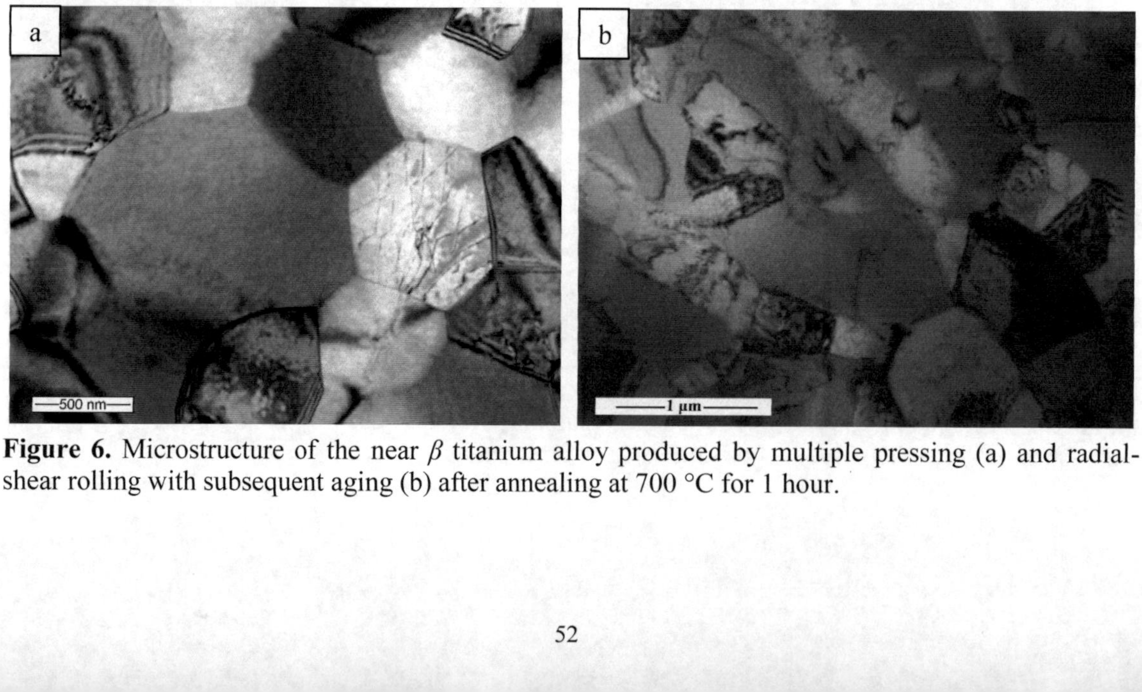


**Figure 6.** Microstructure of the near *β* titanium alloy produced by multiple pressing (a) and radial-shear rolling with subsequent aging (b) after annealing at 700 °C for 1 hour.

## 4. Conclusions

Comparative studies of the structure, mechanical properties, and thermal stability during annealing in the temperature range of 500-800 °C of the ultrafine-grained near β titanium alloy Ti-5Al-5Mo-5V-1Cr-1Fe obtained by the methods of multiple pressing and radial-shear rolling with subsequent aging were carried out. It was shown that as a result of SPD by the method of multiple pressing, the formation of a homogeneous micro-duplex UFG structure with an average grain (phase) size of ~0.2 μm was observed. Radial shear rolling followed by aging led to the formation of an inhomogeneous structure with precipitates of particles of the α phase and a size of subgrains of the β phase of about 0.5 μm, inside which thin-acicular precipitations of the metastable α" phase were observed. Despite the significantly smaller imposed strain and the larger size of the grain-subgrain structure elements, the alloy samples obtained by radial-shear rolling and subsequent aging demonstrated higher strength properties at room temperature compared to the alloy obtained by multiple pressing (the ultimate strength of the alloy after these treatments was 1695 and 1600 MPa, respectively). At the same time, the alloy obtained by multiple pressing showed a higher ductility and thermal stability of the mechanical characteristics due to the micro-duplex UFG structure. At the same time, the drop in strength as a result of annealing in the alloy obtained by pressing was due to the development of recrystallization with conservation of the globular structure, whereas in the alloy after rolling and aging it was associated with the dissolution of strengthening α"-phase precipitates.

**Acknowledgements**

The research is carried out with the support of the Russian Science Foundation (Grant No. 19-19-00033).

**References**

[1] Meyers M A, Mishra A and Benson D J 2006 *Prog. Mater. Sci.* **51** 427
[2] Kolobov Yu R, Valiev R Z, Grabovetskaya G P, Zhilyaev A P, Dudarev E F, Ivanov K V, Ivanov M B, Kashin O A and Naydenkin E V 2007 *Grain Boundary Diffusion and Properties of Nanostructured Materials* (Cambridge: Cambridge Int. Sci. Publ.)
[3] Ovid'ko I A, Valiev R Z and Zhu Y T 2018 *Prog. Mater. Sci.* **94** 462
[4] Sabirov I, Enikeev N A, Murashkin M Yu and Valiev R Z 2015 *Bulk Nanostructured Materials with Multifunctional Properties* (Cham: Springer Intern. Publ)
[5] Divinski S V, Reglitz G, Rösner H, Wilde G, Estrin Yu 2011 *Acta Mater.* **59** 1974
[6] Sauvage X, Wilde G, Divinski S V, Horita Z and Valiev R Z 2012 *Mater. Sci. Eng.* A **540** 1
[7] Naydenkin E V, Ratochka I V, Mishin I P, Lykova O N and Varlamova N V 2017 *J. Mater. Sci.* **52** 4164
[8] Naydenkin E V, Ratochka I V, Mishin I P and Lykova O N 2015 *Rus. Phys. J.* **58** 1068
[9] RF patent No. 2388566, 22.07.2008 [in Russian]
[10] Materials Science and International team, MSIT 2008 *Ternary Alloy Systems. Phase Diagrams, Crystallographic and Thermodynamic Data*, ed G Effenberg and S Ilyenko (Berlin-Heidelberg: Springer)
[11] Zherebtsov S, Kudryavtsev E, Kostjuchenko S, Malysheva S and Salishchev G 2012 *Mater. Sci. Eng.* A **536** 190
[12] Saitova L R, Höppel H W, Göken M, Semenova I P and Valiev R Z 2009 *Int. J. Fatigue* **31** 322

IOP Conf. Series: Materials Science and Engineering **672** (2019) 012010 doi:10.1088/1757-899X/672/1/012010

# Effect of frequency on plasma electrolytic oxidation of zirconium in pulsed unipolar mode

**E V Parfenov[1], V R Mukaeva[1], R G Farrakhov[1], A Saikiran[2], S Hariprasad[2], P Manoj[2], E Lokesh[2] and N Rameshbabu[2]**

[1] Department of Theoretical Basis of Electrical Engineering, Ufa State Aviation Technical University, 12 Karl Marx Street, Ufa, 450008, Russia
[2] Department of Metallurgical and Materials Engineering, National Institute of Technology, Tiruchirappalli 620015, Tamil Nadu, India

E-mail: evparfenov@mail.ru

**Abstract**. Using the method of *in situ* impedance spectroscopy, an optimization of plasma electrolytic oxidation (PEO) was performed, and the influence of the pulse frequency was evaluated for the coating of Zr-1Nb alloy. It was shown that the *in situ* impedance modulus decreased with increasing the pulse frequency in a pulsed unipolar mode. Also, the protective properties of the coating improved with frequency; this was attributed to the increase in the uniform distribution of the microdischarges over the sample surface. As a result, the optimal frequency range of PEO was justified based on the impedance spectroscopy studies.

## 1. Introduction

Zirconium alloys are widely used in the nuclear power industry in the active part of the reactor due to their neutron transparency and good corrosion properties. Also, Zr alloys gain attention as an alternative to Ti alloys for medical implant applications. International research teams focus mainly on two types of zirconium alloys: (i) alloyed with niobium (e.g. Zr-1Nb, the Russian alloy E-110, produced in accordance with the Russian Standard - GOST), (ii) alloyed with tin (e.g. Zr-1.5Sn, Zircaloy-2) [1-3]. These zirconium alloys used for fuel cladding, in superheated water (also containing LiOH) and water vapor, are subject to waterside corrosion, accompanied by oxygen and hydrogen release, which penetrates into the alloy and causes embrittlement, which can lead to the fuel cladding failure [4, 5]. Fretting corrosion is also a serious problem due to excessive vibration in a nuclear reactor [6]. Therefore, severe operating conditions in the reactor require appropriate surface modification by forming protective coatings to improve corrosion and wear resistance of the devices. Plasma electrolytic oxidation (PEO) of zirconium alloys is a promising method for increasing their performance, due to the properties of the formed protective coatings similar to anodizing, and the coatings obtained on the fuel cladding require careful process monitoring and control [7–9].

Recently, the method of impedance spectroscopy was adapted for *in situ* studies of PEO processes using frequency response estimates of the electrolyzer impedance [10, 11]. The methodology of the measurements was developed earlier [12]. After evaluation of the system linearity [13], this approach allows the use of electrochemical impedance spectroscopy (EIS) [14], including the equivalent circuit fitting. Moreover, the frequency response analysis helps to optimize the shape and frequency of the electric pulse by running a small number of experiments. Therefore, this paper is devoted to monitoring the plasma electrolytic oxidation process and optimizing the frequency of the PEO of the Zr-1Nb alloy in the pulsed unipolar mode.

## 2. Experimental

### *2.1. Plasma electrolytic oxidation*

The samples were cut from a sheet of the Zr-1Nb alloy in the form of plates 15×11×0.8 mm in size. PEO was carried out in a pulsed unipolar (PUP) mode under voltage control. The sample was used as the anode, the electrolyzer was connected as the cathode. The processing was carried out in the frequency scanning mode in the range from 28 to 10457 Hz, with 14 experimental points equally separated on a log scale; the exposure at every frequency was 2 s. The PEO process was also carried out at individual frequencies of 62, 713 and 3137 Hz. Other PEO parameters are given in table 1.

**Table 1.** Parameters of plasma electrolytic oxidation of the Zr-1Nb alloy in PUP mode.

| Sample No. | Frequency (Hz) | Electrolyte (g/L) | Temperature (°C) | Voltage pulse amplitude (V) | Duty cycle (%) | Electrolyte volume (L) | Time (min) |
|---|---|---|---|---|---|---|---|
| 1 | 62 | 1 g/L KOH+2 g/L $Na_2SiO_3$+2 g/L $Na_4P_2O_7$ | 20 | 570 | 50 | 5 | 7 |
| 2 | 713 | | | | | | |
| 3 | 3137 | | | | | | |

### *2.2. Surface characterization*

The surface morphology was investigated using a Gemini SEM300 scanning electron microscope (SEM) (Carl Zeiss, Germany). The phase composition of the coating was investigated using an X-ray diffractometer (XRD) Ultima IV, (Rigaku, Japan) equipped with a Cu-Kα X-ray source; scanning was performed from 10 to 80 degrees 2θ, with 0.02 degrees 2θ step, and measurement time at each step of 1 s. The phase composition of the surface was studied for the uncoated alloy, and for all the coatings obtained. Phase identification and semiquantitative analysis was performed using Philips X'Pert HighScore Plus software with the PDF2 databank.

Electrochemical studies were carried out using a P-5X potentiostat (Elins, Russia) in a 0.1 M LiOH solution using an 80-ml three-electrode cell with a silver chloride reference electrode and a platinum counter-electrode. The electrode potential was measured for 2 h until a steady state was achieved. To obtain polarization curves, after establishing the stationary state, a potential sweep was performed in the range from –600 mV to +600 mV with respect to the open circuit potential with a scan rate of 0.25 mV/s. The potential and corrosion current were calculated by the Tafel method from the polarization curves.

## 3. Results and discussion

### *3.1. In situ impedance spectroscopy*

As a result of the in-situ impedance spectroscopy, the characteristics of the frequency response of the PEO process were obtained. Figure 1 shows the impedance modulus during the treatment. Estimates of the frequency response corresponding to different PEO time moments (1, 2, 4, 6 min) are shown in figure 1 by solid lines. The impedance for all the frequencies tends to increase over the PEO time due to the coating growth and the associated growth of coating resistance. For the PUP mode, there is a gradual decrease in the impedance modulus with increasing frequency. This can be attributed to the shunting effect of the coating capacitance connected in parallel with the coating resistance represented during the PEO process by the microdischarge action.

The markers in figure 1 show the impedance modulus during the PEO processing at single frequencies of 62, 713, 3137 Hz. As follows from this figure, the single frequency experiments show similar values of the impedance modulus; this signifies the efficiency of the frequency response method for the process optimization via one step experiment.

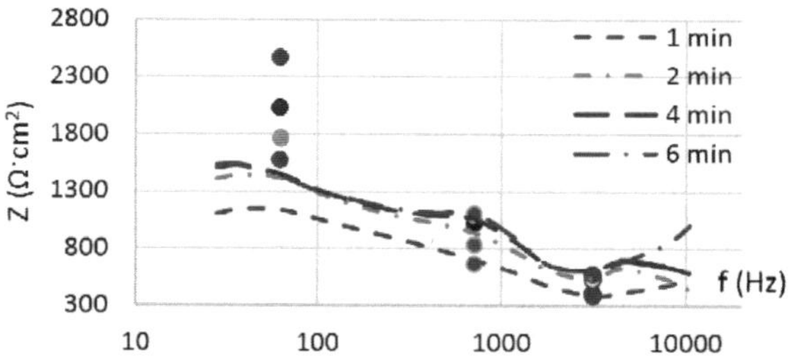


**Figure 1.** In situ frequency response impedance modulus for the PEO of Zr-1Nb in PUP mode.

However, some quantitative changes can be seen at the lowest frequencies: the single frequency impedance is higher than that for the frequency scan. As follows form further discussion, these treatment conditions provide the worst coating; other frequencies contribute to more uniform and protective coatings; therefore, the impact of higher frequencies during the PUP scan provided somewhat better coating than the single frequency treatment. At lower frequencies, the impedance modulus has maximal values, and as a result, it has the minimal current in the system, providing the lowest protective properties to the coating. For the minimal current, the distribution density of microdischarges decreases, which is confirmed by the video recording.

*3.2. Pulse frequency effect on the coating properties*

Figure 2 shows a top view of the PEO coatings obtained at different frequencies. The coatings have a porous structure, which is associated with the effect of microdischarges, and small cracks arising from the mechanical stresses as a result of rapid solidification of the oxide by the surrounding electrolyte during PEO. The coating, formed at a frequency of 3137 Hz, has the most uniform and smooth morphology. The largest coating thickness of 15 μm corresponds to a frequency of 62 Hz, the smallest one of 10 μm corresponds to 3137 Hz.

62 Hz 713 Hz 3137 Hz

**Figure 2.** Morphology (top view) of the coatings obtained at different frequencies of PEO.

As a result of the electrochemical tests, the corrosion properties of the samples were evaluated - free corrosion potential and corrosion current; the results are presented in table 2.

With an frequency increase, the PEO coating more and more passivates the surface, the free corrosion potential increases, and the corrosion current decreases. The highest corrosion current was found for the sample processed at the lowest PUP frequency of 62 Hz; however, it is still lower than that for the uncoated sample. The sample obtained at the frequency of 3137 Hz has the best corrosion resistance.

X-ray diffractometry analysis (table 2) shows that the substrate is characterized by α-Zr lines, and the coating contains crystalline phases of monoclinic and tetragonal modifications of $ZrO_2$. No other crystalline phases were detected. As follows from table 2, the predominant phase in the coating structure is m-$ZrO_2$. The percentage of high temperature modification of the t-$ZrO_2$ phase does not exceed 6%, and it reaches the maximal value at a PEO frequency of 3137 Hz. Therefore, this electrical mode seems to be the most promising to obtain protective PEO coatings.

**Table 2.** Results of electrochemical corrosion tests and XRD analysis.

| Frequency (Hz) | Corrosion properties | | XRD analysis results | | |
|---|---|---|---|---|---|
| | $E_{corr}$ (V) | $i_{corr} \cdot 10^{-4}$ (A/m$^2$) | α-Zr | m-$ZrO_2$ | t-$ZrO_2$ |
| no coating | -0.1497 | 29.0 | 100 % | - | - |
| 62 | -0.1121 | 13.0 | - | 97 % | 3 % |
| 713 | -0.1105 | 4.5 | - | 95 % | 5 % |
| 3137 | -0.0809 | 4.0 | - | 94 % | 6 % |

## 4. Conclusion

This work demonstrates the efficiency of *in situ* impedance spectroscopy application in optimizing the frequency of the pulsed unipolar PEO processing of Zr-1Nb alloy. It has been established that at a frequency of 3137 Hz, the *in situ* impedance of the system is close to minimal value; this leads to the highest average current density due to the uniform distribution of the PEO microdischarges, over the sample surface. This uniform treatment provides coatings with the best protective properties in this study; this is supported by SEM photographs, corrosion tests and X-ray diffractometry analysis.

## Acknowledgments

The authors express their sincere gratitude to Russian Foundation of Basic Research (RFBR), Moscow (Grant No 16-53-48008) and Department of Science and Technology (DST), New Delhi (Grant No INT/RUS/RFBR/IDIR/P-5/2016) under the DST-RFBR inter-disciplinary scientific cooperation program to carry out this work.

## References

[1] Lin J *et al.* 2004 *Materials Science and Engineering* A **381** 104
[2] Cheng Y *et al.* 2012 *Surface and Coatings Technology* **206** 3230
[3] Jeong Y H, Lee K O and Kim H G 2002 *J. Nucl. Mater.* **302** p 9
[4] Motta A T *et al.* 2007 *J. Nucl. Mater.* **37** 61
[5] Theaker J R *et al.* 1994 *Zirconium in the Nuclear Industry (Proceedings of Tenth International Symposium)* (ASTM) pp 221–242
[6] Fisher N J, Weckwerth M K, Grandison D A E and Cotnam B M 2002 *Nuclear Engineering and Design* **213** 79
[7] Borisov A M, Vostrikov V G, Romanovsky E A, Tkachenko N V, Vinogradov A V, Krit B L, Savushkina S V and Polyansky M N 2013 *Journal of Surface Investigation: X-Ray, Synchrotron and Neutron Techniques* **4** 366
[8] Klapkiv M D, Povstyana N Yu and Nykyforchyn H M 2006 *Mater Sci.* **42** 277
[9] Zhang D *et al.* 2003 *Rare Metal Materials and Engineering* **32** 658
[10] Parfenov E V, Yerokhin A L and Matthews A 2007 *Surface and Coatings Technology* **201** 8661
[11] Parfenov E V, Yerokhin A and Matthews A 2009 *Surface and Coatings Technology* **203** 2896
[12] Parfenov E V and Yerokhin A 2011 Methodology of data acquisition and signal processing for frequency response evaluation during plasma electrolytic surface treatments *Process Control: Problems, Techniques and Applications* ed S P Werther (Nova Science Pub. Inc.) pp 63–96
[13] Yerokhin A, Parfenov E V, Liang C J, Mukaeva V R and Matthews A 2013 *Electrochemistry Communications* **27** 137
[14] Barsukov E and Macdonald J R 2005 *Impedance Spectroscopy: Theory, Experiment, and Applications* (Hoboken: John Willey & Sons, Inc.)

# Structure of materials ultrasonically consolidated from coarse-grained and nanostructured plates of commercially pure copper

**N Yu Parkhimovich [1], N R Yusupova [1,2] and A A Nazarov [1,2]**

[1] Institute for Metals Superplasticity Problems of the Russian Academy of Sciences, 39 Khalturin st., Ufa 450001, Russia

[2] Bashkir State University, 32 Z. Validi st., Ufa 450074, Russia

E-mail: pn@imsp.ru, aanazarov@imsp.ru

**Abstract**. Ultrasonic consolidation of commercially pure copper plates with coarse-grained and ultrafine grained structures was carried out. Copper plates in both states were joined by ultrasonic welding using the same route. Consolidated samples were studied in a cross section parallel to the oscillation direction. The compressive strain value and the joint quality in relation to the initial structure of the material were studied. It is shown that the ultrasonic consolidation is a promising method of additive manufacturing using bulk nanostructured materials.

## 1. Introduction

Ultrasonic welding (USW) is a technique of solid state joining based on applying ultrasonic vibrations to sheet materials under a static compressive load. It is successfully used in the technology for joining different materials including metals [1]. Generally, ultrasonic welding has been used for welding of samples such as thin wires and foils. However, due to the availability of high-power USW equipment, this technique is currently becoming promising for additive manufacturing technologies, especially for ultrasonic consolidation of large metal plates and foils into parts [1-3]. The quality of ultrasonically consolidated materials strongly depends on the conditions of USW. Thus, it is necessary to establish regularities of the formation of joints during USW, including the welding of layered structures.

USW can also be considered as a promising way of application of ultra-fine grained (UFG) materials in additive manufacturing technologies, since it is a relatively low temperature process and is expected to allow one to keep the average grain size of materials in the submicron scale as in the case of bulk ultrasonic treatment [4,5]. Improvement and optimization of the USW routes for ultra-fine grained materials may help to increase the area of their applications and increase the efficiency of manufacturing of complex shaped parts from these materials. In the present paper, the consolidation of UFG copper plates is studied and compared with the consolidation of coarse-grained copper plates.

## 2. Materials and methods

For the studies of the ultrasonic consolidation process, samples of coarse-grained and UFG copper were used. Coarse-grained copper was produced by annealing of the copper billet at a temperature of 600 °C in a time interval of 1 h. Then the billet was cut into 0.7 mm thick plates using a spark cutting machine. Samples of UFG copper were processed using severe plastic deformation by high pressure torsion (HPT) applying 5 revolutions at a speed of 5 rpm under a pressure of 5 GPa. After deformation, the samples had an UFG structure and a disc shape with a diameter of 12 mm and a thickness of 0.8 mm. Both slices cut from coarse-grained copper and samples of UFG copper were polished with sandpaper, cleaned with isopropyl alcohol and dried before the consolidation.

The USW process is schematically presented in figure 1. Vibrations are transmitted to the sample by a steel sonotrode, the welding tip of which had a rectangular shape with sizes of 6 × 7 mm and a serrated surface comprised of parallel "teeth". The resonance frequency of the vibrating system was equal to 20.2 kHz. The amplitude of vibrations was controlled by changing the biasing current supplied to the magnetostrictive transducer and measured by the capacity amplitude detector. Consolidated samples were obtained by successive ultrasonic welding of five coarse-grained and UFG plates using the following regimes: static load of P = 7 kN, welding time of t = 1 s for coarse-grained samples and three static loads of P = 3.5, 5 and 7 kN with the same welding time for UFG samples. When welding the UFG samples, the center of the welding tip of the sonotrode was located at a distance of about half the radius from the centers of the discs, where a more uniform and finer structure is formed during the process of HPT.

As a result of the above treatment, two consolidated samples with sizes equal to those of the welding tip were obtained. These samples were cut by spark cutting along the direction of ultrasonic vibrations. For the scanning electron microscopy (SEM) studies, the surface was polished with sandpaper, diamond paste and Struers OP-S colloidal silica suspension in order to reduce the size of abrasive grits. BSE SEM images were acquired on a Tescan Mira electron microscope at an accelerating voltage of 20 kV.

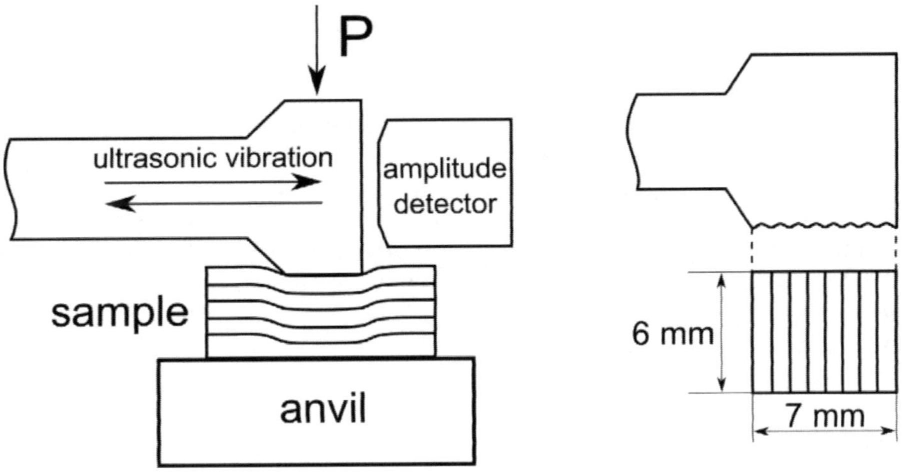


**Figure 1.** Schematics of ultrasonic consolidation: welding of a sample (left side) and the sonotrode tip (right side).

## 3. Results and discussion

### *3.1. Consolidation process*

Some text. During the ultrasonic consolidation of copper plates with the same parameters of the ultrasonic vibration system (frequency and biasing current supplied to the magnetostrictive transducer), the vibration amplitude of the sonotrode tip depended on the numbers of sequentially welded plates both in the cases of coarse-grained and ultrafine grained initial structures. Presented in figure 2 are the dependences of the vibration amplitude on the number of sequential welds. During the consolidation of all coarse-grained plates, the vibration amplitude was fairly constant during the period of welding, while in the case of consolidation of UFG samples, significant changes in the amplitude

IOP Conf. Series: Materials Science and Engineering **672** (2019) 012011 doi:10.1088/1757-899X/672/1/012011

during welding time were observed at all static forces P = 3.5, 5, and 7 kN. The range of these changes are indicated in figure 2 by error bars. This difference in welding behavior can be caused by the difference in the mechanical properties of UFG and coarse-grained copper under ultrasonic vibrations.

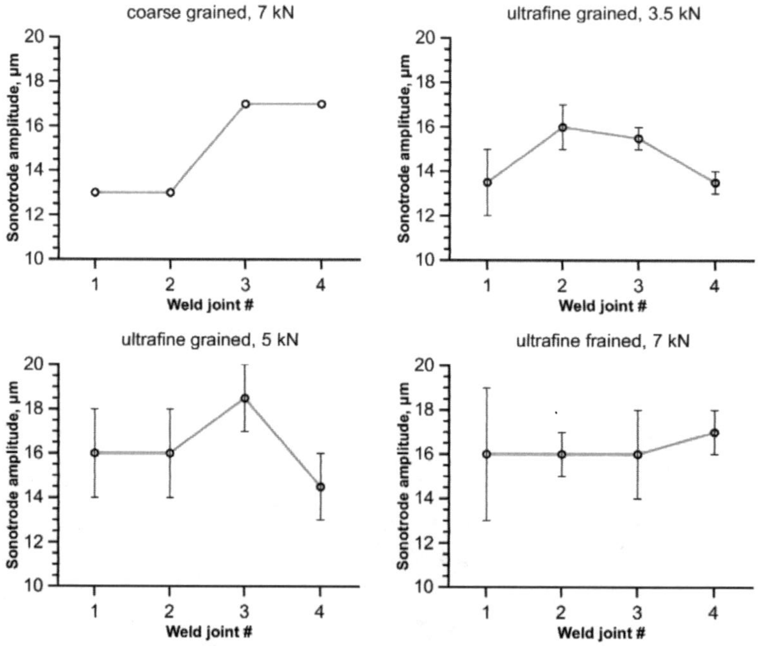


**Figure 2.** Dependence of vibration amplitude of sonotrode on the sequence number of welding joints during ultrasonic consolidation.

The USW led to high compression deformation of the consolidated plates. The compressive strain depends both on the initial structure of the plates and the static compression load. The coarse-grained sample after USW under the static load P = 7 kN underwent a compressive strain of about ε = 30%. The values of the compressive strain of UFG samples welded under P = 3.5, 5 and 7 kN were equal to about ε = 13, 15 and 80%, respectively. It should be pointed out that in the sample welded under P = 3.5 kN, the last plate was not joined properly. Therefore, this static pressure was found to be not sufficient for consolidation with the sonotrode used in the present work.

The higher ε value of the UFG plates is obviously caused by the increased plasticity of the UFG copper samples under ultrasonic vibrations in a comparison to the coarse-grained state [6] and the specific deformation behavior of UFG Cu, which exhibits no strain hardening during compression at room temperature [7].

### *3.2. Welding joint quality of consolidated samples*

Scanning electron microscopic images of the cross sections of the consolidated samples along the direction of ultrasonic vibrations for different conditions of USW and initial structures are presented in figure 3. In most of the samples, the welding joints are clearly identified visually as thin lines with cracks and pores along the samples. However, in the sample consolidated from UFG plates under a static load of P = 7 kN, only one welding joint was identified. Joints between other plates are not visible on the SEM image of the cross-section of the sample. In this sample, there are more areas of well joined material and less macroscopic defects than in the other samples. Thus, USW under the static pressure P = 7 kN resulted in the best consolidation of UFG copper plates.

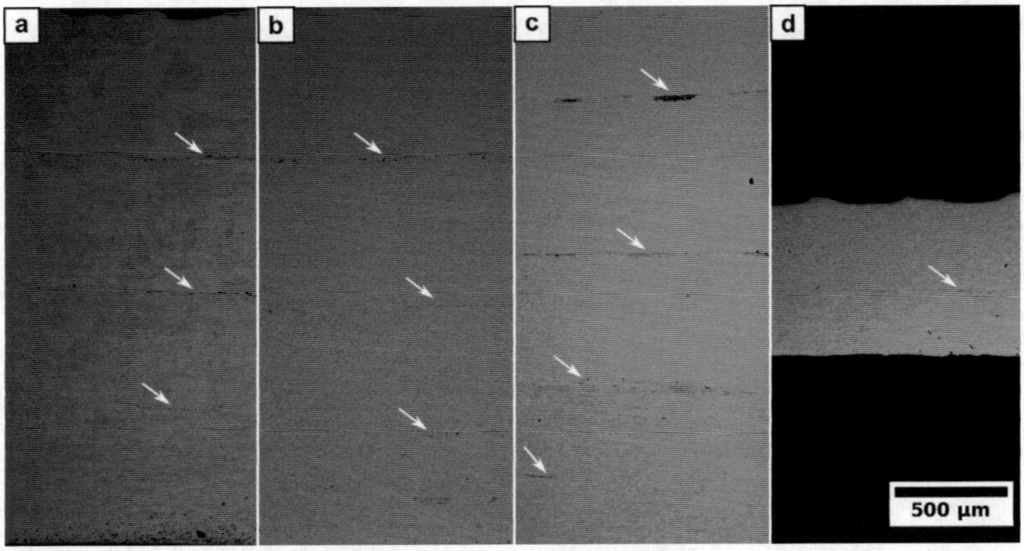


**Figure 3.** Macrostructure of consolidated samples along the direction of ultrasonic vibrations: (a) coarse-grained Cu, P =7 kN; (b) ultrafine grained Cu, P = 3.5 kN; (c) ultrafine grained Cu, P = 5 kN; (d) ultrafine-grained Cu, P = 7 kN. Visible welding joints are marked by arrows.

## 4. Conclusions

The present study shows that the ultrasonic consolidation of UFG copper can be carried out at a lower static pressure than the consolidation of coarse-grained materials. Welding of UFG samples under a static load of P = 7 kN led to the formation of a bulk sample with the minimum amount of visible welding defects. At the same time, the UFG samples undergo higher compressive strain during the consolidation with the same static pressure than the ones with the coarse-grained structure. These differences can be explained by the different deformation behaviors of copper with the two types of structures: the presence of work hardening of coarse-grained copper and its absence for the UFG copper [7]. In order to better understand the advantage of UFG materials in the process of ultrasonic consolidation, detailed studies of the microstructure evolution during welding are needed.

## Acknowledgments

The present work was supported by the Russian Science Foundation (grant # 16-19-10126). Experimental studies were carried out on the facilities of shared services center of IMSP RAS "Structural and Physical-Mechanical Studies of Materials".

## References

[1] 2015 *Power Ultrasonics. Applications of High-Intensity Ultrasound* ed J A Gallego-Juarez and K F Graff (Cambridge: Woodhead Publishing)
[2] White D 2000 *Ultrasonic object consolidation* US patent # 6,519,500
[3] Friel R J and Harris R A 2013 *Procedia CIRP* **6** 35
[4] Mukhametgalina A A, Samigullina A A, Sergeev S N, Zhilyaev A P, Nazarov A A, Zagidullina Y R, Parkhimovich N Y, Rubanik V V and Tsarenko Y V 2017 *Lett. Mater.* **7** 85
[5] Samigullina A A, Mukhametgalina A A, Sergeyev S N, Zhilyaev A P, Nazarov A A, Zagidullina Yu R, Parkhimovich N Yu, Rubanik V V and Tsarenko Yu V 2018 *Ultrasonics* **82** 313
[6] Fang T H, Li W L, Tao N R and Lu K 2011 *Science* **331** 1587
[7] Valiev R Z, Kozlov E V, Ivanov Yu F, Lian J, Nazarov A A and Baudelet B 1994 *Acta Metall. Mater.* **42** 2467

IOP Conf. Series: Materials Science and Engineering **672** (2019) 012012 doi:10.1088/1757-899X/672/1/012012

# Surface modification of CP-Ti metallic implant material by plasma electrolytic oxidation

**N Rameshbabu[1], B Ravisankar[1], A Saikiran[1], E V Parfenov[2] and R Z Valiev[3]**

[1] Department of Metallurgical and Materials Engineering, National Institute of Technology, Tiruchirappalli 620 015, Tamil Nadu, India
[2] Department of Theoretical Basis of Electrical Engineering, Ufa State Aviation Technical University (USATU), 12 Karl Marx Street, Ufa 450008, Russian Federation
[3] Institute of Physics of Advanced Materials, Ufa State Aviation Technical University (USATU), 12 Karl Marx Street, Ufa 450008, Russian Federation

E-mail: rameshrohith@gmail.com

**Abstract**. Plasma electrolytic oxidation (PEO) is a relatively new and environmentally friendly process for development of thick and adherent oxide coatings on conventional and nanostructured metallic implant materials. The PEO process involves anodic oxidation of metals or alloys in aqueous solutions at voltages higher than the breakdown voltage of the oxides present on its surfaces. Currently, the PEO process is gaining increased attention as a novel technique for the fabrication of corrosion resistant, bioactive and functionally modulated composite coatings on commercial pure titanium (Cp-Ti) and Ti based alloys for orthopedic and dental applications. The PEO process coupled with the electrophoretic deposition (EPD) technique is used for the fabrication of $TiO_2$/hydroxyapatite (HA) composite coating on a Cp-Ti implant material in a single step. This paper briefly reviews the properties of the coatings produced on Cp-Ti by PEO and PEO coupled EPD processes.

## 1. Introduction

Metallic materials are the most common synthetic materials used for biomedical implants nowadays. Owing to their high strength and toughness, they are mostly used for the load bearing implant applications [1]. In addition to the strength and toughness, the necessity of the implant material to be non-toxic and corrosion resistant to the body fluid environment, resulted in the usage of stainless steels, Ti and its alloys, Cobalt-chromium alloys [2, 3]. Among the metallic implant materials, Titanium is gaining a lot of attention, due to its low elastic modulus and better biocompatibility, compared to other metallic implant materials [4, 5]. The low elastic modulus (nearly equal to that of the natural bone, i.e., ~ 10 – 30 GPa) of CP-Ti and titanium alloys, results in the lesser stress shielding effects caused due to the mismatch of the elastic moduli of the bone and the implant material [6]. Titanium and titanium alloys are widely used as hard tissue replacements in artificial bones, joints and dental implants [7, 8]. Titanium and its alloys are often used in artificial hip joints, knee joint replacements [9].

Besides artificial bones and joints, titanium and titanium alloys are often used in osteosynthesis, such as maxillofacial implants, bone fracture fixation parts like bone screws and bone plates, etc. [10]. Because of their higher strength, inertness and non-magnetic properties, Ti and its alloys are widely used for cardiovascular implants like heart valves, pacemaker cases [11]. Titanium and its alloys are also used for dental applications such as implants, crowns, bridges and overdentures. Besides having these many advantages, implants made of Ti and its alloys experience corrosion in the body fluid environment, which can be overcome by the surface modification techniques. This article provides a brief review of the usage of CP-Ti as an implant material and the existing ways of improving its corrosion resistance, biocompatibility and antibacterial properties.

## 2. CP-Ti as an implant material

Owing to the higher biocompatibility with the living tissues along with the absence of toxic elements, CP-Ti is a promising material for the implant applications. In spite of the good biocompatibility, CP-Ti lags in the strength perspective compared to other implant materials like Co-Cr alloys and 316L stainless steel, which limits the usage of CP-Ti in load bearing implant applications [12]. The enhancement in the mechanical strength can be achieved either by the alloying or by the grain refinement. The enhancement of the strength by alloying can be attributed to the secondary phase formation or solid solution strengthening by the addition of alloying elements like Al and V, which might lead to the release of the corresponding ions into the body after prolonged exposure to the body fluid, which causes various long-term health problems [13]. The grain refinement of materials by severe plastic deformation (SPD) techniques is gaining attention nowadays, owing to the attainment of fine nanosized grains, thereby enhancing the strength [14]. In addition to the strength, severe plastic deformation enhances the biocompatibility of CP-Ti by enhancing the osseointegration process [15]. Equal channel angular pressing (ECAP) is the most widely used SPD technique for the grain refinement, thereby achieving the strength. But, the plastic deformation of Ti has been a challenge due to its limited plasticity, which can be surmounted by the technique of warm ECAP, in which Ti is worked at an elevated temperature around 400-500 °C [16]. A further increase in strength can be achieved by a combination of techniques like cold rolling, cold extrusion and thermo-mechanical treatments [14, 17, 18]. The production of nanostructured long length rods and sheets suitable for commercial applications can be achieved by a relatively new technique - ECAP-conform. Various researchers reported that nanostructured Ti obtained by ECAP, exhibits an enhanced corrosion resistance, compared to unprocessed Ti, attributing to the rapid formation of strong oxide films and the absence of impurity segregation, as in the case of unprocessed Ti [19, 20]. However, the enhancement in corrosion properties is limited. Further enhancement in the corrosion can be achieved by the surface modification of SPD CP-Ti. This can also improve the cell adhesion, thereby reducing the osseointegration time.

## 3. Surface modification of CP-Ti

CP-Ti is a bio-inert material, the cell adhesion property can be improved by various surface modification techniques. Surface modification techniques like grit blasting followed by acid etching, selective laser melting, 3-D printing of porous implants, electrochemical modification by forming nanotubes, plasma immersion ion implantation, bioabsorbable polymeric coatings have been reported to enhance the osseointegration [21]. Among these techniques, electrochemical modification methods like anodization and plasma electrolytic oxidation are the most effective and easiest surface modification techniques. The porous nature of the oxide coating formed by the PEO process provides a higher surface area for the cell adhesion.

### *3.1. Plasma electrolytic oxidation*

Plasma electrolytic oxidation is a technique used to form hard, corrosion resistant ceramic coatings on valve metals, which can easily form passive oxide films like Al, Ti, Nb, Zr, Mg etc. [22]. PEO is an anodic oxidation process occurring at higher potentials than the breakdown voltage of the formed thin oxide layer [23, 24]. PEO is a 3-stage process as shown in figure 1, in which the valve metal connected to the positive terminal of a power source gets oxidized during the initial stage. The oxidized metal reacts with the anions approaching electrolyte, thereby forming the metal oxide. The oxide layer further grows at the expense of the metal by the diffusion process, and the increased oxide layer thickness resists the current flow and the diffusion of the anions and metal cations will become sluggish. The oxide layer will resist the current flow, till the breakdown of the formed oxide layer is evident. The breakdown of the oxide layer further promotes electrochemical oxidation. The continuous repetition of the breakdown, followed by molten metal oxide ejection and quenching of the molten metal oxide provides rough and porous morphology, which is beneficial for the cell adhesion, along with the better corrosion resistance and increased interlocking between implant material and the tissue. The characteristics of the formed oxide layer will depend on electrical parameters like current density, pulse duty cycle and frequency, along with the electrolyte chemistry.

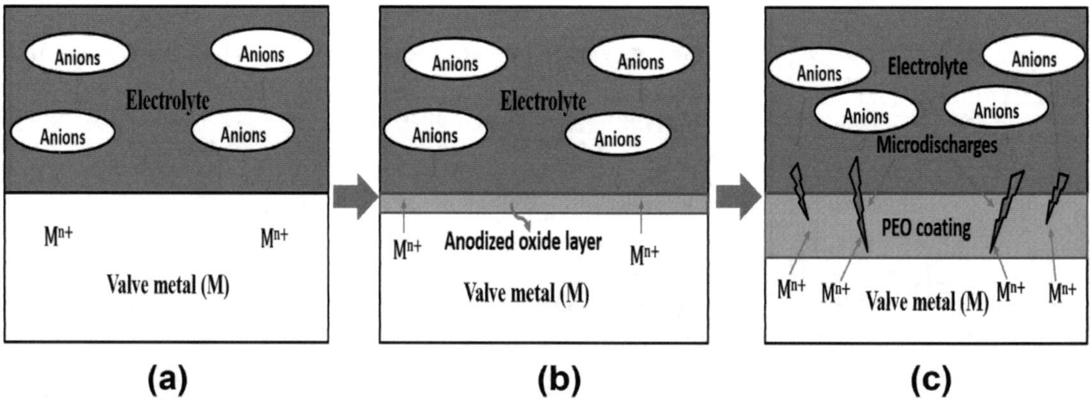


**Figure 1.** (a) Initial stage of PEO process (b) Anodized oxide layer formation (c) PEO coating formation by the microdischarges.

*3.1.1. Effect of electrical parameters.* Duty cycle and frequency have a considerable effect on the PEO coating characteristics developed on CP-Ti as reported by Gowtham et al. [25]. The coating produced at a higher duty cycle (95%) and a higher frequency (1000 Hz) exhibited a higher coating thickness (~ 9 μm). The coatings produced at a higher frequency are relatively less porous than that obtained at a lower frequency (50 Hz), attributed to the short time period available for each cycle. The coating produced at a higher duty cycle and higher frequency exhibited the dense coating with a higher thickness, which provided the better corrosion resistance ($i_{corr}$ =1.9 E−4 mA/cm$^2$) and scratch resistance along with the improved bioactivity. Current density affects the coating characteristics and properties of the coatings as reported by Khan et al. [26]. The PEO coating produced at higher current density (20 Adm$^{-2}$) possess higher coating thickness (~ 9 μm). The amount of rutile phase content is observed to increase with an increase in the current density, along with a higher porosity. The high porosity and the rutile phase content observed in the sample with a higher current density was attributed to the increasing intensity of the discharge events with an increase in the current density.

*3.1.2. Effect of electrolyte chemistry.* Additives such as (Borate ($B_4O_7^{2-}$), Citrate ($C_6H_5O_7^{3-}$) and Silicate ($SiO_3^{2-}$)) show a considerable effect on the coating characteristics in the phosphate electrolyte system as reported by Venkateswarlu et al. [27]. Additives increased the rutile content, porosity, thickness and corrosion resistance. The coatings obtained with the citrate electrolyte system, exhibited highest rutile content (41.6 wt.%), along with the highest thickness (~ 11 μm) and relatively less porous morphology, which resulted in the highest corrosion resistance ($i_{corr}$ = 3.1 E-9 Acm$^{-2}$). Sodium sulfate, Tris (hydroxymethyl) aminomethane, di-ammonium hydrogen phosphate and ammonium acetate additives affect the PEO coating characteristics, developed on CP-Ti as reported by Hari et al. [28]. The additives improved the corrosion resistance, and it is observed that the coating obtained with Tris (hydroxymethyl) aminomethane exhibited a rutile phase content (25.2 wt.%), along with higher corrosion resistance ($i_{corr}$ = 3.71 E−4 μAcm$^{-2}$), attributed to the fine pore morphology and dense inner barrier layer. Higher bioactivity and human osteosarcoma cell adhesion was also observed to be higher in the PEO coating with Tris (hydroxymethyl) aminomethane additive. The zirconate additive can influence the PEO coating characteristics on CP-Ti, as reported by Babaei [29]. The increase in the zirconate level from 0 to 4 g/l, increased the pore size and roughness of the coating on the account of increased conductivity. On the other hand, the thickness of the coatings increased up to 5 μm. The higher (4 g/l) zirconate content, decreased the corrosion protection against the environment at 1000 Hz frequency, indicating the presence of through pores in the coating. The optimized zirconate additive content was reported as 2 g/l for the best corrosion resistant coating. The addition of electrolyte additives, such as, sodium carbonate ($Na_2CO_3$), sodium nitrite ($NaNO_2$) and urea ($CO(NH_2)_2$), resulted in the alteration of the PEO coating characteristics on CP-Ti, as reported by Venkateswarlu et al. [30]. The PEO coatings obtained in the electrolyte systems containing sodium carbonate exhibited the highest rutile content (54.4%), along with higher thickness (~ 55 μm) and least porosity (11.6 %), thereby improving the corrosion resistance ($i_{corr}$ = 2 E−2 $\mu Acm^{-2}$). Fluoride ion also impacts the characteristics of the coatings as reported by Venkateswarlu et al. [31]. The PEO coatings obtained with four different fluorine precursors i.e., $NH_4F$, KF, NaF and $K_2TiF_6$ exhibited superior properties. A higher rutile phase content (37%), along with less porous surface structure was observed in the coating with $K_2TiF_6$ additive, which resulted in the better corrosion resistance ($i_{corr}$ = 8 E−2 $\mu Acm^{-2}$). Thus, the electrical parameters and electrolyte chemistry of the PEO process, greatly influence the coating thickness, morphology and phase composition, thereby increasing the corrosion resistance in the body fluid and making the coated samples more suitable for the bio-implant applications.

### *3.2. Plasma electrolytic oxidation coupled with electrophoretic deposition*

Electrophoretic deposition (EPD) is the mechanism by which the charged nanoparticles suspended in an electrolyte move towards the anode due to its surface charge (zeta potential) and get deposited into the coating formed by the PEO process, thereby producing composite coatings. The suspended nanoparticles should have a sufficiently higher negative zeta potential to get incorporated into the coating [30]. The incorporation of the nanoparticles by the EPD mechanism occurs at higher voltages, which were observed at later stages of the PEO process [32]. The nanoparticle incorporation can be of a reactive or nonreactive type, depending upon the properties of the nanoparticles, such as a particle size, melting point temperature, etc., and the discharge characteristics observed during the PEO process [33, 34].

The electrolyte composition influences the nanosized $ZrO_2$ particle incorporation into the oxide coating developed by the PEO process, as reported by Gowtham et al. [35]. The increase in KOH levels from 0 g/l to 6 g/l, decreased the incorporation of nanoparticles, attributed to the reduction the velocity of the migrating particle in the electrolyte, which in-turn depend on the decreased potential difference attributed to the increase in the conductivity of the electrolyte. The nanoparticles are found to be incorporated partially reactive and partially inert, which was substantiated by the presence of both m-$ZrO_2$ and $ZrTiO_4$ phases. The m-$ZrO_2$ phase content is observed to decrease with the increase in the KOH content, corroborating with the decrease in the particle migration. The thickness and porosity of the coatings are found to be increasing with the KOH levels. The high corrosion resistance ($i_{corr}$ = 5.18 E-8 $mA/cm^2$) is found to be offered by the coating with intermediate KOH level (4 g/l), correlating to its higher thickness along with the thick inner barrier layer. The addition of carbon nanotubes (CNT) is also found to affect the PEO coating characteristics on CP-Ti as reported by Yazici et al. [36]. The increase in the amount of CNT increased the formation of $Al_2TiO_5$, by suppressing the $TiO_2$ rutile phase formation. The CNT content is observed to increase the pore size of the coatings. At 300 V, the coating thickness is found to be higher for the 2 g/l CNT added sample, then the sample without CNT addition, and as the CNT amount increased to 4 g/l, the coating thickness is found to be decreased. On the other hand, at 375 V, the coating thickness found to be decreased by increasing the CNT content from 0 to 4 g/l. Rare earth oxide nanopowders like $CeO_2$ addition to the PEO coatings, has been found to grab the attention of the researchers. The amount of $CeO_2$ addition, influences the coating properties produced on the CP-Ti as reported by Di et al. [37]. The increase in the $CeO_2$ content (0 g/l to 6 g/l), is found to enhance the anatase to rutile phase transformation up to 4g/l, and further increase in the $CeO_2$, lead to the formation of $CeO_2$ phase. The increase in the nanoparticle content decreased the pores size and number by the incorporation of nanoparticles into the pores formed during the PEO process. The thickness of the coatings is found to be increased with the increase in the nanoparticle content. The corrosion resistance of the coatings was also found to be increased by the increase in the $CeO_2$ content, up to 4 g/l and further increase lead to the decrease in corrosion resistance ($i_{corr}$ = 2.295E-6), attributing to the decrease in the rutile content. Doping of Cu nanoparticles into the $TiO_2$ coatings, developed on CP-Ti, results in the enhancement of the antibacterial properties was reported by Yao et al. [38]. Cu nanoparticle addition doesn't significantly influence the coating thickness and morphology along with the phase composition. The antibacterial culture test conducted over a period of 24 h, concluded that Cu doping significantly increased the antibacterial activity by killing E. coli and S. aureus bacteria. The antibacterial activity along with the phosphatase activity of the bone marrow stem cells (BMSC) has found to be increased by the addition of Zn nanoparticles, during the PEO coating formation as reported by Hu et al. [39]. Zn addition does not show any significant changes in the morphology, thickness and phase composition. The antibacterial activity of the Zn doped coatings was attributed to the generation of reactive oxygen species. The Zn nanoparticles played an important role in the adhesion and proliferation of the BMSC's, which were the ideal seed cells for the tissue engineered bone. Ag-HA addition into the $TiO_2$ coatings prepared on CP-Ti results in the coating antibacterial performance as reported by Venkateswarlu et al. [40]. Better corrosion resistance was offered by the Ag-HA/$TiO_2$ composite coating along with the antibacterial property. The Ag-HA/$TiO_2$ coated sample exhibited the zone of bacterial inhibition in the E.coli atmosphere.

## 4. Conclusions

The characteristics of the PEO and PEO-EPD coatings, namely thickness, surface morphology, phase composition, antibacterial properties, cell adhesion, cell growth, corrosion resistance, are found to be influenced by various process parameters like electrical parameters, electrolyte chemistry, treatment time and nanoparticle additions. Precise control of various parameters has to be maintained for obtaining the biocompatible coatings on CP-Ti with better corrosion resistance along with the cell adhesion and proliferation along with the antibacterial properties.

## Acknowledgments

The authors are grateful to the Department of Science & Technology (DST), New Delhi, India through grant No DST/INT/RUS/RSF/P-35 and the Russian Science Foundation (RSF), Moscow, Russia through the grant No. 19-49-02003 to carry out this work.

## References

[1] Valiev R Z, Parfenov E V and Parfenova L V 2019 *Mater. Transac.* **60** 1356
[2] Geetha M, Singh A K, Asokamani R and Gogia A K 2009 *Prog. Mater. Sci.* **54** 397
[3] Long M and Rack H J 1998 *Biomater.* **19** 1621
[4] Okazaki Y, Rao S, Tateishi T and Ito Y 1998 *Mater. Sci. Eng.* A **243** 250
[5] Niinomi M 2003 *Sci. Technol. Adv. Mater.* **4** 445
[6] Niinomi M 2002 *Met. Mater. Trans.* A **32** 477
[7] Liu X, Chu P K and Ding C 2004 *Mater. Sci. Eng.* R **47** 49
[8] Matykina E, Arrabal R, Skeldon P, Thompson G E 2009 *Acta Biomater.* **5** 1356
[9] Taddei E B, Henriques V A R, Silva C R M and Cairob C A A 2004 *Mater. Sci. Eng.* C **24** 683
[10] Giannoudis P V and Schneider E 2006 J. Bone Joint Surg. Am. **88** 1272
[11] Rintoul T C, Bulter K C, Thomas D C, Carriker J W, Maher T R, Kiraly R J, Massiello A, Himely S C, Chen J F and Fukamachi K 1993 *ASAIO J.* **39** 168
[12] Naseri R, Kadkhodayan M and Shariati M 2017 *T. Nonferr. Metal Soc. China* **27** 1964
[13] Nag S, Banerjee R and Fraser H L 2005 *Mater. Sci. Eng.* C **25** 357
[14] Stolyarov V V, Zhu Y T, Lowe T C and Valiev R Z 2001 *Mater. Sci. Eng.* A **303** 82
[15] Estrin Y, Kasper C, Diederichs S and Lapovok R 2009 *J. Biomed. Mater. Res.* A **90** 1239

[16] Rodriguez-Calvillo P and Cabrera J M 2015 *Mater. Sci. Eng.* A **625** 311
[17] Stolyarov V V, Zhu Y T, Alexandrov I V, Lowe T C and Valiev R Z 2003 *Mater. Sci. Eng.* A, **343** 43
[18] Gunderov D V, Polyakov A V, Semenova I P, Raab G I, Churakova A A, Gimaltdinova E I, Sabirov I, Segurado J, Sitdikov V D, Alexandrov I V, Enikeev N A and Valiev R Z 2013 *Mater. Sci. Eng.* A **562** 128
[19] Kim H S, Yoo S J, Ahn J W, Kim D H and Kim W J 2011 *Mater. Sci. Eng.* A **528** 8479
[20] Balyanov A, Kutnyakova J, Amirkhanova N A, Stolyarov V V, Valiev R Z, Liao X Z, Zhao Y H, Jiang Y B, Xu H F, Lowe T C and Zhu Y T 2004 *Scr. Mater.* **51** 225
[21] Souza J C M, Sordi M B, Kanazawa M, Ravindran S, Henriques B, Silva F S, Aparicio C and Cooper L F 2019 *Acta Biomater.* **94** 112
[22] Saikiran A, Hariprasad S, Arun S, Rama Krishna L and Rameshbabu N 2019 *Surf. Coat. Technol.* **372** 239
[23] Yerokhin A, Parfenov E V and Matthews A 2016 *Surf. Coat. Technol.* **301** 54
[24] Arun S, Arunnellaiappan T and Rameshbabu N 2016 *Surf. Coat. Technol.* **305** 264
[25] Gowtham S, Arunnellaiappan T and Rameshbabu N 2016 *Surf. Coat. Technol.* **301** 63
[26] Khan R H U, Yerokhin A L, Li X, Dong H and Matthews A 2014 *Surf. Eng.* **30** 102
[27] Venkateswarlu K, Rameshbabu N, Sreekanth D, Sandhyarani M, Bose A C, Muthupandi V and Subramanian S 2013 *Electrochim. Acta* **105** 468
[28] Hariprasad S, Ashfaq M, Arunnellaiappan T, Harilal M and Rameshbabu N 2016 *Surf. Coat. Technol.* **292** 20
[29] Babaei M, Dehghanian C and Vanaki M 2015 *Appl. Surf. Sci.* **357** 712
[30] Venkateswarlu K, Rameshbabu N, Sreekanth D, Bose A C, Muthupandi V, Babu N K and Subramanian S 2012 *Appl. Surf. Sci.* **258** 6853
[31] Lu X, Mohedano M, Blawert C, Matykina S, Arrabal R, Kainer K U and Zheludkevich M L 2016 *Surf. Coat. Technol.* **307** 1165
[32] Venkateswarlu K, Rameshbabu N, Sreekanth D, Bose A C, Muthupandi V and Subramanian S 2013 *Ceram. Int.* **39** 801
[33] Arun S, Hariprasad S, Ravisankar B, Parfenov E V, Mukaeva V R and Rameshbabu N 2018 *Trans. Indian Inst. Met.* **71** 1699
[34] Arun S, Hariprasad S, Saikiran A, Ravisankar B, Parfenov E V, Mukaeva V R and Rameshbabu N 2019 *Surf. Coat. Technol.* **363** 301
[35] Gowtham S, Hariprasad S, Arunnellaiappan T and Rameshbabu N 2017 *Surf. Coat. Technol.* **313** 263
[36] Yazici S K, Muhaffel F and Baydogan M 2014 *Appl. Surf. Sci.* **318** 10
[37] Di S, Guo Y, Lv H, Yu J and Li Z 2015 *Ceram. Int.* **41** 6178
[38] Yao X, Zhang X, Wu H, Tian L, Ma Y and Tang B 2014 *Appl. Surf. Sci.* **292** 944
[39] Hu H, Zhang W, Qiao Y, Jiang X, Liu X and Ding C 2012 *Acta Biomater.* **8** 904
[40] Venkateswarlu K, Rameshbabu N, Bose A C, Muthupandi V, Subramanian S, MubarakAli, Davoodbasha and Thajuddin N 2012 *Ceram. Int.* **38** 731

# Deformation and current interaction in nanostructured and amorphous Ti based alloys

**V V Stolyarov**[1,2]

[1] Mechanical Engineering Research Institute of Russian Academy of Sciences, Moscow, 101990, Russia
[2] National Research Nuclear University MEPhI (Moscow Engineering Physics Institute), Moscow, 115409, Russia

E-mail: vlstol@mail.ru

**Abstract.** The deformation behavior under tension and pulse current in course-grained, nanostructured and amorphous titanium based alloys is investigated. The influence of grain size, structure state and current mode on shape memory effect and electroplastic effect has been shown. Structure refinement up to nanosize area, occurrence of interphase boundaries, and amorphous state in alloys suppress the electroplastic effect. In TiNi alloys with phase thermoelastic transformation induced by deformation or temperature, current pulses decreases stress jumps downwards induced by manifestation of the electroplastic effect and increases stress jumps upwards connected with the shape memory effect.

## 1. Introduction

The combined application of plastic deformation and an electric current of high density ($10^3$A/ᴍᴍ$^2$) results in the reduction of the applied stresses, called the electroplastic effect (EPE) [1]. It is manifested in stress jumps on the stress-strain tension curve when a single current pulse is introduced to a sample without an essential thermal effect and sample dilatation. The growth of current density and single pulse duration enhances the EPE [2]. When a multipulse current is applied, the temperature of the sample under tension increases thus much that EPE is accompanied by a thermal effect. It is assumed that the primary mechanism of EPE is the electron–dislocation interaction resulting in stress relaxation in the areas of dislocation pile-up in the crystal [3]. However, there are many other possible mechanisms described, for example, for Al or Ti-based alloys [4, 5]. Despite the large number of proposed models and mechanisms of EPE, the direct effect of electropulsing has not been well investigated. A detailed analysis of problems arising in modeling of EPE mechanisms is made in a recent review [6]. Note that such structure elements as grain size, phase morphology during the first experimental or theoretical considerations of EPE were not fully taken into consideration. For instance, the phenomenology of EPE was sufficiently fully investigated only in mono- and coarse-grained (CG) single-phase alloys [2]. It was demonstrated that EPE existed during plastic deformation of a material, and its value in relation to the flow stress varies from a few percent for polycrystals up to ten percent for monocrystals [1]. However, EPE is poorly studied in modern materials, for example, multiphase materials, shape memory effect (SME) alloys, and ultrafine-grained (UFG) and nanostructured (NS) alloys, [7] and for amorphous alloys this information is not available at all. The study of the influence of structural features on EPE is of particular importance. The article makes an attempt to fill in the missing information. The stress-strain curves at tension under the action of a pulse current in alloys with a coarse grain, nanograin and an amorphous structure, as well as in an alloy with phase transformation (MT), are presented.

## 2. Materials and processing

Three kinds of materials were studied. They are commercially pure titanium (CP Ti), Grade 2 (VT1-0), two-phase Ti-6Al-4V (VT6), austenitic SME $Ti_{49.3}Ni_{50.7}$ alloy with CG (up to 50 μm), UFG (less than 1 μm), and NS (less than 100 nm) structures. Fig. 1 shows TEM images of materials in NS states.

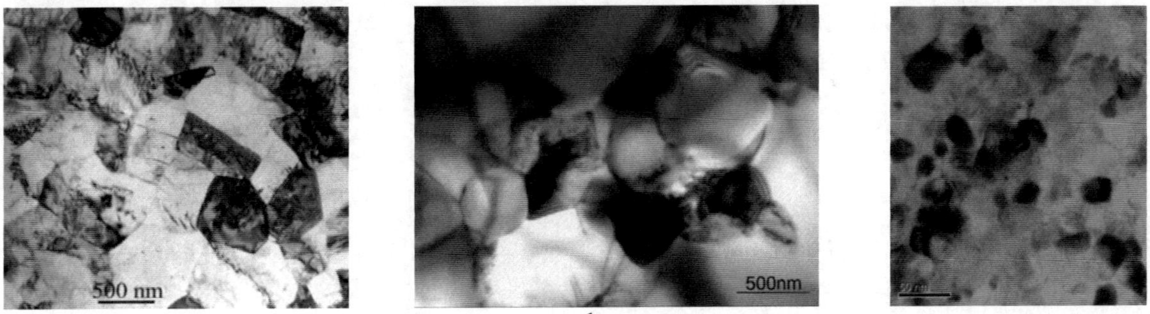


a b c

**Figure 1.** Microstructure of the alloys after: a - ECAE, CP Ti; b - MF, VT6; c - EPR, $Ti_{49.3}Ni_{50.7}$.

The samples were in the form of sheets. Coarse grain structure in the alloys was processed by heat treatment. Structure refinement was achieved by severe plastic deformation, including equal channel angular extrusion (ECAE) for titanium [8], multiple forging (MF) for VT6 [9] and multi-pass cold rolling with electrical current (EPR) [10], accordingly. The processing parameters for EPR included: a strain rate of 4 m/min, a current pulse frequency of $10^3$ Hz, a pulse duration $\tau = 0.8 \times 10^{-4}$ s, a current density $j = 100$ A/mm$^2$. The temperature on the sample-simulator subjected to electropulse current of the maximal density did not exceed 150 °C. The true strain e = ln $h_0 / h_f$ ($h_0$, $h_f$ - start and final thickness) was 1.4. Besides, ribbon-shaped amorphous alloys ($Ti_{50}Ni_{25}Cu_{25}$) processed by the melt-spinning technique were also investigated. Samples of alloys were exposed to tension with a pulse current j =100–1500 A/mm$^2$ and τ = 100 –1000 μs. Dog bone-shaped tensile samples with dimensions of 0.5 x 4 x 20 mm and a test velosity about 1 mm/min were used. The ribbon samples had a cross section of 0.04 x 3 mm and a length of 75 mm.

## 3. Experimental results

It is discovered that the current pulse in the CG CP Ti alloy induces a typical EPE in the form of stress jumps downwards on the stress–strain curve (figure 2 a, b). The amplitude of the stress jumps depends on the density and duration of the current pulse, as well as on the grain size of the alloy structure. For example, the amplitude of stress jumps at $j$ =1500 A/mm$^2$ increases by a factor of four to five with an increase in pulse duration from 100 up to 1000 μs in single-phase CG VT1-0 annealed at 700 °C (figure 2a). In UFG CP Ti, stress jumps are not recorded at the same current density and $\tau$ =100 μs but are only observed at a $\tau$ =1000 μs (figure 2b). It is also visible that the amplitude of stress jumps is sensitive to the structural state. It decreases with a reduction of the grain size from 100–150 MPa in the CG state (figure

IOP Conf. Series: Materials Science and Engineering **672** (2019) 012013 doi:10.1088/1757-899X/672/1/012013

2a) to 25–50 MPa in the UFG state (figure 2b). A similar tendency of decreases in EPE with reduction of the grain size is observed in the (α + β) VT6 alloy (figure 3a, b).

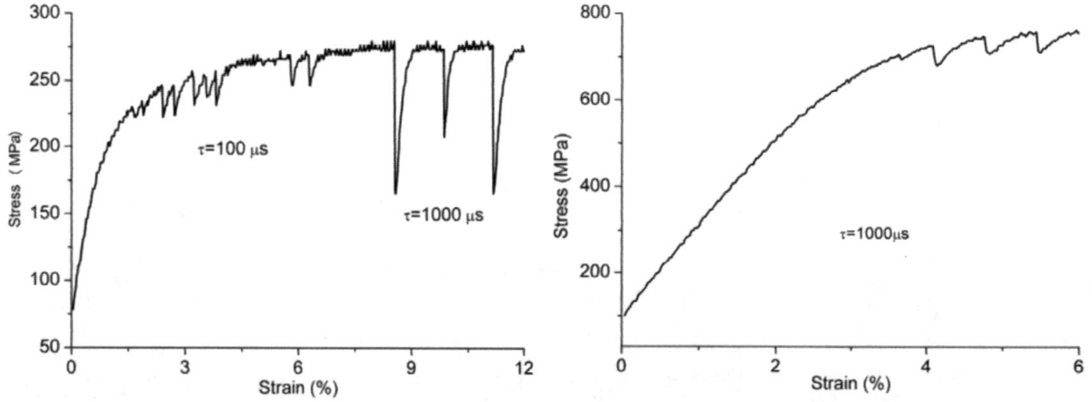


a b

**Figure 2.** Deformation behavior under current (j = 1500 A/mm$^2$) in the CG (a) and UFG (b) titanium.

In a two-phase VT 6 alloy, the amplitude of the stress drops down is much smaller than in pure titanium. They are equal to 15 and 5 MPa in the CG and UFG states, respectively.

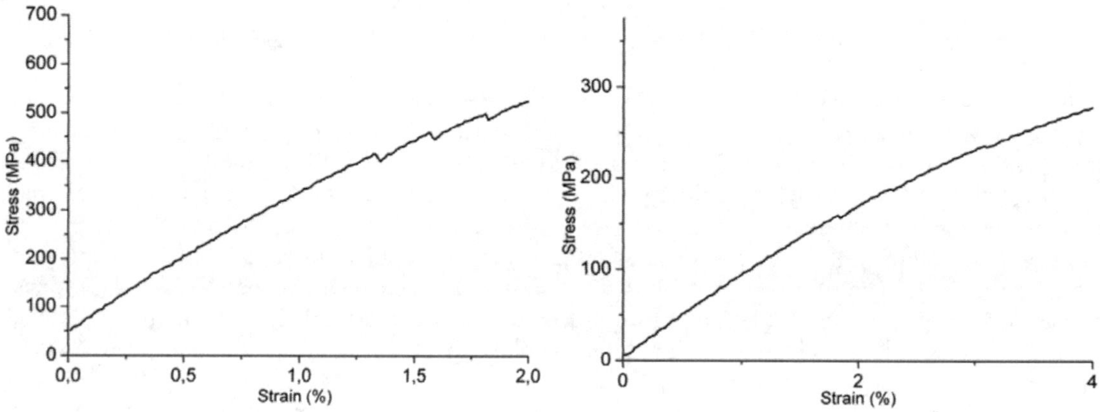


a b

**Figure 3.** Deformation behavior for CG (a) and UFG (b) VT6 alloy under current (τ=1000 μs; j=1500 A/mm$^2$.

Of particular interest are shape memory TiNi alloys with martensitic transformation at low temperature. Depending on the chemical composition, a reverse thermoelastic martensitic transformation occurs at temperatures of –150 to +100 °C. It was first discovered that pulsed current can cause not only traditional stress jumps down, but also up in the CG $Ti_{49.3}Ni_{50.7}$ alloy (figure 4 a, b). It was first discovered that pulsed current can cause not only traditional voltage surges down, but also up. It is due to this that the amplitude of the stress jumps up increases, and the jumps down remains unchanged.

IOP Conf. Series: Materials Science and Engineering **672** (2019) 012013 doi:10.1088/1757-899X/672/1/012013

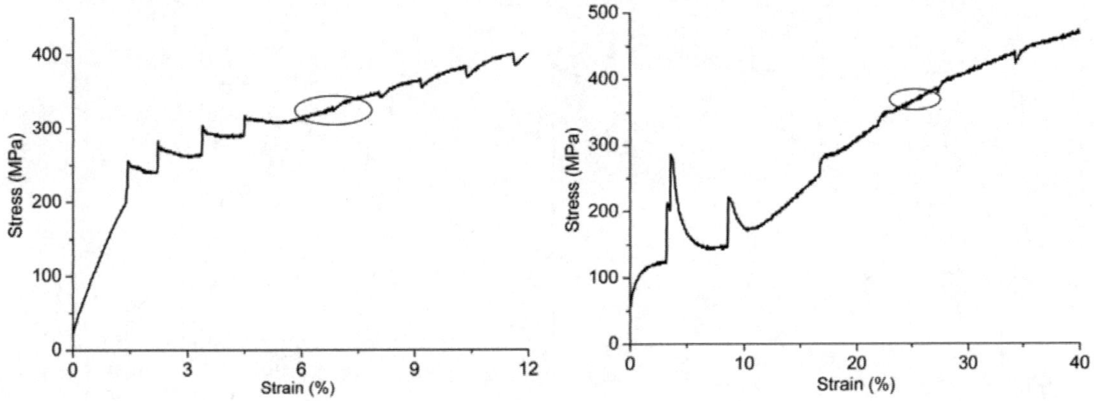


a b

**Figure 4.** Deformation behavior of the CG TiNi alloy under current: a, j=1500 A/mm$^2$; b, j=3000 A/mm$^2$.

Interesting point on a tensile curve for this alloy is an area (indicated in the figures) in which neither kind of stress jump occurs, despite the introduction of a single pulse current (figure 4 a). Up to this area (strain < 6 %), the pulse current leads to occurrence of stress jumps upwards, and after that (strain > 9%) the pulse current initiates stress jumps downwards. Similar situations are observed both at a current density of 1500 A / mm$^2$ and at 3000 A / mm$^2$.

In the nanostructured TiNi alloy, stress jumps upwards can be observed instead of stress jumps downward, which completely disappear (figure 5).

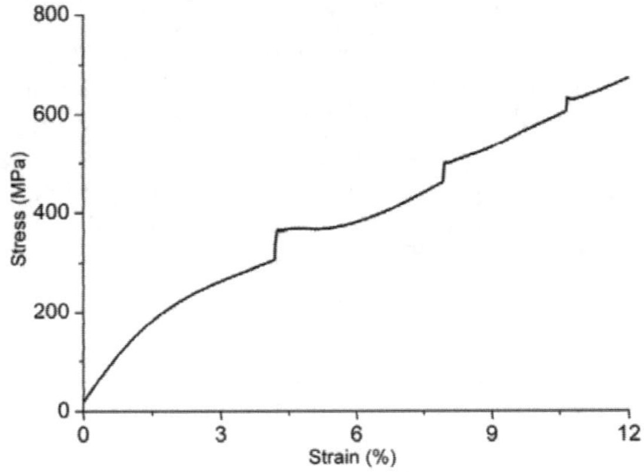


**Figure 5.** Deformation behavior under a current density of 1500 A/mm$^2$ in NS $Ti_{49.3}Ni_{50.7}$.

As for EPE in rapidly quenched $Ti_{50}Ni_{25}Cu_{25}$ alloy, stress jumps at tension are not observed under current with a density of 600 A/mm$^2$ and duration of 800 μs (figure 6a). But, they appear in the ultrafine-grained alloy (grain size about 500 nm) after crystallization annealing at 500 °C for 240 s (figure 6b).

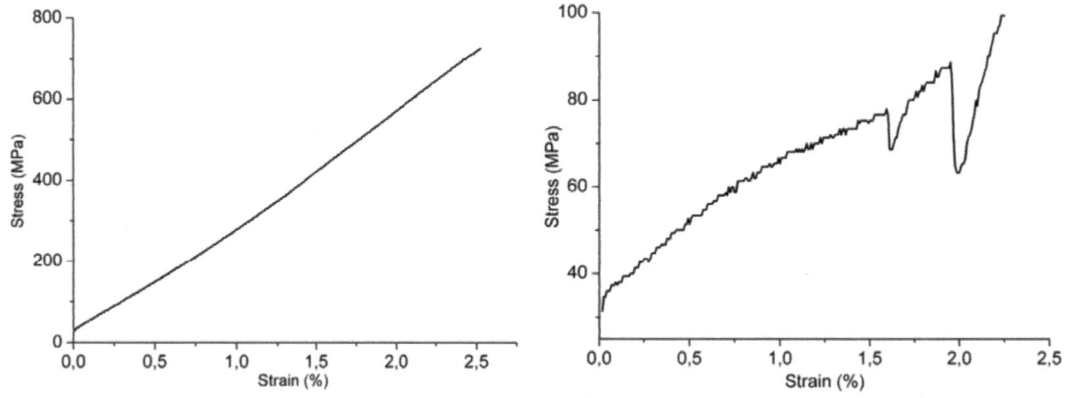


**Figure 6.** Stress–strain curves with a current density of 600 A/mm$^2$ in the initial rapid quenched $Ti_{50}Ni_{25}Cu_{25}$: a) initial state; b) after crystallization at 500 °C.

## 4. Discussion

The results of the study convincingly demonstrate the dependence of EPE on the structural and phase state of the studied alloys. This is especially pronounced in the large-scale effect of grain size on EPE. In titanium-based materials, the transition from a microstructure with a grain size of 20-40 microns to a nanostructure with a grain size of 100-200 nanometers dramatically reduces the amplitude of the jump (figures 2, 3)

Another important structural factor is the presence of interphase boundaries, the role of which is to decelerate the transmission of a mechanical electron pulse to dislocations and reduce the amplitude of the voltage jump. This is clearly seen when comparing EPE in single-phase titanium and two-phase titanium alloy ($\alpha + \beta$) VT6, in which the volume fraction of interphase boundaries can be up to 25%. Interestingly, this is confirmed for coarse-grained (figures 2a, 3a) and ultrafine-grained (figures 2b, 3b) alloys. The mobility of dislocations in the crystal lattices of the hcp and bcc differs and leads to a decrease in the electron-electron diffraction, the mechanism of which is based on electron-dislocation interaction [3]. Both experimental facts related to the size effect and interphase boundaries are in good agreement with the mean free path of dislocations and their ability to multiply in the Frank-Reed source. Both parameters sharply fall upon a structural transition to the region of nanoscale or the appearance of multiphase.

A detailed study of the stress-strain curves in the $Ti_{49.3}Ni_{50.7}$ alloy showed that the parameters of the stress jump are determined by the combined action of two different in nature effects: EPE (downward jump) and SME (upward jump). The SME jump is due to the difference in the stresses of the flow of austenite and martensite coexisting at the initial stage of deformation even with slight heating of the alloy due to the martensitic transformation B19'→B2 [11]. As noted above, EPE is close to zero in the nanostructured state, which leads to the manifestation of only SME and the associated upward voltage jump (figure 5). The decrease in the amplitude of the stress jump in NS $Ti_{49.3}Ni_{50.7}$ depending on the degree of deformation is determined by the change in the ratio of the volume fraction of the martensitic and austenitic phases, as well as the stabilization of one of these phases. In fact, the thermoelastic MT A↔M caused by stress arises in the stress region in which $v_m / v_a$ varies from 0 to 1 [12]. With increasing stress, the volume fraction of unstable austenite decreases, the volume fraction of stable martensite increases. As a result, the amplitude of one of the effects decreases, and the other effect increases, respectively. An interaction of EPE with SME reduces the amplitude of the step up.

In the initial rapid quenched shape memory alloy, EPE is absent (figure 6a), but it appears in the UFG state with a grain size of 500 nm, obtained by crystallization. The experiment confirms the disappearance of EPE with a decrease in size to the nano-region. Indeed, the absence of mobile deformation carriers in amorphous materials does not contribute to the manifestation of EPE. As a rule, crystallization of an alloy contributes to the occurrence of dislocations and the manifestation of EPE. The mode and type of current are also of great importance. An increase in the main current parameters contributes to an increase in the energy and amplitude of stress jumps. This is why an increase in the number of current pulses also increases the amplitude of the jump. It is known that the Joule effect from a multipulse current with duration of up to 1 ms can lead to more significant stress jumps of up to 500 MPa than in the case of the application of single pulses in SME alloy [13].

## 5. Conclusions

The effect of the initial structure-phase state and current regimes on the stress-strain curves was studied in a CG, UFG, NS and an amorphous VT1-0, VT6, TiNi alloys. The pulse current resulted in the appearance of stress jumps up or down on the stress–strain curves due to the manifestation of the electroplastic effect and martensitic transformation, accordingly. An amplitude and direction of the stress jumps depend on structure features (grain size, phase composition), phase transformations in alloys, and current energy.

**Acknowledgments**

The work has been supported by RFBR, grant no. 16-58-48001.

## References

[1] Troitskiy O A 1969 *Pisma Zh. Eksp. Teor. Fiz.* **10** 18
[2] Troitskiy O A, Baranov Yu V, Avraamov Yu S and Shlyapin A D 2004 *Physical Bases and Technologies for Treatment of the Modern Materials* Vols 1, 2 (Moscow–Izhevsk: Inst. Comp. Technol.)
[3] Sprecher A F, Mannan S L and Conrad H 1986 *Acta Metall.* **34** 1145
[4] Kronenberger T J, Johnson D H and Roth J T 2009 *J. Manuf. Sci. Eng.* **131** 031003
[5] Okazaki K, Kagawa M and Conrad H 1980 *Mater. Sci. Eng.* **45** 109
[6] Qin R S 2015 *Mater. Sci. Technology* **31** 203
[7] Stolyarov V 2012 *Rev. Adv. Mater. Sci.* **31** 14
[8] Stolyarov V and Valiev R 2004 *Russ. Metall. (Metally)* **2** 110
[9] Zherebtsov S *et al.* 2005 *Mater. Trans.* **46** 1
[10] Stolyarov V V, Ugurchiev U Kh, Trubitsyna I B, Prokoshkin S D and Prokofiev E A 2006 *Phys. Tech. High Press.* **4** 48
[11] Ugurchiev U *et al.* 2009 *Proc. Conf. Bernstein's Readings on Thermal and Mechanical Treatment of Materials* (Moscow: MISiS) pp 123–124
[12] Brailovski V *et al.* 2003 *Shape Memory Alloys* (Montréal: ETS Publ.)
[13] Stolyarov V V 2010 *Materials Science Forum* **633/634** 595

IOP Conf. Series: Materials Science and Engineering **672** (2019) 012014 doi:10.1088/1757-899X/672/1/012014

# Nanocrystal formation and structure rejuvenation in Al-based alloys

**D Matveev, E Pershina, N Volkov and G Abrosimova**

Institute of Solid State Physics of RAS, Chernogolovka, Moscow distr., Russia

E-mail: gea@issp.ac.ru

**Abstract**. Studies of the structural evolution of the group of Al-based amorphous alloys of Al-Ni-RE (RE = La, Y, Gd) systems showed that, under the combined action of heat and deformation on the structure, one may obtain samples with different crystal sizes and different fractions of the nanocrystalline phase. Under crystallization of a heterogeneous amorphous phase, the size of the nanocrystals is smaller, and their fraction is larger than under crystallization of a homogeneous amorphous phase. The use of cryogenic cycling method allows rejuvenating the amorphous phase in partially crystalline alloys.

## 1. Introduction

Amorphous and nanocrystalline alloys belong to the group of materials being widely studied due to a combination of splendid physical properties: magnetic, strength, corrosion-resistant, and others [1-3]. Fe-based alloys have excellent magnetic properties: low coercivity (0.5 – 1 A/m) and high saturation magnetization, which exceeds 1.4 T. Superior hysteresis properties were obtained in Co-based alloys [3], which have almost zero magnetostriction. Good properties of the material are determined by its structure. One of the examples of promising amorphous-nanocrystalline materials are light Al-based alloys. They are high-strength, and the values of yield stress in Al-based alloys containing 6-15% of a transition metal (Fe or Ni) and a few percent of a rare-earth metal reach 1.6 GPa under sufficiently high plasticity [4-7]. The properties are structure-sensitive and may change significantly under a change in the structure [8, 9]. Structural changes may occur both in the amorphous state and under the formation of crystalline phases. It was established that the changes under heat treatment depend considerably on temperature: above or below the glass transition temperature [10]. Significant changes are observed also under plastic deformation: the formation of shear bands or the reduced density regions occurs in the amorphous phase. In these regions, the processes of diffusion mass transfer are accelerated, and a region with a thickness to 100 μm is formed near the shear band, the mechanical properties of which differ from the properties of the remaining amorphous matrix [11-13].

It is important to note that structural changes result in the formation of materials with a novel combination of properties. There are a lot of works devoted to the study of the structure and properties of amorphous and amorphous-nanocrystalline materials [14-18]. The present paper considers the possibilities of both creation of a partially crystalline structure with different structural parameters, as well as rejuvenation of an amorphous structure in partially crystalline samples.

## 2. Experimental

Al-Ni-RE (RE = La, Y, Gd) amorphous alloys were obtained in the form of ribbons with a thickness of about 30 μm and a width of about 1 cm by melt quenching. The obtained samples were subjected to heat treatment and deformation in a different order. Isothermal annealing in a furnace, heating in a calorimeter at a constant rate of 20°/min and thermal cycling in the temperature range of 77-373 K were used for heat treatment. The method of multiple rolling was used for deformation. After each stage of treatment, the structure was investigated with a Siemens D500 X-ray diffractometer using $CoK_\alpha$ radiation and JEM 2100 and JEM-100CX electron microscopes. The microhardness was measured with a Micro-Vickers hardness tester with a load of 10 g. The nanocrystal size was determined by XRD analysis as well as by dark-field electron microscopy images. The nanocrystal size was estimated using the Scherrer formula [19]. The fraction of the nanocrystalline phase was determined by the ratio of the integral phase intensities according to [20-22].

## 3. Results and discussion

### *3.1. Formation of a nanocrystalline structure under combined heat treatment*

One of the methods of exposure on the amorphous structure was carrying out combined treatments. It was established earlier [23, 24] that the use of the combined treatment "annealing + deformation" and "deformation + annealing" allows changing significantly the structural parameters and properties of the material. Figure 1 demonstrates an X-ray diffraction pattern of the $Al_{88}Ni_6Y_6$ alloy heated in a calorimeter up to a temperature corresponding to the maximum of the first peak on the DSC curve after preliminary (before DSC heating) 50% deformation (1) and preliminary (before DSC heating) annealing at 100 °C during 3 h (2). The structure of the samples is two-phase and consists of Al nanocrystals and an amorphous matrix. Although the X-ray diffraction patterns are similar, the structural parameters turn to be different. After heating, the average nanocrystal size in the preliminarily annealed sample is 11 nm, and the fraction of the nanocrystalline phase is 10%; the same values for the preliminary deformed sample are 9 nm and 12%, respectively. Similar values were obtained for other alloys. For instance, in an alloy containing La, the average nanocrystal size is 30 nm in the case of preliminary annealing at 150 °C during 3 h and then heating in the calorimeter up to the temperature corresponding to the maximum of the first peak on the DSC curve, and it is 26 nm in the case of preliminary 50% deformation and then heating in the calorimeter up to the same temperature; the fraction of the nanocrystalline phase is also greater in the case of preliminary deformation. In the case of crystallization of the samples without preliminary treatment the average nanocrystal size turns to be significantly larger and is 32 nm for an alloy containing La, 23 nm for an alloy containing Y, and 26 nm for an alloy containing Gd. The observed differences are related to the fact that the formation of a heterogeneous amorphous structure occurs under preliminary treatment [25-32].

The properties of the samples depend considerably on the structural state. The studies have shown that the microhardness of this alloy is 270 MPa in the initial amorphous state and 440 MPa in the partially crystalline state. Figure 2 demonstrates the X-ray diffraction pattern of the alloy of the same system, but with a different ratio of Ni and Y. It is seen that with an increase in the annealing temperature, the fraction of the nanocrystalline phase increases along with an increase in the microhardness of the alloy: 280 MPa after annealing at 100 °C and 310 MPa after annealing at 170 °C (in both cases, the duration of annealing was 30 h). In all the investigated samples, the microhardness also increases with an increase in the fraction of the nanocrystalline phase. However, a significant decrease in microhardness is observed under the formation of a multiphase structure.

Thus, the use of combined treatments enables creating amorphous-nanocrystalline materials with different structural parameters (nanocrystal size, fraction of the nanocrystalline phase) and, consequently, with different properties.

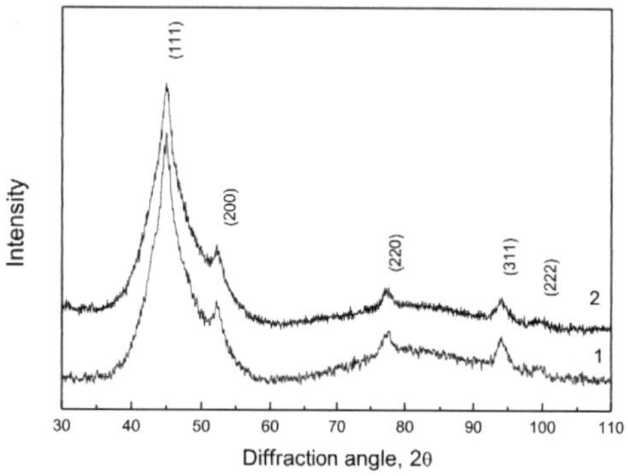


**Figure 1.** X-ray diffraction pattern of the $Al_{88}Ni_6Y_6$ alloy heated in the calorimeter after preliminary deformation (1) and preliminary annealing (2).

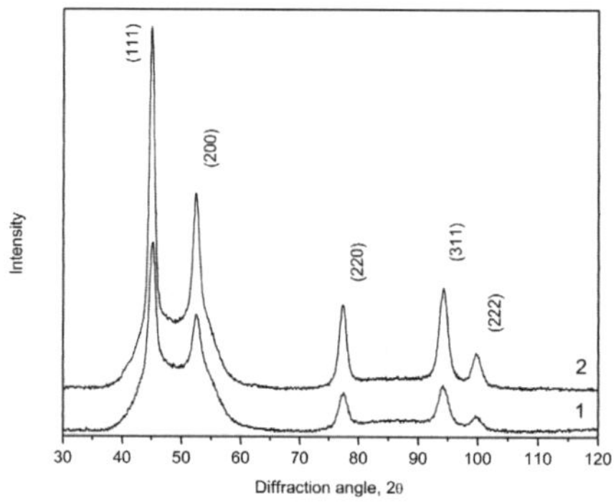


**Figure 2.** X-ray diffraction pattern of the $Al_{88}Ni_{10}Y_2$ alloy annealed during 30 h at 100 °C (1) and 170 °C (2).

*3.2. Rejuvenation of the amorphous structure under cryogenic thermal cycling*

Another way is the possibility of rejuvenation of the amorphous structure. It is well known that under crystallization of some amorphous alloys embrittlement occurs, and in this case, the revival of the amorphous structure is necessary for the plasticity recovery. To study the possibility of rejuvenation of the amorphous structure, the samples in the partially crystalline state were subjected to cryogenic cycling in the temperature range of 77 - 373 K. The idea of this treatment is based on the following assumption: since the coefficients of thermal expansion of the amorphous phase and nanocrystals are different, stresses will be accumulated in the structure under the multiple cycles "heating-cooling", which should lead to structural changes. This idea was tested earlier for heterogeneous amorphous alloys [33, 34]. Its testing on amorphous-crystalline samples (with a greater difference in values of the coefficients of thermal expansion) should lead to larger structural changes [35].

The mentioned treatment was carried out on partially crystalline samples, where the amorphous-nanocrystalline structure was created by annealing or deformation. Figure 3 shows X-ray diffraction patterns of the $Al_{87}Ni_8Gd_5$ alloy before (1) and after (2) cryogenic thermal cycling. Al-reflections are marked with crosses. The inset illustrates the decomposition of the X-ray diffraction pattern in amorphous (a) and crystalline (c) components. It is seen that the fraction of the nanocrystalline phase decreases during cryogenic thermal cycling. No signs of a crystalline phase were observed in the X-ray diffraction patterns with an increase in the number of cycles to more than 100.

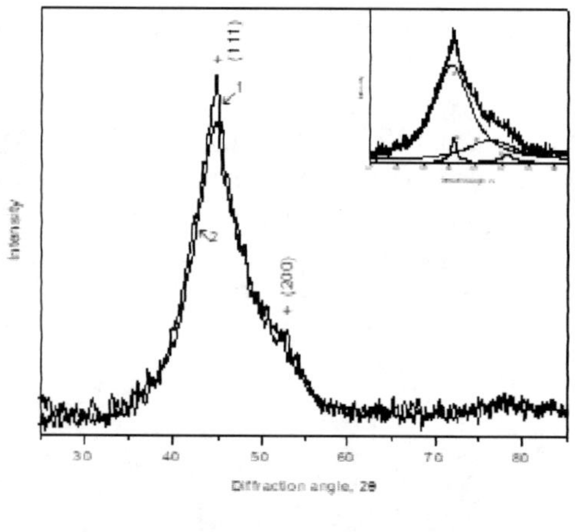


**Figure 3.** X-ray diffraction patterns of the Al87Ni8Gd5 alloy before (1) and after (2) cryogenic thermal cycling.

IOP Conf. Series: Materials Science and Engineering **672** (2019) 012014 doi:10.1088/1757-899X/672/1/012014

Thus, the application of cryogenic cycling to partially crystalline alloys allows decreasing the fraction of the nanocrystalline structure component and, under certain conditions, rejuvenating completely the amorphous phase.

## 4. Conclusions

Studies of the structural evolution of a group of Al-based alloys of Al-Ni-RE (RE = La, Y, Gd) systems have shown that using the combined heat and deformation influence on the structure, it is possible to obtain samples with different crystal sizes and different fractions of the nanocrystalline phase. The application of the cryogenic cyclic method enables rejuvenation of the amorphous phase in partially crystalline alloys.

**Acknowledgments**

The research is carried out within the state task of ISSP RAS and partially supported by RFBR (grant no. 19-03-00355).

## References

[1] Yoshizawa Y, Yamauchi K, Yamane T and Sugihara H 1988 *J. Appl. Phys.* **64** 6047
[2] Kim Y H 1991 *Mater. Trans. JIM* **32** 331
[3] He J, Zhou L, Zhao D L and Wang X L 2009 *J. Mater. Res.* **24** 1607
[4] Gleiter H 1989 *Prog. Mater. Sci.* **33** 223
[5] Birringer R 1989 *Mater. Sci. Eng.* A **117** 33
[6] Abrosimova G E *et al.* 1999 *Nanostructured Materials* **12** 617
[7] Abrosimova G E 2011 *Physics-Uspekhi* **54** 1227
[8] Cremanshi V *et al.* 2000 *J. Mater. Res.* **15** 1936
[9] Abrosimova G *et al.* Slipenyuk A 2001 *J. Mater. Sci.* **36** 3933
[10] Abrosimova G, Aronin A and Ignatieva A 2007 *Mater. Sci. Eng.* A **449-451** 485
[11] Pan J, Chen Q, Liu L and Li Y 2011 *Acta Mater.* **59** 5146
[12] Maaß R, Samwer K, Arnold W and Volkert C A 2014 *Appl. Phys. Letters* **105** 171902
[13] Louzguine D V and Inoue A 2006 *J. Non-Cryst. Solids* **352** 3903
[14] Henits P, Kovács Zs and Révész Á 2008 *Rev. Adv. Mater. Sci.* **18** 597
[15] Abrosimova G E and Aronin A S 2009 *Physics of Solid State* **51** 1765
[16] Tkatch V I *et al.* 2011 *J. Non-Cryst. Solids* **357** 1628
[17] Pershina E, Abrosimova G, Aronin A, Matveev D and Tkatch V 2014 *Mater. Letters* **134** 60
[18] Louzguine-Luzgin D V, Bazlov A I, Ketov S V and Inoue A 2015 *Mater. Chem. Phys.* **162** 197
[19] Guinier A 1956 *Theorie et Technique de la Radiocristallographie* (Paris: Dumond)
[20] Gloriant T, Gich M, Suriñach S, Baró M D and Greer A L 2000 *Mater. Sci. Forum* **343-346** 365
[21] Yang H W, Wen J, Quan M X and Wang JQ 2009 *J. Non-Cryst. Solids* **355** 235
[22] Abrosimova G E, Aronin A S and Kholstinina N N 2010 *Physics of the Solid State* **52** 445
[23] Abrosimova G and Aronin A 2017 *Mater. Letters* **206** 64
[24] Aronin A, Matveev D, Pershina E, Tkatch V and Abrosimova G 2017 *J. All. Comp.* **715** 176
[25] Yavari A R 1988 *Acta Metall.* **36** 1863
[26] Abrosimova G E, Aronin A S, Ignat'eva E Yu and Molokanov V V 1999 *JMMM* **203** 169
[27] Gleiter H 2008 *Mater. Sci. Forum* **584-586** 41
[28] Mattern N *et al.* 2009 *Acta Mater.* **57** 903
[29] Şopu D, Albe K, Ritter Y and Gleiter H 2009 *Appl. Phys. Letters* **94** 191911
[30] Ghafari M *et al.* 2012 *Appl. Phys. Letters* **100** 133111
[31] Chen N *et al.* 2013 *Nanotechnology* **24** 045610
[32] Abrosimova G, Aronin A and Budchenko A 2015 *Mater. Letters* **139** 194
[33] Ketov S V *et al.* 2015 *Nature* **524** 200
[34] Hufnagel T C 2015 *Nature Materials* **14** 867
[35] Abrosimova G, Volkov N, Tran V T, Pershina E, Aronin A 2019 *Mater. Letters* **240** 150

IOP Conf. Series: Materials Science and Engineering **672** (2019) 012015 doi:10.1088/1757-899X/672/1/012015

# Post-dynamic recrystallization of the Cu-0.6Cr alloy

**D A Aksenov[1,2], R N Asfandiyarov[1,2], G I Raab[1] and A G Raab[1]**

[1] Ufa State Aviation Technical University, 12 K. Marx str., Ufa 450008 Russia
[2] Institute of Molecule and Crystal Physics UFRC RAS, 71 pr. Oktyabrya, Ufa 450054 Russia

E-mail: AksyonovDA@mail.ru

**Abstract.** The results of the studying the transformation of structural parameters in the samples of the Cu-0.6Cr alloy during high-temperature free upsetting and subsequent cooling at different rates with the use of different media: liquid nitrogen, water and air are presented. Using EBSD and X-ray diffraction (XRD) analysis, we demonstrate that a decrease in the cooling rate of the samples of the Cu-0.6Cr alloy after a large deformation of e~2 at a temperature of 800-850 °C promotes the formation of smaller structural fragments, an increase in the dislocation density and a reduction in the fraction of high-angle boundaries.

## 1. Introduction

The processes of dynamic polygonization and recrystallization during high-temperature deformation processing have a decisive effect on the type of structure being formed. They participate in structural transformations and the formation of properties in different classes of functional materials. There are a number of papers investigating the kinetics of these processes in various materials and under different deformation procedures [1-9]. Analysis of these papers shows that the predominant condition for the nucleation of recrystallized grains is the non-equilibrium state of grain boundaries, and recrystallization occurs via the mechanism of splitting of high-angle boundaries [1,2]. As a result, in the process of high-temperature deformation, the fraction of recrystallized grains increases and, correspondingly, the structure becomes refined. There are publications investigating the effects of deformation temperature and strain rate in a low-alloyed Cu-Cr-Zr alloy for electrical engineering applications on the polygonization and recrystallization processes in the structure. Using EBSD and TEM, it was found that an increase in strain rate leads to an increase in the volume of recrystallized grains [10].

With the development of modern techniques of severe plastic deformation (SPD), which provide active refinement of the structure to an ultrafine range under large strains, an important factor is the study of the effect of deformation and post-deformation processing parameters on the recrystallization processes, structural changes and a set of properties in metals. Therefore, the aim of the present work is to study the effect of large deformation (e~2) by free upsetting of the samples of the Cu-0.6Cr alloy at a temperature of 800-850 °C and subsequent cooling at different rates on the transformation of the initial structure.

## 2. Material and experimental procedures

The Cu-0.6%Cr alloy in the form of cylindrical billets with a diameter of 20 mm and a height of 40 mm was subjected to upsetting by 85-90% at a temperature of 800-850 °C. Preliminary, the billets of chromium bronze had been exposed to a temperature of 1000 °C for 1 hour with subsequent water quenching. The deformation was performed on a hydraulic press with a force of 6 GPa and a strain rate of 4 mm/s. Temperature was controlled using a Testo830-T1 pyrometer. At the beginning of the deformation, the temperature of the samples was 850±5 °C, and after the deformation was finished, the temperature was not less than 800 °C. As a result of the deformation, round samples with a diameter of 56 mm and a height of 5 mm were produced. In order to study the effect of the cooling rate of the samples on the changes in structural parameters, cooling was performed in three media: in liquid nitrogen, in water and in air.

Structural studies of the cross-section of the samples were performed using EBSD and XRD analysis. Scanning electron microscopy with electron backscatter diffraction (EBSD) was performed employing a TESCAN MIRA 3 LMH field emission microscope equipped with an Oxford Instruments HKL Channel 5 system. The scanning steps were 0.1 and 0.4 μm. Grain misorientation maps were constructed, grain and subgrain size were estimated, a quantitative analysis of the low- and high-angle boundaries was conducted. Diffraction patterns were taken with a Rigaku Ultima IV diffractometer in the Bragg-Brentano geometry using CuKα radiation generated at a voltage of 40 kV and a current of 40 mA. X-ray patterns were obtained within a range from 35° to 145° in the continuous scanning regime with a rate of 0.25 °/sec.

## 3. Results

As a result of the upsetting at a high temperature and subsequent quenching in different media, 3 states with different types of structures were obtained. To characterize the structural state and determine the grain misorientations, we performed an EBSD analysis with identification of the size of grains and subgrains. It can be seen from the bar charts of grain misorientation distribution (figure 1) that after quenching in nitrogen, grains with misorientation angles above 15° prevail.

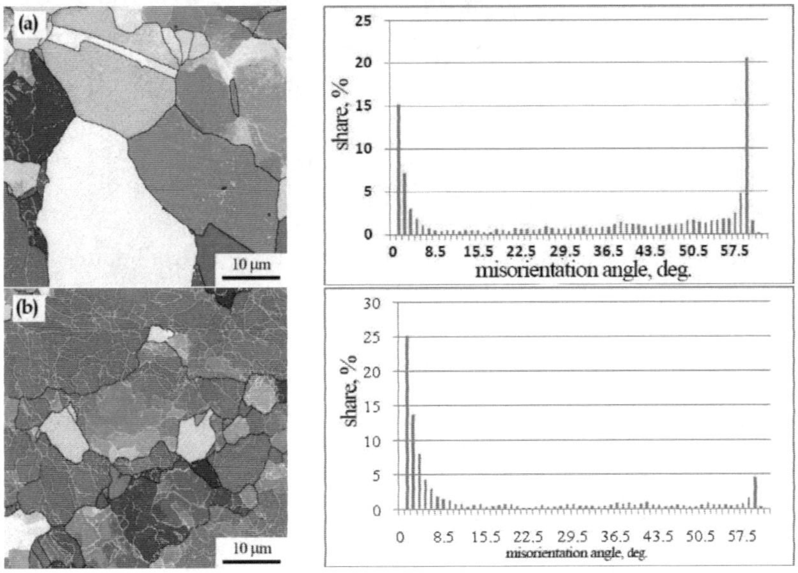

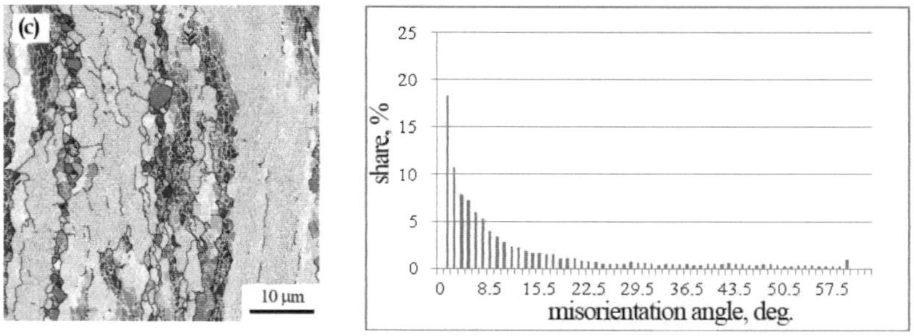


**Figure 1.** Maps and bar charts of grain misorientation distribution: (a) upsetting 800 °C + cooling in liquid nitrogen, (b) upsetting 800 °C + cooling in water, (c) upsetting 800°C + cooling in air.

A high fraction of boundaries, over 20%, refers to misorientation angles of 58-60°, which correspond to twin boundaries. The fraction of twin boundaries decreases with decreasing cooling rate, as shown by the results of the study of the water-quenched condition. At the same time, the fraction of medium-angle boundaries increases and the fraction of high-angle boundaries decreases with decreasing cooling rate (figure 2).

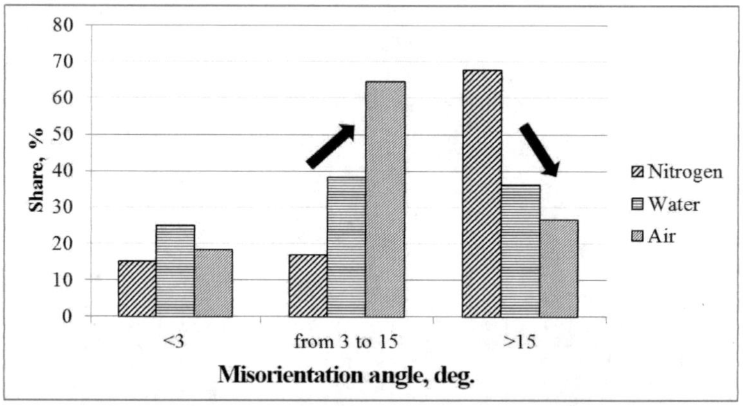


**Figure 2.** Variation in the fraction of low-, medium- and high-angle boundaries.

The data on the size of grains and subgrains forming in the process of hot deformation and subsequent cooling are presented in table 1.

**Table 1.** Average sizes of grains and subgrains in the Cu-0.6Cr alloy.

| | Average subgrain size, µm | Average grain size, µm |
|---|---|---|
| Upsetting 800 °C quenching in nitrogen | 5.0±0.3 | 7.5±0.4 |
| Upsetting 800 °C quenching in water | 3.1±0.3 | 6.5±0.4 |
| Upsetting 800 °C quenching in air | 1.1±0.2 | 1.7±0.2 |

The XRD results indicate that with an increase in the cooling time (cooling in air), the lattice parameter decreases (table 2), which is related to the process of decomposition of the supersaturated solid solution. The lattice parameter after quenching in nitrogen is 3.618(4) Å, and after cooling in air the lattice parameter is restored to a value of 3.616 (4) Å. The character of solid solution decomposition affects the dislocation density, since dispersed particles are effective stoppers for dislocation propagation.

**Table 2.** XRD analysis data for the Cu-0.6Cr alloy.

| | a, Å | CRS, nm | e, % | Dislocation density, $m^{-2}$ |
|---|---|---|---|---|
| Upsetting 800 °C quenching in nitrogen | 3.618(4) | 142±7 | 0.020 | $1.9*10^{13}$ |
| Upsetting 800 °C quenching in water | 3.617(6) | 155±7 | 0.023 | $2.1*10^{13}$ |
| Upsetting 800 °C quenching in air | 3.616(4) | 83±6 | 0.042 | $6.8*10^{13}$ |

## 4. Conclusions

1. Large deformation (e~2) by free upsetting of the samples of the Cu-0.6Cr alloy at a temperature of 800-850 °C and subsequent cooling in different media leads to the formation of the following structural parameters:
- During cooling in liquid nitrogen, a grain type of structure is formed with an average grain size of 7.5±0.4μm, a low dislocation density ($1.9*10^{13}$ $m^{-2}$) and a high fraction of growth twins. This condition corresponds to the stage of finished dynamic recrystallization, the fraction of high-angle boundaries is over 60%.
- During cooling in water, the fractions of medium- and high-angle boundaries are practically the same – 36% and 38%. The fraction of twin boundaries decreases. However, the dislocation density is close to the level of dislocation density after cooling in liquid nitrogen –$2.1*10^{13}m^{-2}$. This condition corresponds to the stage of unfinished dynamic recrystallization. The average grain size is 6.5±0.4 μm.
- Cooling in air leads to the formation of a predominantly subgrain type of structure. In this case, the process of post-dynamic recrystallization occurs. The average size of the grain-subgrain structure is 1.7±0.2 μm, the fraction of high-angle boundaries is not more than 23%. Active decomposition of the supersaturated solid solution takes place. The dispersed particles of the second phases, formed as a result of the exposure to high temperature, are efficient barriers to the dislocation propagation in the grain interiors. As a consequence, the dislocation density grows to $6.8*10^{13}$ $m^{-2}$.

2. A decrease in the cooling rate of the Cu-0.6Cr alloy samples after large deformation at a temperature of 800-850 °C promotes the formation of smaller structural fragments, an increase in the dislocation density, and a reduction in the fraction of high-angle boundaries.

## Acknowledgments

The study was supported by the Russian Science Foundation (project №19-19-00432).

## References

[1] Beltran O, Huang K and Loge R E 2015 *Comp. Mater. Sci.* **102** 293
[2] Huang C, Jia X and Zhang Z 2018 *Metals* **8** 585
[3] Zhang Y and Tian B 2016 *J. Mater. Res.* **31** 1275
[4] Chashchukhina T I, Voronova L M, Degtyarev M V and Pokryshkina D K 2011 *Phys. Met. Metallogr.* **111** 304
[5] Gao W, Belyakov A, Miura H and Sakai T 1999 *Mater. Si. Eng.* A **265** 233
[6] Chen M-S, Lina Y C and Lia K-K 2017 *Procedia Engineering* **207** 2125
[7] Zhang Y, Sun H-L, Volinsky A A, Tian B-H, Chai Z, Liu P and Liu Y 2016 *Acta Metall. Sin. (Engl. Lett.)* **29** 422
[8] Dong L, Zhong Y, Ma Q, Yuan C and Ma L 2008 *Tsinghua Science and Technology* **13** 609
[9] Zhang Y, Huili S, Volinsky A A, Tian B, Chai Z, Liu P and Liu Y 2016 *JMEPEG* **25** 1150
[10] Morozova A, Borodin E, Bratov V, Zherebtsov S, Belyakov A and Kaibyshev R 2017 *Materials* **10** 1394

# Structural features and mechanical properties of Grade 4 titanium from VSMPO-AVISMA (Russia) and Grade 4 titanium from Carpenter Technology Corporation (USA), subjected to ECAP-Conform

**I S Kodirov, R Z Valiev, G I Raab, G N Aleshin and A G Raab**

Institute of Physics of Advanced Materials, Ufa State Aviation Technical University, 12 K. Marx str., Ufa 450008 Russia

E-mail: galioshin@mail.ru

**Abstract**. In this paper, we analyze the microstructural features and mechanical properties of the Grade 4 titanium produced in Russia (VSMPO-AVISMA) and Grade 4 titanium from Carpenter Technology Corporation (USA). It is shown that during the ECAP-Conform processing and subsequent drawing, a homogeneous UFG structure is formed in both Ti materials, providing an enhanced level of mechanical properties with an ultimate tensile strength $\sigma_B > 1200$ MPa and an elongation to failure $\delta \sim 12\%$.

## 1. Introduction

Commercially pure (CP) Ti and its alloys are widely applied in the chemical, aircraft-building and medical industries. CP Ti for medical applications is used in the production of implants for traumatology, orthopedics and maxillofacial surgery [1]. The development of modern processing methods, such as severe plastic deformation (SPD), makes it possible to enhance considerably the mechanical properties of Ti through nanostructuring and provide a level of properties superior to that of the Grade 5 Ti alloy (analogue of the Russian VT6 alloy) widely used in medicine [1-6]. Of special interest is the SPD technique of ECAP-Conform (ECAP-C) which enables the fabrication of products that are in most demand, i.e. long-length rods [5, 7-8]. Considering the fact that the Grade 5 alloy is toxic due to the presence of Al and V additions, replacing this alloy with the high-strength nanostructured Grade 4 Ti in applied practice is highly relevant [9]. Therefore, in this work we studied the structure and mechanical properties of nanostructured CP Grade 4 Ti produced in Russia (VSMPO-AVISMA, ASTM F 67-13, ASTM B 348-13) it terms of its applicability for medical products.

## 2. Material and experimental procedure

In this work, we studied the CP Grade 4 Ti produced by VSMPO-AVISMA Corporation (Verkhnyaya Salda, Russia). The material's chemical composition is given in table 1. The chemical composition of the foreign-made Grade 4 Ti (Carpenter Technology Corporation, USA) is also listed for comparison.

**Table 1.** Chemical composition of the alloys under study.

| Material, manufacturer | Chemical elements, wt.% | | | | | | | |
|---|---|---|---|---|---|---|---|---|
| | Ti | Mo | W | Fe | Sn | C | Al | Co |
| Grade 4, VSMPO-AVISMA | 98.96 | <0.5 | 0.073 | 0.315 | 0.0076 | <0.0015 | 0.049 | 0.014 |
| Grade 4, Carpenter | 99.08 | <0.5 | 0.186 | 0.122 | 0.0190 | 0.0180 | 0.012 | 0.011 |

With a view to relieve the stresses and stabilize the structure of the alloy, we annealed the as-received alloy billets in the shape of rods with a diameter of 10 mm and a length of 1000 mm in a Nabertherm furnace at a temperature of 680 °C for 1 hour [4].

Microstructural studies were performed on a Jeol JEM 2100 transmission electron microscope, and X-ray diffraction analysis was performed on a RIGAKU ALTIMA-4 X-ray diffractometer.

The resulting structural state was further regarded as the initial one. To produce a UFG structure with a view to increase the strength characteristics of Grade 4 Ti, we processed the rods by ECAP-Conform (ECAP-C) at a temperature of 250 °C for the purpose of structure refinement. 6 ECAP-C passes were implemented, and subsequently the deformed rods were subjected to drawing to a diameter of 8 mm at a temperature of 250 °C. The produced billets were subjected to tensile mechanical tests.

## 3. Results and discussion

The structural studies of the initial state of the alloys, both by light microscopy and by transmission electron microscopy, showed that both materials consist of the α-phase grains with sizes of ~18±3 μm having arbitrary, mainly high-angle, boundaries (figure 1). The X-ray studies confirmed the presence of the α-phase in the whole bulk of the alloys in the initial state and the absence of precipitates of any other phase. The results of the interpretation of the X-ray diffraction patterns for both materials are presented in table 2. According to these data, the lattice parameters of the alloys, their dislocation densities, the sizes of coherent scattering regions do not differ much.

**Table 2.** XRD analysis results of Grade 4 Ti in different conditions.

| | CSR size, nm | Dislocation density, $m^{-2}$ | Lattice parameter, Å (a/c) | δ % | $\sigma_{0.2}$ МПа | $\sigma_{в}$ МПа |
|---|---|---|---|---|---|---|
| (VSMPO-AVISMA) annealing 1 h 680 °C | 114.1±2.0 | 4.78E+13 | 2.9521/ 4.6895 | 18 | 640 | 769 |
| (Carpenter) annealing 1 h 680 °C | 112.7±2.5 | 9.22E+13 | 2.9524/ 4.6906 | 17.6 | 643 | 758 |
| (VSMPO-AVISMA) UFG Ti* | 73.6±0.3 | 3.34E+14 | 2.9542/ 4.6885 | 12 | 1190 | 1200 |
| (Carpenter) UFG Ti* | 72.3±0.7 | 3.28E+14 | 2.9542/ 4.6890 | 12 | 1170 | 1180 |

*UFG Ti: ECAP-C + drawing

The tensile mechanical tests (table 2) demonstrated very close results for the initial states, i.e. the states after annealing at 680 °C for 1 hour. The ultimate tensile strength (UTS) of the Grade 4 Ti (VSMPO-AVISMA) is $\sigma_в \approx$ 769±6 MPa, the elongation to failure is δ≈18.0±0.5%; for Grade 4 Ti (Carpenter), the UTS is $\sigma_в \approx$ 758±10 MPa, the elongation to failure is δ ≈ 17.6±0.5%.

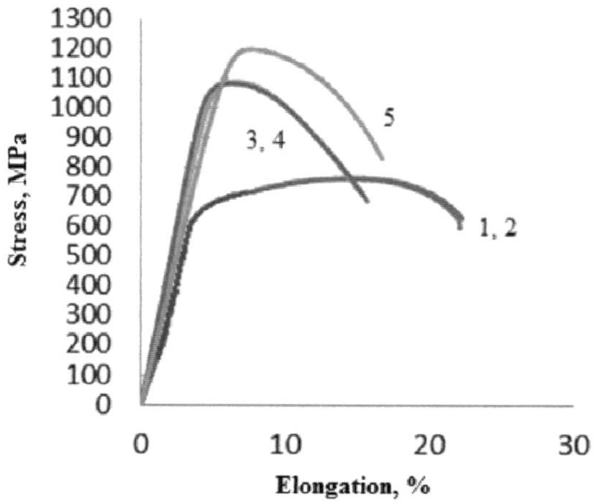


**Figure 1.** σ–δ curves obtained during the tension of Grade 4 Ti in different conditions: 1 – initial condition (VSMPO-AVISMA); 2 – initial condition (Carpenter); 3 – after 6 ECAP-C passes at 250°C (VSMPO-AVISMA); 4 – after 6 ECAP-C passes at 250°C (Carpenter); 5 – after ECAP-C + drawing at 250 °C to ø 8 mm (VSMPO-AVISMA).

As it can be seen from the σ–δ curves based on the results of the tensile mechanical tests of the Ti rods processed by ECAP-C, the tensile strength of the rods increases to 1100 MPa. The uniform elongation during tension for both materials reaches $\delta_u \sim 2.2\%$, and the elongation to failure reaches ~12%.

Figure 2 shows the microstructure of Grade 4 Ti from both manufacturers after ECAP-Conform. As revealed by the microstructure images and electron-diffraction patterns, an ultrafine-grained structure is observed in both materials, having structural elements (grains, fragments) with sizes about 0.1-0.25 μm. In the electron-diffraction patterns, the presence of azimuthal blurring of spots located on concentric circles indicates a high density of crystalline defects (figure 2a). The observed grain/subgrain microstructure has boundaries of deformation origin with a predominantly high-angle misorientation. On the whole, no differences are observed in the fine structure of the two materials. Thus, the results of the structural studies by transmission electron microscopy of Ti after ECAP-C processing indicate that the USA Grade 4 Ti (Carpenter) and the Russian-made Grade 4 Ti (VSMPO-AVISMA) are practically identical in terms of structure.

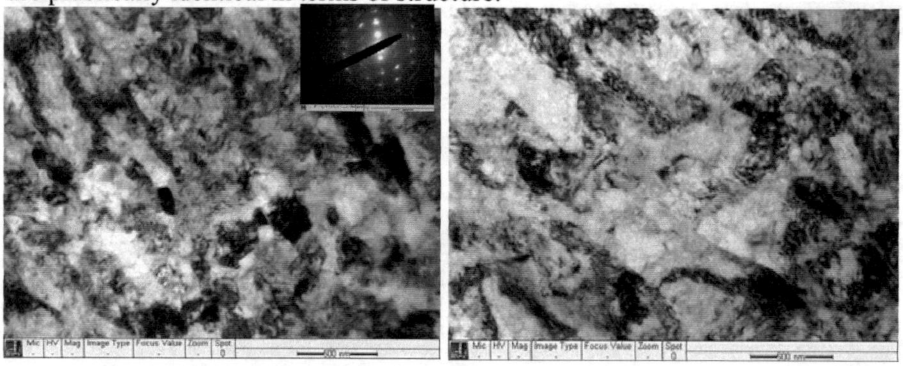

a b

**Figure 2.** Microstructure of Grade 4 Ti after ECAP-C at 250 °C: a – VSMPO-AVISMA, b – Carpenter. TEM.

With a view to further increase the strength of the Ti rods processed by ECAP-C for 6 passes at 250 °C, they were subjected to drawing at a temperature of 250°C to a diameter of 8 mm. The conducted tensile mechanical tests of the Ti rods after drawing to ø 8 mm (figure 1) show that the ultimate tensile strength $\sigma_в$ increased to ~ 1200 MPa.

Therefore, the performed studies into the microstructure and mechanical properties of nanostructured CP Grade 4 Ti of Russian manufacture (VSMPO-AVISMA + IPAM, Ufa) and of U.S. manufacture (Carpenter) have demonstrated that both materials are practically identical in terms of their deformation behavior, structural state and level of achieved properties that is superior to that of the analogues used in current practice. The latter is an important condition for their use in the fabrication of nanostructured rods intended for the manufacture of medical products.

**Acknowledgements**
This research was supported by the Russian Foundation for Basic Research under project No. 17-08-00720/17 and by the Russian Science Foundation under project No. 19-49-02003.

**References**

[1] 2018 *Titanium in Medical and Dental Applications* ed Froes F H and Ma Q (Duxford: Woodhead Publishing)
[2] Valiev R 2004 *Nature Mater.* **3** 511
[3] Stolyarov V V, Zhu Y T, Alexandrov I V, Lowe T C and Valiev R Z 2003 *Mater. Sci. Eng.* A **343** 43
[4] Zherebtsov S V, Salishchev G A and Galeyev R M 2002 *Defect and Diffusion Forum* **208–209** 237
[5] Semenova I P, Polyakov A V, Raab G I, Lowe T C and Valiev R Z 2012 *J. Mater. Sci.* **47** 7777
[6] Langdon T G 2013 *Acta Mater.* **61** 7035
[7] Valiev R Z, Zhilyaev A P and Langdon T G 2014 *Bulk Nanostructured Materials: Fundamentals and Applications* (Hoboken: John Wiley & Sons, Inc.)
[8] Utyashev F Z and Raab G I 2013 *Deformation Methods for the Fabrication and Processing of Ultrafine-Grained and Nanostructured Materials* (Ufa: Gilem) [in Russian]
[9] Polyakov A V, Semenova I P, Bobruk E V, Baek S M, Kim H S and Valiev R Z 2018 *Adv. Eng. Mater.* **20** 1700863

IOP Conf. Series: Materials Science and Engineering **672** (2019) 012017 doi:10.1088/1757-899X/672/1/012017

# Structure of deformed Al-based amorphous alloys

**D Matveev[1], E Pershina[1], E Postnova[1], D Fokin[2] and A Aronin[1]**

[1] Institute of Solid State Physics of RAS, Chernogolovka, Moscow distr., Russia
[2] OPTEC Moscow, Russia

E-mail: aronin@issp.ac.ru

**Abstract**. The structure of deformed Al-based amorphous alloys was studied by X-ray diffraction, scanning, transmission and probe electron microscopy. It has been found that the material in the shear bands has lower values of Young's modulus and density. The shear band width is about 50 nm. In the band, no redistribution of the alloy components typical for phase separation is observed. The size of the regions with reduced Young's modulus around the shear band is in the range from several tenths of a micron to 1 μm.

## 1. Introduction

Today metal amorphous and partially crystalline alloys are objects under extensive study. The size of samples varies depending on the material composition and the cooling rate. These materials have an unusual structure and demonstrate high physical and chemical properties. Fe-based partially crystalline alloys (FINEMET) have splendid hysteresis properties. Zr-based amorphous alloys in the form of rods and plates (BMG) show good mechanical properties. The value of elastic deformation in them can reach a few per cent [1]. Al-based amorphous-nanocrystalline alloys with rare-earth metals have superior strength properties with a low specific weight [2,3]. These materials can be formed under crystallization as a result of heating or deformation. Deformation of amorphous alloys usually occurs by the formation of shear bands (deformation bands). The amorphous phase in the shear band has characteristics different from those of the surrounding amorphous matrix. In the shear band, the amorphous phase is less ordered and has reduced density [4,5]. The value of differences in the density of the amorphous phase in deformation bands can range from a few to tens of per cent [6, 7]. In the deformation band, the processes of mass transfer are simplified, and the diffusion coefficient at room temperature can exceed that in the surrounding matrix by 5-6 orders of magnitude [8]. This significant change in the kinetic characteristics in these distinguished regions leads to the simplified formation of nanocrystals in them [9,10]. Nanocrystals are not always formed in shear bands [11,12], and the question remains open, what conditions are necessary for their formation. Deformation bands come out to the surface of the sample, and steps are formed. The thickness of the shear bands ranges from several tens to hundreds of nanometers [13]. There are several models which describe the processes resulting in the formation of shear bands. They use the concept of free volume and its evolution. Thus, according to the model described in [4,13], both the production of free volume and its annihilation occur under the development of the flow in the shear band. In principle, the composition of the amorphous phase in the shear bands can differ from that of the surrounding matrix. Then phase separation will be observed, which is found in these materials under deformation [14-16]. Shear bands determine largely the mechanical properties of amorphous materials, being their “weak point”. The amorphous phase around the shear band is weakened [17,18]. Microhardness decreases in the shear band and in the matrix surrounding the bands. The size of the region where the microhardness decreases significantly exceeds the thickness of the shear band and can reach 150 μm [19, 20]. Such a size of the reduced microhardness region was observed in a Zr-based alloy under compression. The question arises to what extent this size of the region with reduced strength properties around the band is typical of other deformation methods, as well as if the composition of the band differs from that of the matrix.

The present paper is devoted to clarification of these questions. Al-based alloys with transition and rare-earth metals were chosen as the research objects. The present work aims at determining the local distribution of regions with different Young's moduli in the samples containing numerous deformation bands. The samples of the $Al_{87}Ni_8La_5$ amorphous alloy were selected as the objects; this alloy is deformed at room temperature with the formation of shear bands.

## 2. Experimental

$Al_{87}Ni_8La_5$ and $Al_{90}Y_{10}$ amorphous alloys were obtained by rapid melt quenching. The ribbons had a thickness of about 40 μm and a width of 10 mm. The obtained samples were subjected to deformation by multiple rolling and high-pressure torsion (HPT). The structure and composition were investigated by X-ray diffraction methods using Co Kα and Cu Kα radiation and by transmission, scanning and scanning transmission electron microscopy. Deformation of the $Al_{87}Ni_8La_5$ alloy was carried out by multiple rolling (at room temperature); the decrease in thickness was 90%. The $Al_{90}Y_{10}$alloy was deformed by high-pressure torsion (5GPa). Maps of the distribution of mechanical properties were plotted by probe microscopy methods (PeakForceQNM). To plot the maps, the samples were preliminarily polished so that no steps remained on the surface, which arise when deformation bands come out to the surface.

## 3. Results and discussion

Figure 1 shows the image of the surface of the $Al_{87}Ni_8La_5$ alloy after deformation by multiple rolling. One can see numerous steps, which are formed as a result of deformation bands coming out to the surface of the sample. The steps form a net, the distance between the bands varies from several tenths of a micrometer to several micrometers. Nanocrystallization (formation of Al nanocrystals) does not occur; there is only a diffuse halo in the corresponding X-ray diffraction pattern (figure 1b).

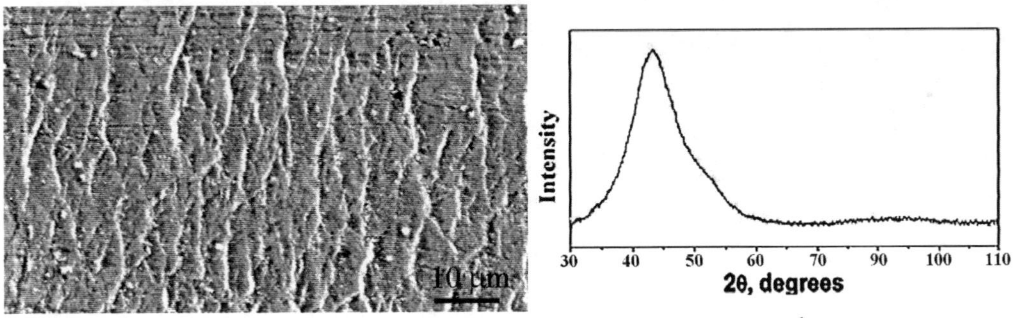


a b

**Figure 1.** Steps of shear bands on the surface of the deformed $Al_{87}Ni_8La_5$ alloy (a); X-ray diffraction pattern of the deformed $Al_{87}Ni_8La_5$ alloy (b).

No traces of a change in the properties related to the deformations bands are seen on the initial map of distribution of the Young's modulus over the surface of this sample (under a load of 0.3 μN). Figure 2 illustrates a map of Young's modulus distribution obtained under a load of 0.85 μN. One can see that the regions of reduced Young's modulus appear. These areas appear with increasing load. The width of these regions is tenths of a micrometer, and the location correlates with the width and location of the steps on the sample surface (figure 1a). Note that the regions of reduced Young's modulus have a lateral size of less than 1 μm. These sizes of the regions of reduced Young's modulus do not correspond to the regions of reduced microhardness observed by nanoindentation, the width of which was more than 100 μm [20]. Such a large difference in the size of the region is probably caused by different types of deformation or the other alloy composition.

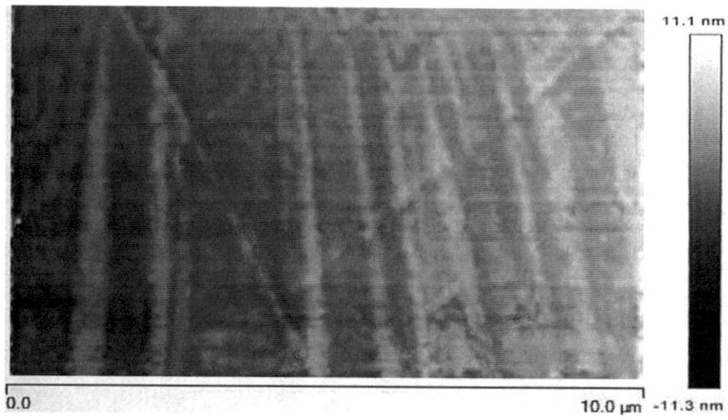


**Figure 2.** Maps of distribution of the Young's modulus in the deformed $Al_{87}Ni_8La_5$ alloy (a – PeakForce = 0.85 μN).

The authors of [20] deformed the sample by compression and studied the distribution of microhardness around a separate (single) band, whereas in our samples there were numerous bands formed under multiple rolling. An amorphous-nanocrystalline structure (figure 3a) was observed in the $Al_{90}Y_{10}$ sample after deformation by the HPT method at 0.5 revolution. The average size of Al nanocrystals was 8 nm. Typical peaks from nanocrystals are present against the background of the halo in the X-ray diffraction pattern (figure 3b).

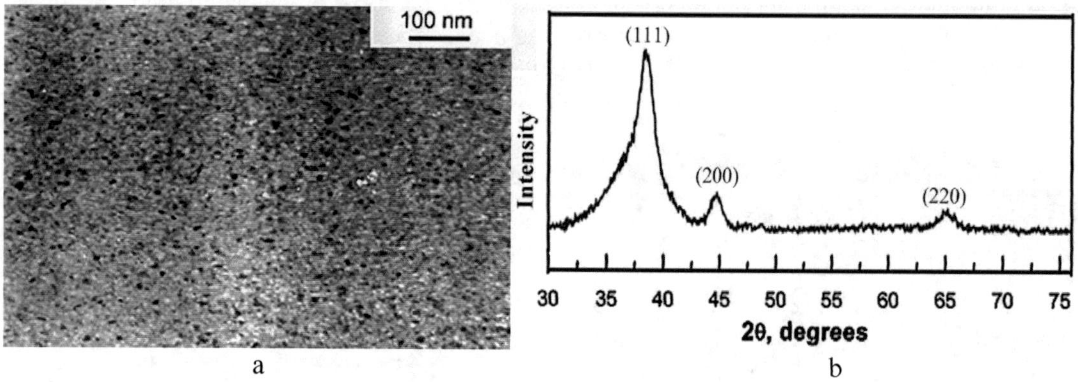


**Figure 3.** TEM image (a) and X-ray diffraction pattern (b) of the $Al_{90}Y_{10}$ amorphous sample after deformation by the HPT method (0.5 revolution).

Scanning across the shear bands (as demonstrated in figure 4a) allowed obtaining the distribution of Al and Y concentrations in the sample along the marked scanning line. A significant decrease in the intensity of the signal from the shear band compared to the signal from the surrounding sample is seen in the obtained plot (figure 4b). Most probably, this is related to a reduced material density in the shear band. It is also possible that the thickness of the electron microscopic foil in the region of the shear band was less, as well as the simultaneous influence of several factors on intensity of the signal. The width of the region correlates with that of the shear band (figure 4a) and is about 50 nm. Note that the ratio of the amplitudes of Al and Y signals inside and outside the shear band remains constant.

Consequently, redistribution of the concentrations under deformation and band formation does not occur.

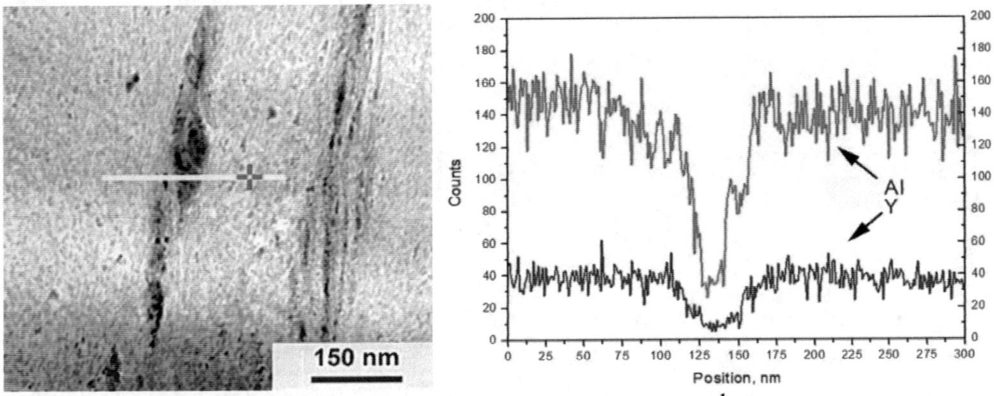


a b

**Figure 4.** STEM image (a) (the size of the electron beam was about 2 nm) and EDS spectra (b) (along the line in the image) of the $Al_{90}Y_{10}$ sample after deformation by HPT method (0.5 revolution).

## 4. Conclusions

Studies of the structural evolution of Al-based amorphous alloys of Al-Ni-RE (RE = La, Y) systems under deformation have demonstrated the following:

- no redistribution of element concentration occurs in the shear bands, i.e., the composition of the band and the matrix is similar;
- the sizes of regions with reduced Young's modulus around the shear band are in the range from several tenths of a micron to 1 μm.

**Acknowledgments**

The research is carried out within the state task of ISSP RAS and partially supported by RFBR (grant no. 19-03-00355).

## References

[1] Trexler M and Thadhani N 2010 *Prog. Mater. Sci.* **55** 759
[2] Eckert J *et al.* 2008 *Rev. Adv. Mater. Sci.* **18** 169
[3] Kim H *et al.* 1991 *Mater. Trans. JIM* **32** 331
[4] Cao A *et al.* 2009 *Acta Mater.* **57** 5146
[5] Packard C *et al.* 2007 *Acta Mater.* **55** 5348
[6] Stolpe M *et al.* 2014 *Acta Mater.* **64** 231
[7] Klaumunzer D *et al.* 2011 *Phys. Rev. Lett.* **107** 185502
[8] Aronin A and Louzguine-Luzgin D 2017 *Mechanics of Mater.* **113** 19
[9] Louzguine-Luzgin D and Inoue A 2006 *J. Non-Crystall. Sol.* **352** 3903
[10] Abrosimova G *et al.* 2011 *Physics of the Solid State* **53** 229
[11] Abrosimova G and Aronin A 2018 *J. Alloys Compd.* **747** 26
[12] Pershina E *et al.* 2015 *Mater. Lett.* **134** 60
[13] Greer A *et al.* 2013 *Mater. Sci. Eng.* R **74** 71
[14] Abrosimova G *et al.* 2015 *Mater. Lett.* **139** 194
[15] Louzguine-Luzgin D *et al.* 2019 *Scripta Mater.* **167** 31
[16] Abrosimova G *et al.* 1999 *JMMM* **203** 169
[17] Ketov S *et al.* 2016 *J. Alloys Compd.* 2016 **687** 221
[18] Nasu T *et al.* 2015 *Mater. Trans.* **56** 249
[19] Yoo B-G *et al.* 2009 *Scripta Mater.* **61** 951
[20] Pan J *et al.* 2015 *Acta Mater.***59** 5146

IOP Conf. Series: Materials Science and Engineering **672** (2019) 012018 doi:10.1088/1757-899X/672/1/012018

# Study of the processing of long-length billets by continuous free bending

**R N Asfandiyarov[1,2], D A Aksenov[1,2] and G I Raab[1]**

[1] Ufa State Aviation Technical University, 12 K. Marx str., Ufa 450008 Russia
[2] Institute of Molecule and Crystal Physics UFRC RAS, 71 pr. Oktyabrya, Ufa 450054 Russia

E-mail: a.r.n@list.ru

**Abstract.** This paper investigates the features of deformation processing by continuous free bending in rollers, aimed at enhancing the mechanical properties of long-length metallic billets without any change in their cross section. As the material for the investigation, we selected a chromium-zirconium bronze for electrical engineering applications, used, for example, for the production of contact wires that require high mechanical characteristics. Computer simulation reveals that during the processing by continuous free bending, a gradient field of accumulated strain is formed, with maximum values in the peripheral region and with minimum values in the center of the billet. It is found that during bending, by the 6th cycle the peripheral layers of the billet accumulate a strain of e~2, and the damage reaches a value of ~1. In the process of verification by means of a physical experiment, cracks emerge on the billet surface during the 6th cycle, confirming the simulation data. Structural analysis shows that during multi-cycle processing by this method, a gradient type of structure is formed. An important advantage of the processing by continuous free bending is the enhancement of the mechanical properties of long-length samples without any reduction in the cross-sectional area: e.g., after 4 processing cycles the ultimate tensile strength increased from 250±20 to 380±20 MPa, and the variation of the cross-sectional area was ~5-7%. The obtained results are in agreement with some principles of severe plastic deformation processing.

## 1. Introduction

The results of studies of the severe plastic deformation (SPD) techniques demonstrate the efficiency of their application for the enhancement of mechanical properties [1-3], however, attention should be paid to the difficulty of implementation of SPD technologies, primarily related to the insufficient technological efficiency of these processes. For instance, the techniques based on equal-channel angular pressing (ECAP) involve some tribological problems that need to be solved. Conventional methods of deformation processing, such as rolling and drawing, enable enhancing the mechanical properties of metallic materials, but at the same time reduce significantly the cross-sectional area.

## 2. Method description

An alternative to the above-mentioned techniques could be deformation by bending that implements shear. Pioneer studies of bending as a deformation procedure demonstrate the positive role of this approach for the refinement of the initial structure in sheet billets [4-6].

In the present study, we investigated the deformation processing of long-length rod-shaped billets by ***continuous free bending technique*** (figure 1) in round rotating rollers with open passes, without any lubricants. This design enables a visual inspection of the process.

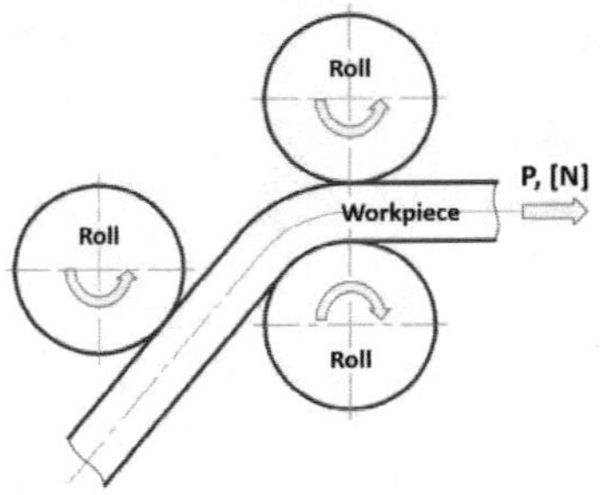


**Figure 1.** Principle of continuous free bending.

## 3. Material and experimental procedures

As the material for the study, we selected the low-alloyed alloyCu-1Cr-0.08Zr (wt.%). This alloy is used for the manufacture of electrodes for resistance spot welding and resistance seam welding, railway contact wires, as well as other parts that require a combination of high electrical conductivity and high hot strength [7-9]. In the study, we used rod-shaped samples with a diameter of 10 mm and a length of 600 mm. In the process of research using the Deform-3D software system, we performed the finite-element computer simulation to study the strain state and the thermal effect of deformation. Further, we conducted a physical experiment – the samples were subjected to thermomechanical treatment comprising two stages:

- at the first stage, a high-temperature treatment – exposure for 1 hour at a temperature of 1000 °C, followed by water quenching;
- at the second stage, the samples were subjected to deformation by continuous free bending in rollers. The number of processing cycles was 1, 4 and 6 (until the emergence of cracks on the surface of a sample). After each cycle, the billet was rotated by 90°. The processing temperature was room temperature, and no lubricants were used.

After the processing of the samples, we performed mechanical tensile tests and structural analysis using light and scanning electron microscopy.

## 4. Results

### *4.1. Computer simulation of the continuous free bending process*

At this stage of research, the main attention was given to the process of strain and damage accumulation (figure 2).

It follows from the analysis of the accumulated strain distribution fields that in the process of bending a gradient strain field is formed: higher strains are observed at the periphery, while lower strains are found in the central region of the deformed billets (figure 2 a, b). Correspondingly, both the structural changes (refinement) and strengthening in the surface peripheral regions will proceed more actively than in the central region of the billets. The accumulated strain after 1 cycle varies from 0.2 in the central region to 0.5 at the surface. To ensure a symmetrical strain distribution, it is necessary to implement at least 4 processing cycles, resulting in strains varying from e=0.75 in the center to e=1.8 at the surface of the billet. After 6 cycles, the values of accumulated strain are e=1.7-2.0 at the periphery and e=0.8-1.0 in the central region (figure 2 a, b). It should be noted that for products experiencing bending and wear during operation the enhanced strength of surface layers provides an increase in operational life.

The results of our study also show that by the 6th processing cycle, the level of damage reaches ~1, which indicates that the technological plasticity is exhausted. The increment in damage in one processing cycle is ~0.2. This result is confirmed by the physical experiment: after the 6th cycle, cracks with a depth of about 200-300 μm emerge on the billet surface.

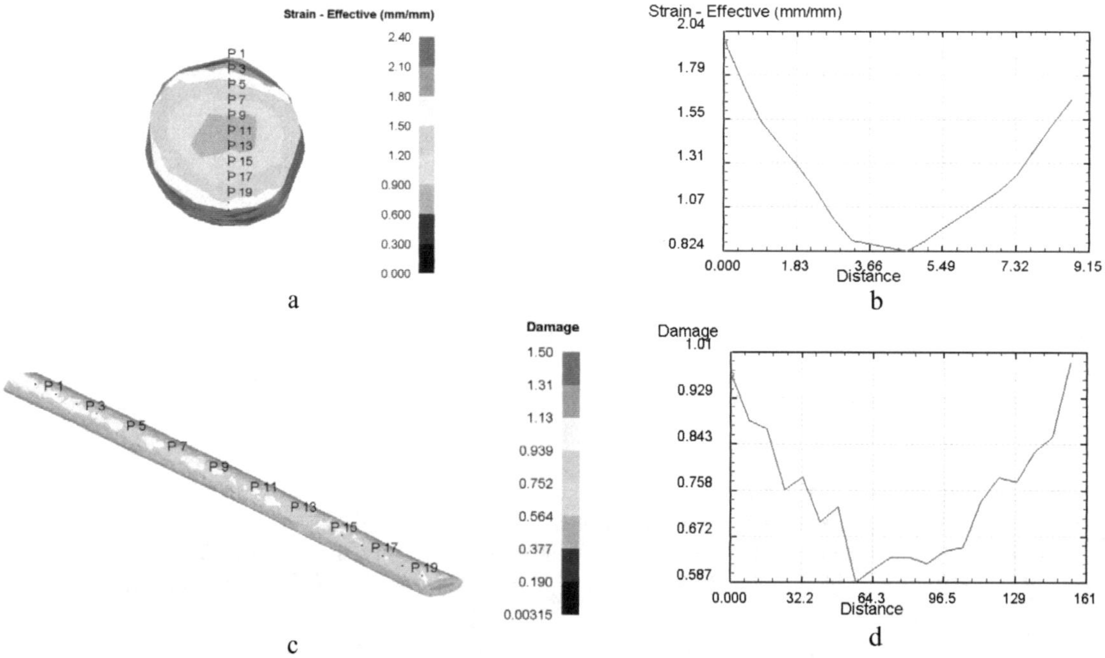


**Figure 2.** Billet after 6 processing cycles: accumulated strain distribution fields (cross section, a, b) and accumulated damage (surface, c, d).

*4.2. Microstructural analysis and mechanical properties*

The microstructure in the sample after 6 cycles of continuous free bending is shown in figure 3. As a result of the processing, a gradient type of structure is formed. In the central region of the sample, coarse grains with growth twins are observed, the mean grain size is 42±3 μm. In the near-surface layer, grains close to equiaxed ones are observed, with a mean size of 15.0±0.8 μm. In the surface layer, a refined structure with a mean grain size of 7.1±0.5 μm is found. This distribution corresponds to the results of the computer simulation of the strain state. In the bulk of the billet, large particles of elongated second-phase particles with an average cross size of 0.90 ± 0.07 μm are observed. Mechanical fragmentation of particles is not observed.

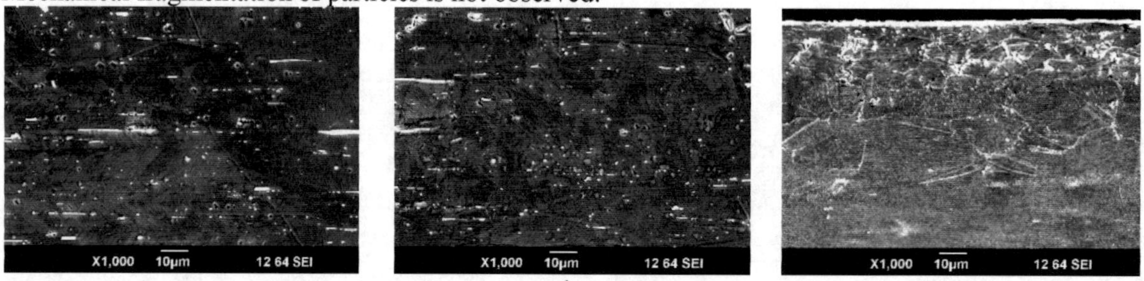


a b c

**Figure 3.** Microstructure in the sample after 6 cycles of continuous free bending (longitudinal section, SEM): a – central region, b – near-surface layer, c – surface layer.

After 4 processing cycles, in the Cu-1Cr-0.08Zr alloy, the ultimate tensile strength increased from 250±20 MPa in the initial as-quenched state to 380±20 MPa in the deformed state.

## 5. Conclusions

The following conclusions can be drawn from the conducted study:

1. The computer study of the stress-strain state in the process of continuous free bending in rollers shows that:

- During bending, a gradient strain field is formed, irrespective of the number of processing cycles. Higher strains are accumulated in the peripheral region. For example, after 6 processing cycles, the difference in the values of accumulated strain in the central and peripheral regions reaches ~1, the maximum value being e=2.
- The increment of the accumulated strain in one processing cycle is up to 0.2 in the central region and up to 0.5 on the surface, and the increment values decrease with increasing number of cycles.
- The damage increment in one processing cycle is ~0.2. By the 6th processing cycle, the damage level reaches values exceeding 1, which indicates that the technological plasticity of the peripheral regions of the billet has been exhausted.

2. Verification of the computer simulation of continuous free bending in rollers by means of a physical experiment shows that:

- A gradient type of structure is formed in the sample, with a mean grain size of 42±3 μm in the central region, 15.0±0.8 μm in the near-surface layer, 7.1±0.5 μm in the surface layer.
- The processing by continuous free bending in rollers ensures the enhancement of mechanical properties of long-length samples without a considerable reduction in the cross-sectional area: after 4 cycles of processing of the Cu-1Cr-0.08Zr alloy, its ultimate tensile strength increased from 250±20 to 380±20 MPa, and the change in the cross-sectional area was only ~5-7%.

## Acknowledgments

The authors are grateful for the support from the Ministry of Science and Higher Education of the Russian Federation under Grant agreement No. 14.586.21.0059 (UIN: RFMEFI58618X0059).

## References

[1] Valiev R Z, Zhilyaev A P and Langdon T G 2014 *Bulk Nanostructured Materials: Fundamentals and Applications* (Hoboken: Wiley STM)
[2] Utyashev F Z and Raab G I 2013 *Deformation Methods for the Fabrication and Processing of Ultrafine-Grained and Nanostructured Materials* (Ufa: Gilem) [in Russian]
[3] Segal V M, Beyerlein I J, Tome C N, Chuvil'deev V N and Kopylov V I 2010 *Fundamentals and Engineering of Severe Plastic Deformation* (New York: Nova Science Publishers)
[4] Method for producing ultrafine-grained materials using repetitive corrugation and straightening. US patent No. 20020088506, application date 05.03.2001, publication date 11.07.2002.
[5] Huang J Y, Zhu Y T, Jiang H and Lowe T C 2001 *Acta Mater.* **49** 1497
[6] Sunil B R *Materials and Manufacturing Processes* **30** 1262
[7] Osintsev O E 2004 *Copper and Copper Alloys. Domestic and Foreign Grades: Reference Book* (Moscow: Mashinostroenie) [in Russian]
[8] Vinogradov A, Patlan V, Suzuki Y, Kitagawa K and Kopylov V 2005 *Acta Mater.* **50** 1639
[9] Nikolaev A K 2001 *Tsvetnye Metally* **5** 84 [in Russian]

IOP Conf. Series: Materials Science and Engineering **672** (2019) 012019 doi:10.1088/1757-899X/672/1/012019

# SEM and AFM analysis of the shear bands in Zr-based BMG after HPT

**V Astanin[1], D Gunderov[1,3], E Boltynjuk[2], E Ubyivovk[2], A Churakova[1,3] and S Gunderova[1]**

[1] Ufa State Aviation Technical University, 12 K. Marx st., Ufa 450008 Russia
[2] Saint Petersburg State University, 7-9 Universitetskaya nab., Saint Petersburg 199034 Russia
[3] Institute of Molecule and Crystal Physics UFRC RAS, 71 pr. Oktyabrya, Ufa, 450054 Russia

E-mail: v.astanin@gmail.com

**Abstract**. The surface relief formed by shear bands in bulk metallic glasses (BMG) under high-pressure torsion (HPT) has been investigated by the method of scanning electron microscopy (SEM) and atomic force microscopy (AFM). For this purpose, two halves of disks of the bulk metallic glass were jointed together and processed by HPT. The SEM examination of the internal surfaces of two joint halves of an HPT-processed disk allowed to study the formation and accumulation of shear bands under an increased imposed strain. The maximum density of the shear bands is observed at the edges of the HPT samples and in areas adjacent to the upper anvils. The observed minimum shear band spacing is equal to 0.5 μm after HPT processing for 5 revolutions.

## 1. Introduction

Bulk metallic glass (BMG) [1, 2] attract significant scientific and practical interest. However, during their deformation, the plastic flow is localized in thin shear bands (SB), which leads to brittle fracture during deformation by tension. One of the effective ways to transform the structure and properties of amorphous materials is deformation using high-pressure torsion. Studies [3-6] showed that under the effect of HPT on initially amorphous alloys, partial nanocrystallization and the formation of a nanocluster structure can occur, a growth of free volume and an increase in speed sensitivity of deformation is noted. The urgent task is to study the formation of shear bands in BMG with HPT. However, during TEM studies of BMG after HPT, shear bands are often not detected [6], due to the small band thickness (10 nm) and since the difference in the amorphous structure of the bands and the matrix is not sufficient to form a visible contrast. When deformed by pressing or rolling, the SBs emerge onto the surface of the BMG in the form of steps, which are well observed by SEM and AFM [7]. However, with HPT, the sample is clutched between the anvils, and there are no “free surfaces” for the SB exit in this scheme. In [8], a new method of HPT of BMGs was proposed, which made it possible to conduct SEM study of shear bands on a free surface. In the current article we continue to use the AFM method.

IOP Conf. Series: Materials Science and Engineering **672** (2019) 012019 doi:10.1088/1757-899X/672/1/012019

## 2. Experimental results and discussion

The $Zr_{52.5}Cu_{17.9}Ni_{14.6}Al_{10}Ti_5$ (at. %) bulk metallic glass rods 10 mm in diameter and 40 mm in length (hereinafter referred to as Vit105) were used in this study. Disks with a thickness of 1 mm were cut out of the initial BMG ingot. Each disk was cut into two halves (figure 1a), and the internal surfaces of both halves were polished. Then the halves of the disks were joined together with a thin layer of a lacquer in order to prevent the adhesion of the two parts, and processed by HPT, as shown in figure 1b, under a pressure of 6 GPa at room temperature.

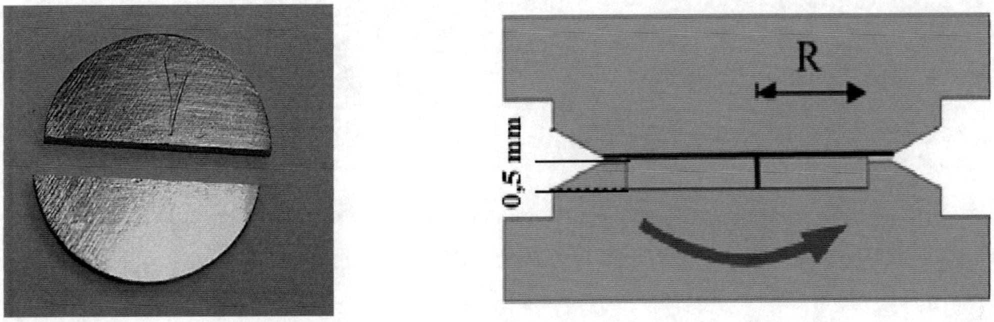


(a) (b)

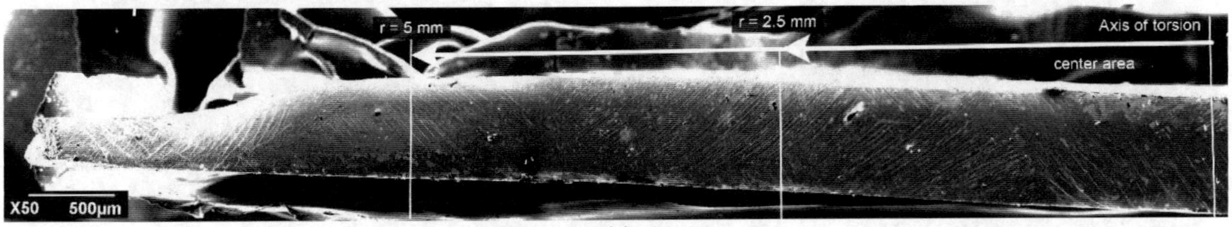


(c)

**Figure 1.** (a) Scheme of the two halves of the as-cast HPT disk. (b) Scheme of constrained HPT processing: the lower anvil has a groove with a depth of 0.5 mm, the upper anvil is flat. (c) SEM image of the internal surface of the sample HPT 30°.

The observed structure obtained by HPT with various regimes is inhomogeneous across the internal surface of the sample. Shear bands with a much higher density are formed on the side of the sample (figure 1c). The observed bands can be conventionally divided into primary shear bands, less intense secondary shear bands that intersect primary ones, and even thinner tertiary shear bands that are visible at high magnifications (figure 3). Such a division of SB in BMGs samples after rolling was also observed [7]. In the main area of the sample subjected to HPT with a rotation of 30° (HPT 30°) a system of parallel primary SB is formed, directed at an angle of 45° to the upper surface (figures 1, 2), with an average distance between the lines about 10 μm. In the marginal regions of the sample, the SB density is noticeably higher. This is attributed to the larger strain in this area.

An increase in the number of revolutions of HPT leads to an increase in the density of SBs. The distance between the bands in the areas of $r = 2.5$ mm was about 0.3 μm after HPT for turns $n = 5$. In the samples after HPT $n = 5$, the SB of the first type have an angle of about 45° to the upper surface of the sample, and the shear bands of the second type are nearly parallel to the surface of the sample. These SB thus divide the sample volume into numerous units (figure 3).

IOP Conf. Series: Materials Science and Engineering **672** (2019) 012019 doi:10.1088/1757-899X/672/1/012019

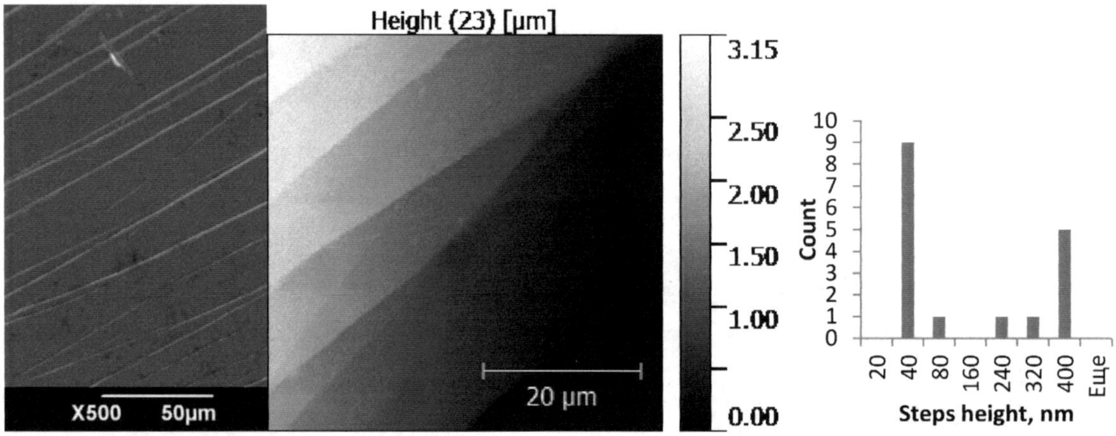


**Figure 2.** SEM and AFM images of the internal surface of the Vit105 BMG subjected to HPT 30°, and the step height distribution at the exit of the shear bands (AFM).

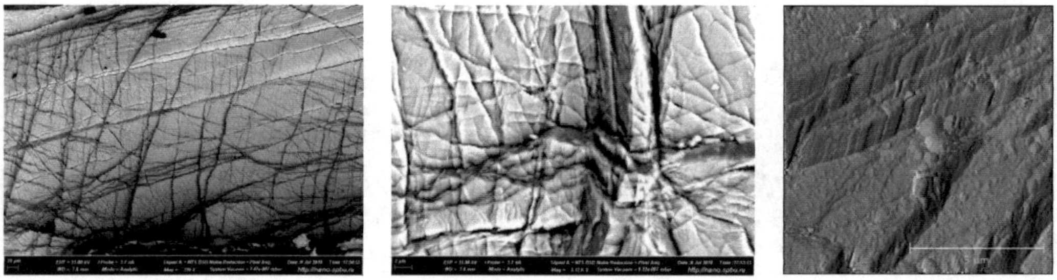

**Figure 3.** SEM and AFM (lateral forces) images of the internal surface of the Vit105 BMG subjected to HPT n=5.

The density and distribution of the shear bands, stated by AFM, correspond to the pattern observed by SEM. However, the use of AFM made it possible to estimate the height of the shear bands, which in the sample HPT 30° for the primary bands were about 400 nm, the secondary ones - about 50 nm. In the HPT sample deformed to n = 5, it is more difficult to analyze the AFM pattern due to the high density of the bands and a rougher surface profile of the sample. According to the first estimates, the height of the primary bands after HPT n = 5 reached 1000 nm. Hence, with an increase in the number of revolutions (shear angle) of HPT, both the density of the shear bands and their intensity (the height of the steps) increase.

## 3. Conclusions

Two halves of disks of the BMG were processed by HPT, and the parameters of SB evolution with deformation were examined using the SEM and AFM imaging of the internal surfaces of two joint halves of the HPT-processed disk. HPT processing leads to the formation of a high density of the SB in the BMG. An increase in HPT strain leads to a significant increase in the SB density, and the spacing between SB after HPT n = 5 is 0.5 µm. The maximum density of the SB is observed at the edges of the HPT samples. The use of AFM made it possible to estimate the height of the shear bands, which in the sample HPT 30° for primary bands was about 400 nm, after HPT n = 5 - about 1000 nm, for secondary bands - about 50 nm.

IOP Conf. Series: Materials Science and Engineering **672** (2019) 012019 doi:10.1088/1757-899X/672/1/012019

**Acknowledgements**
The authors are grateful to Saint-Petersburg State University for research grant No. 6.65.43.2017. for the research grant from the DFG under contract No. HA1344/30-1. The authors acknowledge the support of RFBR grant No. 17-08-00974. AFM images were processed using Gwyddion software (http://gwyddion.net/). Part of the work was carried out using the equipment of the "Nanotech" Engineering center (http://nanotech.ugatu.ac.ru).

**References**
[1] Louzguine-Luzgin D V and Inoue A 2013 Bulk metallic glasses: formation, structure, properties, and applications *Handbook of Magnetic Materials* (vol 21) ed K H J Buschow (Amsterdam: Elsevier) chapter 3 pp 131–171
[2] Abrosimova G E 2011 Physics Uspekhi **181** 1265
[3] Valiev R, Gunderov D, Zhilyaev A P, Popov A G and Pushin V 2004 *Journal of Metastable and Nanocrystalline Materials* **22** 21
[4] Glezer A M, Sundeev R V and Shalimova A V 2011 *Doklady Physics* **56** 476
[5] Ubyivovk E V, Boltynjuk E V, Gunderov D V, Churakova A A, Kilmametov A R and Valiev R Z 2017 *Materials Letters* **209** 327
[6] Boltynjuk E V, Gunderov D V, Ubyivovk E V, Monclús M A, Yang L W, Molina-Aldareguia J M, Tyurin A I, Kilmametov A R, Churakova A A, Churyumov A Y and Valiev R Z 2018 *Journal of Alloys and Compounds* **747** 595
[7] Liu J W, Cao Q P, Chen L Y, Wang X D, Jiang J Z 2010 *Acta Mater.* **58** 4827
[8] Gunderov D V, Churakova A A, Boltynjuk E V, Ubyivovk E V, Astanin V, Asfandiyarov R N, Valiev R Z, Xioang W, Wang J T 2019 *Journal of Alloys and Compounds* **800** 58

IOP Conf. Series: Materials Science and Engineering **672** (2019) 012020 doi:10.1088/1757-899X/672/1/012020

# Plasma electrolytic polishing of nitrided steel under force convection condition

**P N Belkin, S A Silkin, I G Dyakov, I V Tambovskiy, S S Korableva and S A Kusmanov**

Department of Physical, Mathematical and Natural Science, Kostroma State University, 17 Dzerzhinskogo Street, Kostroma 156005, Russia

E-mail: sakusmanov@yandex.ru

**Abstract**. The effect of plasma electrolytic polishing in an ammonium chloride electrolyte under force convection condition on the characteristics of medium carbon steel after anodic plasma electrolytic nitriding was investigated. The morphology of the steel surface was studied using a metallographic microscope. The phase composition of the surface layers was studied using an X-ray diffractometer. Treatment specimens were tested under dry friction conditions with a counter body made of bearing steel. Corrosion studies were performed using potentiodynamic polarization curves in a 3.5 % sodium chloride solution. Nitriding increased the surface hardness up to 1130 HV due to the formation of a nitride-martensite layer with an outer oxide layer. After polishing of the nitrided surface, the loose outer part of the oxide layer containing microcracks and pores is removed, surface lightening has occurred. The reduction of the wear rate of 45-fold and the corrosion current density of 4.8-fold of nitrided steel by plasma polishing without changing the structure of the diffusion layers was shown.

## 1. Introduction

Plasma electrolytic technologies have a wide potential usage in various industries. These include the aerospace and oil industries, tool and electrical ones [1], as well as biomedical applications in surgery [2], disinfection of solutions [3], etc. Plasma electrolytic nitriding (PEN) increases significantly the wear and corrosion resistance of some structural steels due to the rise in the hardness of the surface layer and the protective action of nitrides and iron oxides. There are examples of PEN of S0050A medium carbon steel in a carbamide-based electrolyte [4], 316L steel in a solution of ammonium carbonate [5], and low-alloy steel 40Cr in electrolyte contained ammonia and ammonium chloride [6]. The disadvantage of the cathode PEN is the erosion of the surface due to electrical discharges. Anodic PEN is accompanied by anodic dissolution [3], which reduces the surface roughness of the treated steel. In addition, the anodic process leads to the formation of an oxide layer with protective properties under the conditions of atmospheric corrosion [7], but not always having good behavior in friction pairs [8]. The likely reason is the presence of a loose part of the oxide layer.

An effective method of the enhancement of metal surfaces is plasma electrolytic polishing (PEP), also implemented in aqueous electrolytes, but with use of other current passing modes. PEP is characterized by high processing speed, low surface roughness of the polished layer, the absence of aggressive and toxic reagents. This method has been successfully used for polishing knee caps, titanium micro implants [9], etc. This process reduces the surface roughness of nitrided steel without removing the hardened layer [10]. Therefore, the purpose of this work is to increase the wear and corrosion resistance of nitrided steel by means of its additional PEP, which can be performed in the same electrolyzer.

## 2. Experimental

Cylindrical samples (a diameter of 10 mm, a length of 15 mm) of medium carbon steel (0.45 wt.% C) were subjected to plasma electrolytic nitriding in a cylindrical electrolyzer under force convection condition. There was an axially symmetric electrolyte flow supplied through a nozzle located at the bottom of the electrolyzer. In the upper part of the electrolyzer, the electrolyte was overflowing into the sump and was further pumped through a heat exchanger at a rate of 2.6 l/min. The solution temperature was 20±2 °C. The samples were connected as the positive output, and the electrolyzer was connected as the negative output of the 15 kW DC power supply. An aqueous solution of ammonium nitrate $NH_4NO_3$ (5 wt.%) with ammonium chloride $NH_4Cl$ (15 wt.%) was used as the working electrolyte. After switching the voltage to 200 V, the samples were immersed in the electrolyte at a speed of 1–2 mm/s. Once the sample was immersed at a depth equal to its height, the voltage was decreased to 135 V in order to reach the prescribed temperature of the sample – 850 °C, the current was 8.2±0.2 A. The PEN time was 5 min and after that, the samples were quenched in the electrolyte (hardening). PEP was carried out in an aqueous solution of ammonium chloride (3 wt.%) with added glycerol or oxalic acid (5 wt.% each) at a voltage of 275 and 300 V. The immersion depth of the upper end was 5 mm. The electrolyte was pumped through a heat exchanger at a flow rate of 1 l/min. The electrolyte temperature was kept at 80±2 °C. In this case, the sample temperature was known to be 110–120 °C while the PEP time was 1 min.

The phase composition of the surface layers after PEN was investigated with the use of an X-ray diffractometer with Co Kα radiation by a simple scanning mechanism in the theta-2theta-mode with a step of 0.1° and a scanning rate of 1.25°/min. The morphology of the steel surface was studied using a metallographic microscope. Microhardness of the surface layer of the sample after PEN+PEP was measured on a Falcon 503 apparatus at a loading of 50 g. The corrosion resistance of the samples was determined by the potentiodynamic polarization curves using an SP-150 Biologic potentiostat-galvanostat with a standard three-electrode cell in a 3.5% solution of sodium chloride at a sweep rate of 1 mV/s. Dry wear testing of the samples was carried out against a bearing steel disk with a normal load of 10.4 N, a sliding speed of 1 m/s, and a sliding distance of 1000 m at room temperature using a laboratory unit.

## 3. Results and discussion

The anodic PEN results in the formation of a surface layer of oxides (FeO and $Fe_3O_4$) and a modified layer underneath containing nitrides ($Fe_4N$), martensite and retained austenite. The morphology study of the steel surface after PEN shows the presence of microcracks and peeling of the outer part of the oxide layer (figure 1a). This morphology changes after PEP of the nitride surface. The loose outer part of the oxide layer containing microcracks and pores is removed (figure 1b). Table 1 shows the weight loss of the samples on the processing PEP. Moreover, X-ray analysis confirms the decrease in the intensity of the peaks of oxides after PEP, while the peaks of nitrides are maintained. $Fe_2O_3$ is not detected after PEP in the chloride electrolyte with the addition of glycerol at 275 V and oxalic acid at 300 V. In the latter case, FeO is not also found. At the same time, these additives lead to a decrease in the weight loss of the samples and the current yield (Table 1). This is due to the surface passivation by the interaction products of functional groups of organic molecules with the metal, which leads to a current consumption. This process can be carried out according to the following scheme:

$$(n+1)Fe^{3+} + mH_2O_{ads} + xOH^-(\bullet OH)_{ads} \rightarrow [Fe(Fe_nO_m)\cdot(OH)_x]_{ads} + 3(n+1)H^+$$

$$[Fe(Fe_nO_m)\cdot(OH)_x]_{ads} \rightarrow Fe_nO_m + Fe(OH)_x$$

The microhardness of the diffusion layers reduces slightly when these additives are used, except for PEP with glycerol at 275 V (figure 2), which makes it possible to use these additives in the electrolyte for polishing.

**Figure 1.** Morphology of steel surface after PEN (a) and PEP (b) in an electrolyte containing $NH_4Cl$.

**Table 1.** The characteristics of PEP and surface properties

| Electrolyte composition | Voltage (V) | Sample weight loss (mg) | Current yield | Corrosion current density (μA/cm²) | Wear rate (ng/N·m) | Friction coefficient |
|---|---|---|---|---|---|---|
| Not treated | – | – | – | 39.5 | 1327 | 0.61 |
| $NH_4Cl$ | 300 | 38.3 | 0.68 | 30.2 | 48 | 0.58 |
| $NH_4Cl$ + Gl | 275 | 29.3 | 0.47 | 8.2 | 29 | 0.57 |
| $NH_4Cl$ + Gl | 300 | 27.7 | 0.49 | 13.7 | 38 | 0.54 |
| $NH_4Cl$ + Ox | 275 | 27.5 | 0.54 | 25.5 | 38 | 0.54 |
| $NH_4Cl$ + Ox | 300 | 28.9 | 0.56 | 22.7 | 29 | 0.52 |

Gl is glycerol and Ox is oxalic acid.

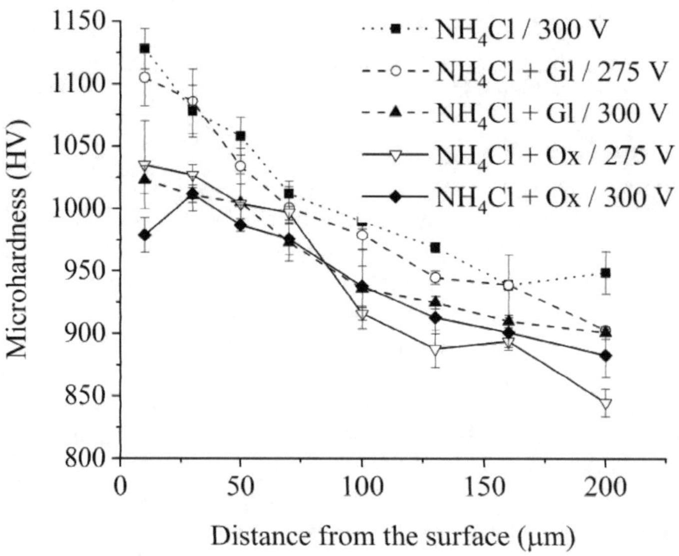


**Figure 2.** Microhardness distribution in the surface layer of steel after PEN+PEP.

Corrosion tests show that the PEP of PEN-treated steel can improve the corrosion resistance of the steel (Table 1). The contact of the surface with the corrosion media occurs through a modified layer enriched with iron oxides and nitrides, which play a positive role in increasing the corrosion resistance. The maximal decrease in the corrosion current density occurs after PEP in an electrolyte containing glycerol and oxalic acid owing to the additional surface passivation.

Tribological tests show that friction coefficient of samples treated is always lower than that of untreated sample (Table 1). PEN provides the formation of a hardened sublayer in which mainly elastic deformations occur. The outer layer containing iron oxides and retained austenite, has good behavior, since plastic deformations are localized in this layer. PEP slightly affects the hardness of the layer, but it reduces the wear rate (Table 1). The maximal reduction of the friction coefficient and wear rate is observed with use of glycerol and oxalic acid as the electrolyte components for PEP. This may also be associated with additional surface passivation and improved plasticity of the contact of friction pairs.

## 4. Conclusions

An increase in the wear and corrosion resistance of medium carbon steel can be achieved by means of PEN, followed by PEP in the same electrolyzer. Anodic PEN is carried out in an aqueous solution of ammonium nitrate (5 wt.%) and ammonium chloride (15 wt.%) with the following quenching. The temperature of the sample is 850 °C and the electrolyte temperature is 20 °C during PEN. PEP is advantageous in a 3% solution of ammonium chloride with glycerol or oxalic acid (5 wt.% each) at electrolyte temperature of 80 °C and a voltage of 275 and 300 V for 1 min. The corrosion current density in a 3.5% sodium chloride solution reduces from 39.5 $\mu A/cm^2$ in the untreated sample to 8.2 $\mu A/cm^2$ in steel after PEN and PEP in an ammonium chloride electrolyte with glycerol due to the protective action of the nitride zone and the dense part of the oxide layer. The dry friction coefficient with a counter-body of bearing steel slightly decreases from 0.61 in the untreated sample to 0.52 after PEN and PEP in an ammonium chloride electrolyte with oxalic acid. The weight wear rate being 1327 ng/(N·m) in the untreated sample is reduced to 29 ng/(N·m) after PEN and PEP in an ammonium chloride electrolyte with glycerol or oxalic acid. The structure of diffusion layers does not change.

**Acknowledgments**
This work was financially supported by the Russian Science Foundation (Contract No. 18-79-10094) for Kostroma State University.

**References**

[1] Aliofkhazraei M, Rouhaghdam A S and Gupta P 2011 *Crit. Rev. Solid State Mater. Sci.* **36** 174
[2] Vankov A and Palanker D 2007 *J. Appl. Phys.* **101** 124701
[3] Sakiyama Y, Tomai T, Miyano M and Graves D B 2009 *Appl. Phys. Lett.* **94** 161501
[4] Nie X, Wang L, Yao Z C, Zhang L and Cheng F 2005 *Surf. Coat. Technol.* **200** 1745
[5] Roy A, Tewari R K, Sharma R C and Sherhar R 2007 *Surf. Eng.* **23** 243
[6] Kusmanov S A, Smirnov A A, Silkin S A and Belkin P N 2016 *Surf. Coat. Technol.* **307** 1350
[7] Parshutin V V and Pasinkovslii E A 2007 *Surf. Eng. Appl. Electrochem.* **43** 431
[8] Kusmanov S A, Silkin S A, Smirnov A A and Belkin P N 2017 *Wear* 386-387 239
[9] Nestler K, Böttger-Hiller F, Adamitzki W, Glowa G, Zeidler H and Schubert A 2016 *Procedia CIRP* **42** 503
[10] Kusmanov S, Zhirov A, Kusmanova I and Belkin P 2019 *Surf. Eng.* **35** 507

IOP Conf. Series: Materials Science and Engineering **672** (2019) 012021 doi:10.1088/1757-899X/672/1/012021

# On strengthening of ultrafine grained austenitic steels subjected to large strain deformation

**A Belyakov, M Odnobokova, Z Yanushkevich, M Nazarova and R Kaibyshev**

Belgorod State University, Pobeda 85, Belgorod 308015, Russia

E-mail: belyakov@bsu.edu.ru

**Abstract**. Our current results on the structure-property relationship in ultrafine grained (UFG) austenitic steels, subjected to large strain deformation under conditions of cold to warm working are critically reviewed. Commonly, the grain size and dislocation density that evolve in UFG steels and alloys during large strain deformation depends sensitively on the processing conditions, i.e., temperature and strain rate. The grain size decreases, and the dislocation density increases with a decrease in deformation temperature. Therefore, a unique power-law relationship with a grain size exponent of -0.6 holds between the deformation grain size and the dislocation density. A power-law function between the grain size and dislocation density is obtained for UFG steels even after annealing. Thus, the yield strength of UFG steels can be expressed by a modified Hall-Petch-type relationship, which includes the dislocation strengthening, by using the grain size or dislocation density as a unique variable.

## 1. Introduction

Structural steels and alloys are commonly processed by various thermo-mechanical treatments [1]. Depending on the processing conditions, the thermo-mechanical treatment can involve dynamic and post-dynamic recovery and/or recrystallization. Both recovery and recrystallization are intended for controlled microstructure evolution during deformation and heat treatment. The most important microstructural parameters affecting mechanical behavior and properties of structural metallic materials are the grain size and dislocation density. Effect of hot working conditions on dynamic recrystallization (DRX) and the dynamic grain size evolved after sufficiently large strains is fairly clarified [2-4]. The power law relationships between the grain size and flow stress or temperature-compensated strain rate have been revealed for various metals and alloys. Studying the dislocation density in DRX grains evolved during hot working is complicated by heterogeneous dislocation distribution in DRX microstructures [5], although an inverse proportional stress dependence of the substructural parameters has been suggested [6].

The revealed dependencies for the microstructural parameters and the deformation conditions imply a power law function between the grain size and the dislocation density evolved in DRX microstructures. The aim of the present work, therefore, is to clarify the structural relationships in DRX microstructures evolved during large strain warm deformation, and to elucidate the strengthening mechanisms aroused in DRX microstructures. In contrast to DRX microstructures, the structural/substructural relationships in ultrafine grained (UFG) microstructures developed by large strain cold deformation followed by an annealing have not been elaborated. Thus, another aim of this study is to reveal a possible relationship between the grain size and dislocation density and their effect on the yield strength of UFG materials processed by severe plastic deformation and annealing.

## 2. Experimental

Two widely used austenitic stainless steels, 304L (0.05%C – 18.2%Cr – 8.8%Ni – 1.65%Mn – 0.43%Si – 0.05%P – 0.04%S) and 316L (0.04%C – 17.3%Cr – 10.7%Ni – 2%Mo – 1.7%Mn – 0.4%Si – 0.04%P – 0.05%S) were used in the present study as typical representatives of the classical DRX materials during hot working and susceptible to rapid grain refinement during severe plastic deformation. The steel samples were subjected to plate rolling at room temperature and a temperature of 573 K to a total strain of 3 and bar rolling at temperatures of 773-1173 K to a total true strain of 2. Then, the samples subjected to plate rolling were annealed at 873-1073 K. Structural investigations were carried out by using a JEM – 2100 transmission electron microscope (TEM) and a Nova Nanosem 450 scanning electron microscope, equipped with an orientation image microscopy (OIM) system based on an electron back-scattered diffraction pattern analyzer. The dislocation density was evaluated as the number of dislocations crossing an arbitrary selected area on at least six representative TEM images. Tensile tests were carried out with dog-bone specimens having a gauge length of 16 mm and a cross section of $3\times1.5$ mm$^2$ using an Instron 5882 universal testing machine.

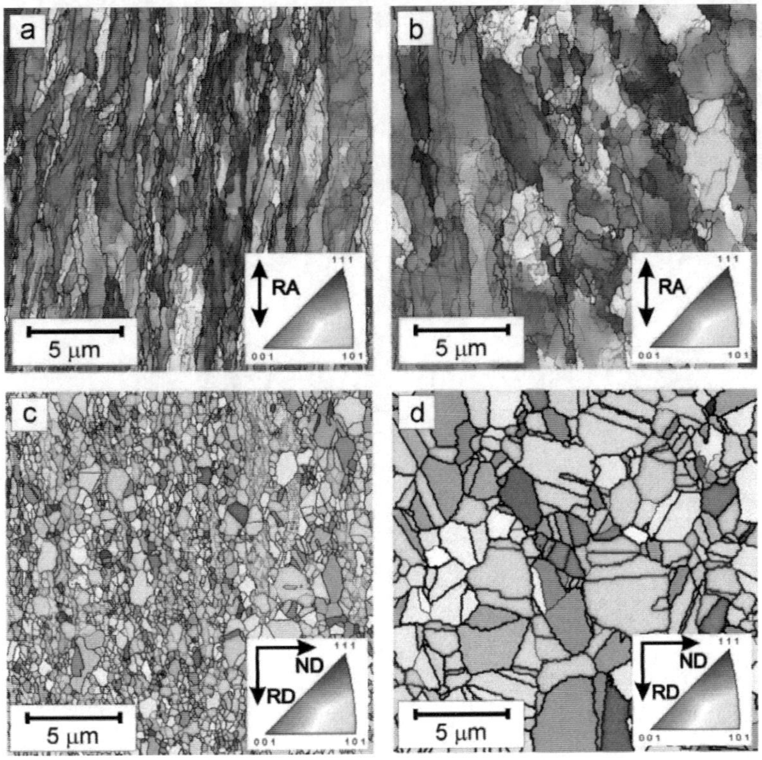


**Figure 1**. Typical microstructures evolved in 304L stainless steel after warm rolling at 773 K (a) and 973 K (b) and those after cold rolling followed by annealing at 973 K (c) and 1073 K (d). Colours in (a, b) and (c, d) correspond to the rolling axis (RA) and the normal direction (ND), respectively.

## 3. Results and discussion

Typical microstructures that evolve in austenitic stainless steels during warm rolling or cold rolling followed by annealing are shown in figure 1. The elongated grains containing numerous dislocation subboundaries testify to continuous DRX as the mechanism of microstructure evolution during warm

IOP Conf. Series: Materials Science and Engineering **672** (2019) 012021 doi:10.1088/1757-899X/672/1/012021

working. The transverse grain size, measured along the normal direction, increases with an increase in rolling temperature. Correspondingly, the dislocation density in the grain interiors increases with a decrease in the rolling temperature (figure 2a). Both the grain size and the dislocation density are characterized by weakening temperature dependencies as the rolling temperature decreases. Such behavior leads to a power law relationship between the grain size and the dislocation density with a grain size exponent of -0.6 in the warm rolled steel samples, namely, $\rho = 18\ D^{-0.6}$ in figure 2b.

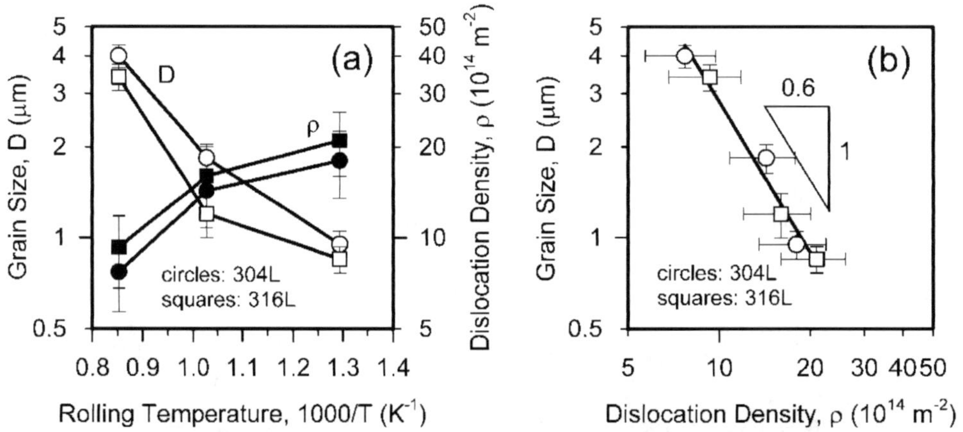


**Figure 2**. Effect of the rolling temperature on the grain size and the dislocation density (a) and the relationship between the grain size and the dislocation density evolved by warm rolling (b).

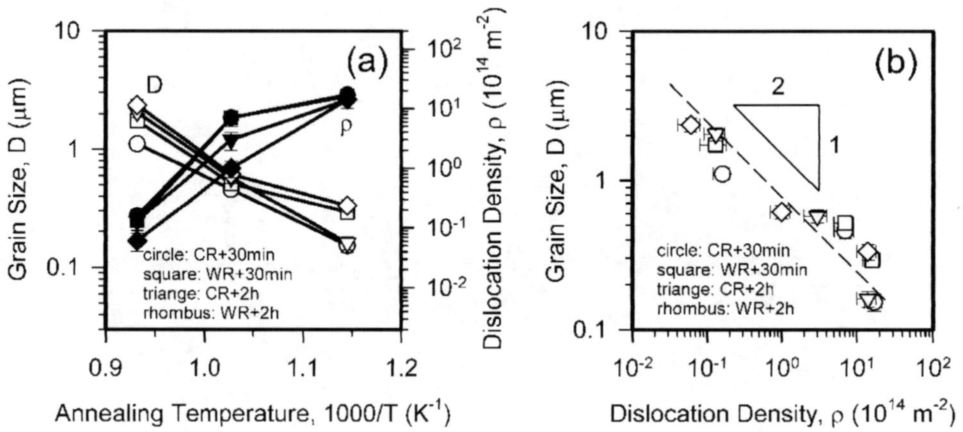


**Figure 3**. Effect of annealing for 30 min and 2 h on the grain size and the dislocation density (a) and the relationship between the grain size and the dislocation density (b) in steel samples subjected to rolling at room temperature (CR) and 573 K (WR).

Annealing of cold rolled steel samples results in the development of an almost equiaxed grain microstructure (figures 1c and 1d). An increase in the annealing temperature leads to a growth of the mean grain size, while the dislocation density decreases (figure 3a). Both the annealed grain size and the dislocation density within the grains can be expressed by power law functions of inverse temperature. Similar to the warm worked microstructures, the dislocation density in the cold rolled and annealed steel samples can be expressed by a unique power law function of the annealed grain size with an exponent of 2, i.e., $\rho = 0.61\ D^{-2}$ in figure 3b. It can be concluded, therefore, that deformation microstructures that evolved at sufficiently large strains under conditions of warm working and annealed microstructures developed in largely strained samples are characterized by unique relationships, which link the grain size and the dislocation density through power law functions.

The fine grained samples processed by warm rolling and cold rolling, followed by annealing, exhibit enhanced strength properties. The yield strength ($\sigma_{0.2}$) of structural steels depends on their mean grain size ($D$) and is commonly expressed by the Hall-Petch relationship [7, 8]. In the case of high dislocation density ($\rho$) in grain interiors, the modified Hall-Petch relationship is used for yield strength calculation taking into account substructural strengthening as follows [9, 10].

$$\sigma_{0.2} = \sigma_0 + kD^{-0.5} + \alpha G b \rho^{0.5} \quad (1)$$

Here $\sigma_0$ is the strength of the same material with an infinite grain size, $k$ and $\alpha$ are strengthening factors ($\alpha$ involves Taylor factor), $G$ is the shear modulus, and $b$ is the Burgers vector. It is interesting to note that the yield strength can be expressed by using one microstructural/substructural parameter, i.e., either grain size or dislocation density, using equation (1), if there is a unique relationship between these parameters. Thus, the yield strength was calculated using equation (1), substituting the corresponding power law functions, $\rho \sim D^{-n}$ indicated above, for the grain size and the dislocation density. Figure 4 illustrates a good correspondence between the experimentally measured values of the yield strength and those calculated using the grain size, $\sigma(D)$, or the dislocation density $\sigma(\rho)$ taking $k$ = 300 MPa μm$^{-0.5}$ and $\alpha$ = 0.5.

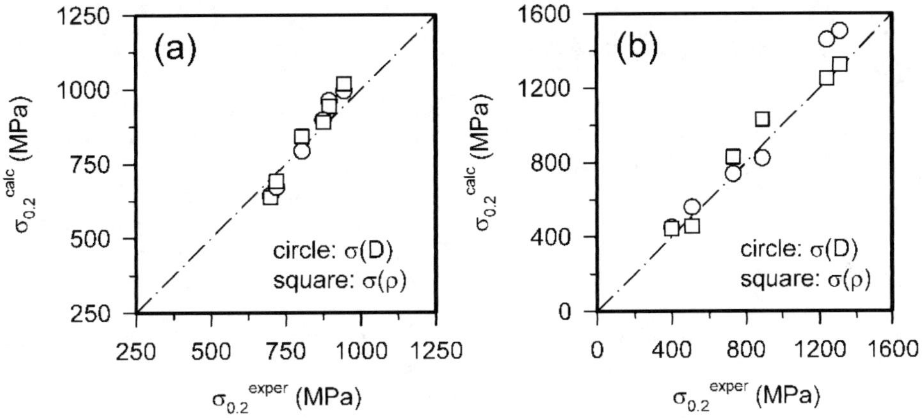


**Figure 4**. Relationship between the experimental and calculated yield strength for UFG stainless steels processed by warm rolling and annealing (a) or cold rolling and annealing (b).

## References

[1] Waters T F 1996 *Fundamentals of Manufactoring for Engineers* (London: UCL Press)
[2] McQueen H J and Jonas J J 1975 Recovery and Recrystallization during High Temperature Deformation *Treatise on Materials Science and Technology* (vol. 6) ed R J Arsenault (New York: Academic Press) pp 393-493.
[3] Humphreys F J and Hatherly M 1996 *Recrystallization and Related Annealing Phenomena* (Oxford: Elsevier)
[4] Sakai T, Belyakov A, Kaibyshev R, Miura H and Jonas J J 2014 *Prog. Mater. Sci.* **60** 130.
[5] Sakai T and Ohashi M 1990 *Mater. Sci. Technol.* **6** 1251.
[6] Tikhonova M, Kaibyshev R and Belyakov A 2018 *Adv. Eng. Mater.* **20** 1700960
[7] Hall E O 1951 *Proc. Phys. Soc.* **B64** 742
[8] Petch N J 1953 *J. Iron Steel Inst.* **174** 25
[9] Hughes D A and Hansen N 2000 *Acta Mater.* **48** 2985
[10] Yanushkevich Z, Dobatkin S V, Belyakov A and Kaibyshev R 2017 *Acta Mater.* **136** 39

IOP Conf. Series: Materials Science and Engineering **672** (2019) 012022 doi:10.1088/1757-899X/672/1/012022

# Influence of severe plastic deformation on the structure and mechanical properties of eutectic Al-Zn-Mg-Fe-Ni alloy

**I G Brodova[1,2], A N Petrova[1] and T K Akopyan[3]**

[1] M.N. Mikheev Institute of Metal Physics of Ural Branch of Russian Academy of Sciences, 620108, 18 S. Kovalevskoy Street, Ekaterinburg, Russia
[2] Ural Federal University named after the first President of Russia B. N. Yeltsin, 620002, 19 Mira Street, Ekaterinburg, Russia
[3] National University of Science and Technology MISiS, 119049, 4 Leninsky av., Moscow, Russia

E-mail: brodova@imp.uran.ru

**Abstract**. The structure and phase evolution of the new eutectic high strength aluminium alloy Nikalin were investigated under high pressure torsion (HPT) by several numbers of the revolution of the anvil. The chemical composition of the investigated Nikalin alloy was as follows: Al (base)–7.22Zn–2.95 Mg–0.52 Fe–0.57 Ni–0.2 Zr. (wt.%). The initial material was a coarse grained cast ingot after homogenization. It was established that the HPT process resulted in the deformation dissolution of the nanosized T-phase precipitates and the formation of a supersaturated aluminum solid solution simultaneously with the strong refinement of the structure to the grain-subgrain size of 130-150 nm. Due to that, the yield stress of the HPT alloy increased by a factor of 1.5, the ultimate tensile strength increased by a factor of 1.4, while preserving good ductility of 6-7%. The observed effect of the additional supersaturated solution upon HPT relative to the homogenized state appeared upon post-deformation annealing at 140 °C. An increase in the microhardness of the HPT alloy due to the $MgZn_2$ phase precipitation was observed at 0.5- hours of annealing.

## 1. Introduction

The Nikalins is the name of a new class of eutectic aluminum based alloys. The structure and properties of the castings, rolled sheets and radial-shear rolling rods of the high-strength alloy (Al-7Zn-3Mg-Fe-Ni) were recently investigated [1-3]. The results showed the potential of the thermo-deformation treatment for producing high strength semi-finished products of good quality.

It is well known that high pressure torsion (HPT) is an effective technique for improving the structure of aluminum-based materials and their mechanical properties [4-6]. An analysis of the published papers has shown that the structure and deformation behavior of eutectic alloys produced by the HPT process have not been adequately studied, and have not been studied earlier for alloys of the Nikalin type.

The purpose of the present work is to investigate the peculiarities of the structural-phase transformations during HPT, to determine the hardening mechanisms during deformation and post deformation annealing, and to determine the mechanical properties of an alloy of the Al-Zn-Mg-Fe-Ni system.

## 2. Experimental

The chemical composition of the investigated Nikalin alloy was as follows: Al (base)–7.22Zn–2.95 Mg–0.52 Fe–0.57 Ni–0.2 Zr. (wt. %). The initial material was a cast ingot after two- stage homogenizing annealing at 450 °C and at 540 °C for 3 hours at each stage. The choosing of the homogenization regime is described in [2]. The material had an initial static yield stress $\sigma_y$= 220 MPa, a tensile strength $\sigma_u$= 350 MPa, and a microhardness $H_v$=1400 MPa.

Cross-sectional discs with a thickness of 1.2 mm were cut from a billet with a diameter of 20 mm. The discs were processed by HPT at room temperature. The accumulated true strain varied between e=6.0, 6.7, and 7.1, corresponding to the number of revolutions of the anvil n=5, 10 and 15. The accumulated true strain was calculated for the half radius of the HPT samples, considering pre-stressing to a thickness of 1 mm. The produced samples were used for structural investigations, static tensile tests and post deformation annealing.

Structural investigations were performed by means of a Tecnai G2 30 Twin transmission electron microscope (TEM) and a Quanta-200 scanning electron microscope (SEM). X-ray diffraction (XRD) analysis was carried out using an Empyrean (PANalytical) diffractometer at room temperature and CuKα radiation. Based on the X-ray diffraction data, the lattice parameter of the aluminium matrix and t the mean-square microstrain of the crystal lattice were calculated. Hardness $H_v$ was measured using a PMT-3 device at a load of 0.2 N (the error did not exceed 10%).

For static mechanical tests, two tensile specimens with a gauge length of 5.7 mm, a width of 2.0 mm, and a thickness of 1 mm were machined from the HPT disc using electrical discharge machining. Microspecimens were pulled to failure using a Shimadzu AGX-50 Plus universal testing machine. Tensile tests were conducted at room temperature and at a strain rate of $\dot{\varepsilon}=2\times10^{-3}$ $s^{-1}$. Structural investigations and hardness measurements were conducted at half the radius of the HPT samples. The shape of the tensile microsamples and the scheme of a cut-off appear to provide the measurements of static mechanical properties at half the radius area of the HPT discs.

Tensile tests of the homogenized cast alloy were conducted on standard cylindrical specimens at a strain rate $\dot{\varepsilon}=2.5\times10^{-3}$ $s^{-1}$ by means of a Zwick Z250 testing machine according to ASTM E 8/E 8M-08.

## 3. Results

### *3.1. Mechanical properties*

The obtained experimental σ-ε diagrams of HPT alloys are shown in a figure 1a. Figures 1b and 1c demonstrate the yield stress $\sigma_y$ and microhardness depending on the number of revolutions n. It is shown that the yield stress value changes consistently with an increase in the accumulated strain as well as microhardness. The highest values of $\sigma_y$=521 MPa and $H_v$=2900 MPa correspond to the samples after HPT, n=10. With increasing n up to 15, these mechanical characteristics decrease by 100 and 340 MPa. The tensile strength of HPT Nikalin is higher by a factor of 1.4 compared to that of the initial homogenized alloy. The plasticity of HPT samples is δ=6-7% at various n.

### *3.2. Structure and phase composition*

To explain the observed regularities of the mechanical behavior of Nikalin and the hardening mechanisms during HPT, the evolution of the structure and phase composition depending on the accumulated strain has been studied.

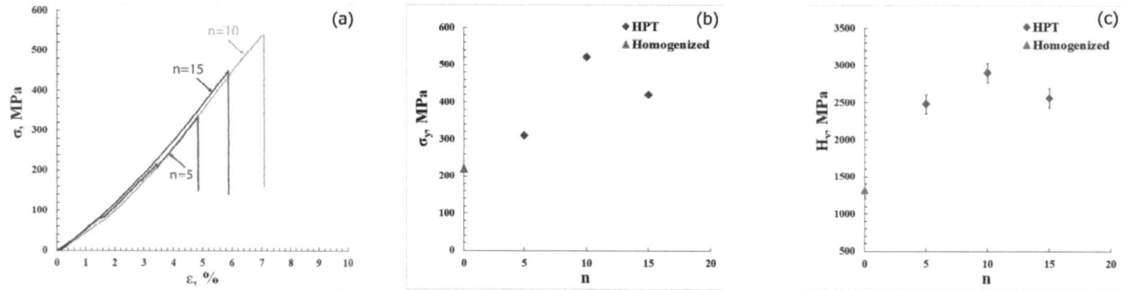


**Figure 1.** Mechanical properties of the HPT alloy: a - stress-strain diagrams, b - yield stress and c - microhardness changes depending on the number of revolutions.

The phase composition of the alloy after homogenization is represented by the Al (Zn, Mg) solid solution, the eutectic (Al +$Al_9FeNi$), and the T phase - $Al_2Zn_3Mg_3$.

The quantitative analysis of the structure has shown that the average grain size was 380 μm (figure 2a). The eutectic $Al_9FeNi$ aluminides are arranged in groups along the grain boundaries. The thickness of the eutectic layer is 2 μm (figure 2b). The secondary T phase precipitated during cooling after homogenization. The T phase particles with a size of 15-20 nm are shown up predominantly within the grains (figure 2c).

The evolution of each structural component under deformation was investigated considering the complex phase composition of the homogenized alloy. According to SEM data, the morphology and the size of the $Al_9FeNi$ aluminides slightly change upon HPT compared to the initial state. Because of the almost zero solubility of iron and nickel in the Al solid solution, the content of the eutectic phase retained unchanged before and after HPT. The eutectic $Al_9FeNi$ aluminides are arranged in groups after HPT (n=5, 10). An increase in strain leads to a uniform distribution of the particles (n=15).

The results of structural investigations of the alloy after HPT treatment (n=5 and n=15) are shown in figure 3. After HPT (n=5), a shear band structure was formed (figure 3a). The TEM image represents a slightly misoriented fine subgrained structure within the bands (figure 3b). An increase in *n* leads to the development of crystals fragmentation, rotations modes results in the formation of HABs and their predominance (figure 3d). An augmentation of point reflections on the circle type MD pattern at n=15 indicates an increase in the the fraction of HABs (figure 3d). One can see deformed grains-subgrains with a strong deformation contrast as well as small (20-30 nm) defect-free grains that testify the onset of dynamic recrystallization (figure 3d). Thus, the mixed structure of deformed and recrystallized crystallites was formed after HPT, n=15.

The grain-subgrain size distributions are constructed depending on the accumulative strain. A slight decrease in the size of the Al-matrix fragments from 155±75 nm to 130±75 nm was established with an increase in n. Thus, the HPT treatment results in significant grain refinement as compared with the grain size of the initial alloy. An increase in the accumulated strain leads to a transition from a fragmented structure with a developed LABs network to a mixed submicrocrystalline structure with a predominance of HABs.

Figure 4 shows the segments of the x-rays patterns for the initial homogenized alloy and for the alloy after HPT. For these patterns, all reflection lines refer to the Al solid solution and to the $Al_9FeNi$ aluminides. An increase in the mean-square microstrain from 0.13% up to 0.19% was observed with an increase in the accumulated strain *n*. The lattice parameter was a=0.4059 nm for the homogenized alloy and did not change upon HPT (a=0.4058-04060 nm, Δa = 0.0001-0.0002 nm). The constant lattice parameter values cannot indicate the absence of diffusion of the atoms of the dissolved elements into the solid solution, because Mg atoms increase the aluminum lattice parameter, and Zn atoms decrease it.

According to the TEM data, the nanosized T-phase particles were presented in the structure of HPT alloys (figure 3c). The absence of reflections from this phase on the x-rays patterns of the deformed samples is due to its small volume fraction. The volume fraction of the T phase particles decreases with increasing accumulated strain and only single particles of the T phase are precipitated within the grain-subgrains of samples after the HPT, n = 10 and n= 15. Thus, with an increase in the accumulated strain, the deformation dissolution of aluminides and the formation of a supersaturated Al solid solution occur, but given the data of the works [6,7], we can assume that Zn and Mg atoms from the disappeared aluminide precipitates were used for the formation of adsorption layers in GBs. To confirm these phenomena, additional studies of structural phase transformations in HPT Nikalin are planned using STEM and Atom probe tomography.

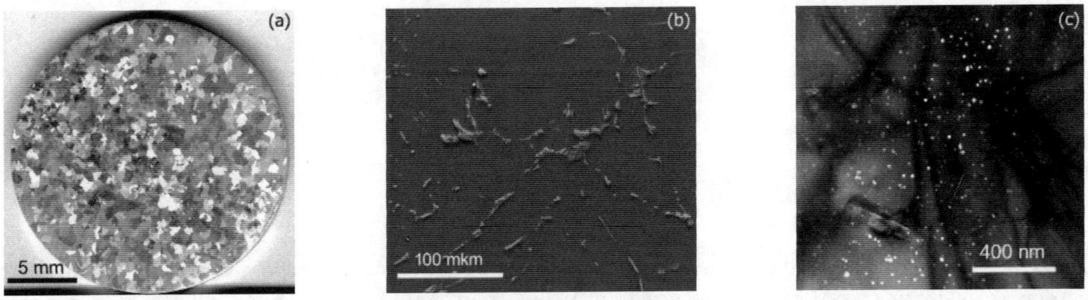


**Figure 2.** Structure of the homogenized alloy: a - grains, b - SEM image of $Al_9FeNi$ particles, c - DF TEM image of the $Al_2Mg_3Zn_3$ hardening phase.

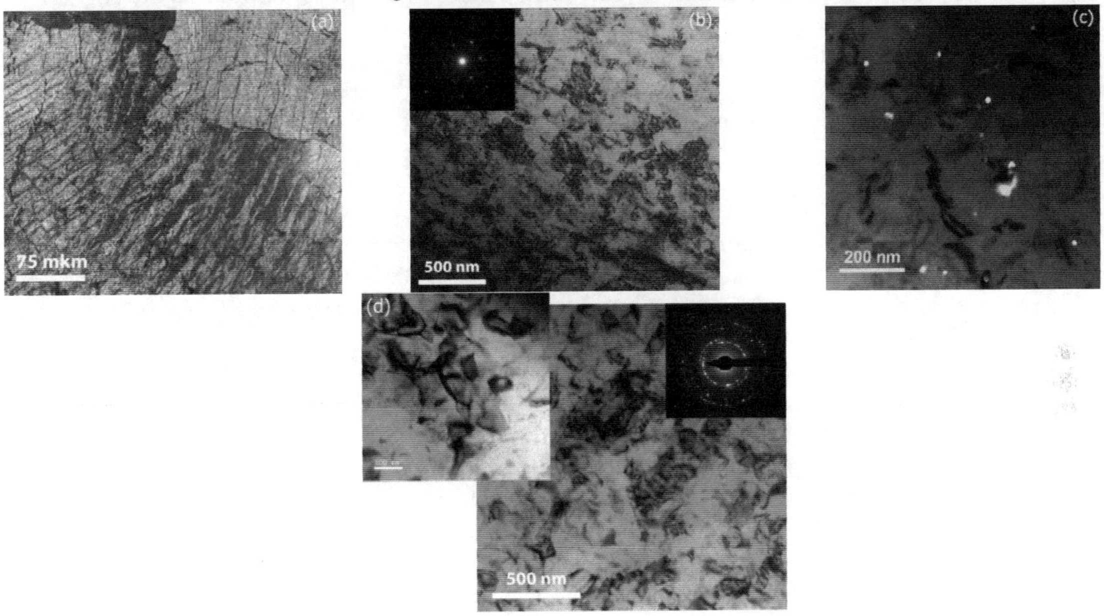


**Figure 3.** Structure of the HPT alloy: a - band structure after n=5, b - submicrocrystalline structure after n=5, c- $Al_2Mg_3Zn_3$ phase particles (DF TEM), d - submicrocrystalline structure after n=15.

The observed effect of an additional supersaturation solution relative to the homogenized state appears upon post-deformation annealing of HPT samples at 140 °C for 6 hours. X-rays analysis shows that this treatment results in the reduction of the concentration of defects and in decreasing lattice parameter to a=0.4046 nm. The mean-square microstrain after annealing is reduced to 0.12%-0.08%. On the X-rays patterns of the aged HPT samples, the additional reflection lines were determined. These lines correspond to the stable $MgZn_2$ phase (figure 4). The effect of the supersaturated solid solution decomposition leads to non-monotonic microhardness-time dependences. In spite of the general tendency toward the microhardness decrease due to the recovery, an increase in the microhardness at short exposures of about 0.5 and 1 hours was observed. The effect of dispersion hardening in samples with greater accumulated strain appears at shorter annealing times.

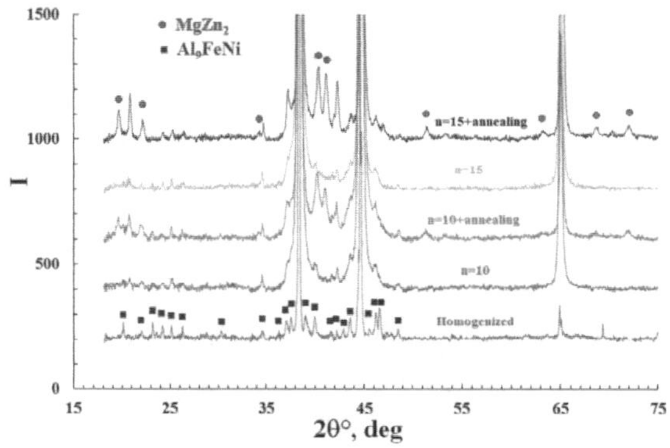


**Figure 4.** X-rays diffraction patterns for the homogenized, HPT alloy and HPT alloy after annealing.

After 6 hours of annealing, the HPT samples have high hardness of 1700–1800 MPa. Thus, the decrease in hardening is compensated by the preservation of the grain boundary hardening according to the Hall-Patch law and the dispersion hardening with the formation of a stable nano-sized $MgZn_2$ phase. A similar increase in the hardness of the quenched alloy occurs at the same temperature of artificial aging only after a long exposure for more than 3 hours [7]. Thus, the formation of the submicrocrystalline structure changes the precipitation kinetics and the type of precipitates of the hardening phases. The $MgZn_2$ phase particles precipitated from the supersaturated aluminum solid solution after HPT treatment. Aging of the quenched alloy leads to the formation of T-phase precipitates.

## 4. Conclusions

The HPT experiment showed that severe plastic deformation had a strong influence on the structure and properties of Nikalin. It was established that a submicrocrystalline deformed structure with an average size of grain-subgrains of 130-150 nm was formed after HPT. The growth of the stress state of the HPT alloy (n=15) activated the process of dynamic recrystallization, and a mixed structure was formed consisting of deformed fragments and recrystallized grains.

It was established that in the HPT process, the accumulation of structural defects (vacancies and dislocations), the deformation dissolution of nanosized T-phase precipitates, and the formation of a supersaturated aluminum solid solution occur simultaneously with the refinement of the structure.

The alloy with the submicrocrystalline structure had high mechanical properties as compared with the initial homogenized state. The yield stress of the HPT samples increased by a factor of 1.5, the ultimate tensile strength increased by a factor of 1.4, while preserving the same ductility of 6-7%.

X-ray diffraction and microhardness measurements were used to established structural and phase transformations with post-deformation annealing at 140 °C from 0.25 to 6 hours. It was found that the preliminary HPT deformation significantly accelerated the decomposition of the supersaturated aluminum solid solution. The effect of dispersion hardening is observed already after 0.5-1 hours of annealing of the submicrocrystalline alloy at 140 °C. In addition, the formation of the submicrocrystalline structure changes the precipitation kinetics and type of precipitates of the hardening phases. It was determined that 6-hour annealing at 140 °C did not cause degradation of the submicrocrystalline structure and HPT Nikalin retained high strength.

## Acknowledgments

The research was carried out within the state assignment of the Ministry of Science and Higher Education of the Russian Federation (theme "Structura", № AAAA-A18-118020190116-6) supported in part by RFBR (project № 18-03-00102). The electron microscopy investigations were performed at the Center of Collaborative Access "Testing Center of Nanotechnologies and Advanced Materials" of the M.N. Mikheev Institute of Metal Physics of Ural Branch of Russian Academy of Sciences.

IOP Conf. Series: Materials Science and Engineering **672** (2019) 012022 doi:10.1088/1757-899X/672/1/012022

**References**

[1] Belov N, Akopyan T and Naumova E 2017 *Mater. Sci. Technol.* **33** 656
[2] Shurkin P K, Belov N A, Akopyan T K, Aleshchenko A S, Avxentieva N N and Alabin A N 2017 *Phys. Met. Metallogr.* **118** 896
[3] Shurkin P K, Akopyan T K, Galkin S P and Aleshchenko A S 2019 *Metal Science and Heat Treatment* **60** 764
[4] Sabirov I, Murashkin M Y and Valiev R Z 2013 *Mater. Sci. Eng.* A **560** 1
[5] Zhang Y D, Jin S B, Trimby P, Liao X Z, Murashkin M Y, Valiev R Z and Sha G 2019 *Mater. Sci. Eng.* A **752** 223
[6] Sauvage X, Murashkin M Yu, Straumal B B, Bobruk E V, Valiev R Z 2015 *Adv. Eng. Mater.* **17** 1821
[7] Nguyen Q C, Csanádi T, Györi T, Valiev R Z, Straumal B B, Kawasaki M, Langdon T G 2012 *Mater. Sci. Eng.* A **543** 117

IOP Conf. Series: Materials Science and Engineering **672** (2019) 012023 doi:10.1088/1757-899X/672/1/012023

# Contribution of structural factors to the specific electrical resistance of nanostructured Cu-Cr-Zr alloy

**R G Chembarisova**[1], **E A Sarkeeva**[1,2] **and I V Alexandrov**[1,2]

[1] Ufa State Aviation Technical University, Ufa, 450008, Russia
[2] Sino-Russian Joint Laboratory of Functional Nanostructured Materials, Changzhou University, Changzhou 213164, Peoples Republic of China

E-mail: chroza@yandex.ru

**Abstract**. The specific resistance of the nanostructured Cu-1.8wt.%Cr-0.8wt.%Zr alloy has been analyzed by means of analytical modeling. The contribution of different structural parameters to the specific resistance of the material has been estimated.

## 1. Introduction

The energy industry demands the conductors with high strength and high electrical conductivity [1]. Reducing grain size allows increasing the strength characteristics of such materials. At the same time, this leads to an increase in electrical resistance. In this regard, the issue of the contribution of structural parameters to the formation of a state with optimal strength and electrical conductivity is relevant [2, 3]. In this paper, analytical modeling methods are used to analyze the contribution of different structural parameters into the specific electrical resistance of the Cu-1.8 wt.% Cr-0.8 wt.% Zr alloy in various structural states.

## 2. Experimental and theoretical approaches

A detailed description of the experiments is presented in [3]. Samples of the Cu-1.8 wt.% Cr-0.8 wt.% Zr alloy, were kindly provided by Prof. Wei Wei in the initial coarse-grained (CG) state and in the state after equal-channel angular pressing (ECAP). The initial state (IS) was formed as a result of high-temperature heat treatment at 1000 °C for 0.5 h, followed by quenching in 5% NaCl solution. Samples of the alloy in the IS were deformed by 4 ECAP passes using route *A* at room temperature (RT). As a result, a nanostructured (NS) state of the alloy was formed (4ECAP state). The samples of the alloy in the 4ECAP state were subjected to subsequent aging at a temperature of 450 °C for 1 hour, which led to the formation of the 4ECAP+A state. The microstructure was studied using an OLIMPUS GX51 optical microscope, a JEOL JSM-6490LV scanning electron microscope and a JEOL 2100 transmission electron microscope. X-ray diffraction analysis of the samples was carried out on a Rigaku Ultima IV diffractometer. The measurements of the specific electrical resistance were carried out at RT by the eddy current method. The experimental values of the sizes of coherent scattering domains $D_{XRD}$, the magnitude of the mean-square lattice of elastic microdistortions $\langle \varepsilon^2 \rangle^{1/2}$, lattice parameter $a$, average particle size $d_{av}$, grain size $d$, and average distance between the particles $L_{av}$ are presented in table 1.

The analysis of the specific electrical resistance of various structural states was carried out taking into account the contribution of alloying atoms, dislocations, particles of secondary phases, grain boundaries (GBs) and thermal vibrations of the crystal lattice. The contribution of the dissolved atoms of the alloying elements was estimated by the formula presented in [4]:

$$\rho_{\mathrm{a}} = k^3\hbar l^4 n_{\mathrm{a}} C_{\mathrm{sa}}\left(1-\frac{\sin 2kl}{2kl}\right)\Big/\left(9e^2\pi\, n_{\mathrm{e}}\right) \tag{1}$$

where $n_{\mathrm{a}}$ is the volume concentration of the matrix atoms, $k=2\pi/\lambda$ is the wave number, $\lambda=h/m^*\upsilon_{\mathrm{F}}$ is the electron wavelength, $m^*$ is the effective electron mass, $\upsilon_{\mathrm{F}}$ is the Fermi velocity, $\hbar$ is Planck's constant, $C_{\mathrm{sa}}$ - atomic fraction of atoms $a$ in solution, $e$ is the electron charge value, $n_{\mathrm{e}}$ is the density of free electrons, $l$ is the size of the region of interaction of an electron with an alien atom. The presence of dislocations increases the specific electrical resistance by the value of $\rho_{\mathrm{disl}}$ [4]:

$$\rho_{\mathrm{disl}} = \hbar k^2 l^2 \rho_{\mathrm{tot}}\left(1-\frac{\sin 2kl}{2kl}\right)\Big/\left(9\, n_{\mathrm{e}} e^2\right). \tag{2}$$

The contribution of large and small particles of secondary phases to resistivity was estimated according to the formula presented in [5]:

$$\rho_{\mathrm{seca}} \approx \frac{m_{\mathrm{e}}\upsilon_{\mathrm{F}} f_{\mathrm{seca}}}{n_{\mathrm{a}} e^2}\frac{\pi}{V_{\mathrm{pa}}}\left(R+r\right)^2 \tag{3}$$

where $R$ is the radius of the interaction region of the electron with a particle ($R$ – particle radius), $r$ is the screening length, $r\approx 3\cdot 0.554\cdot 10^{-10}$ m, $m_{\mathrm{e}}$ is the electron mass. $V_{\mathrm{pa}}$ is the volume of spherical particles of a secondary phase. In the case of large misorientation angles, the average value of the contribution of GBs was estimated according to the formula $\rho_{\mathrm{gb}} = 5.6\cdot 10^{-16} N_{\mathrm{gb}}$, $N_{\mathrm{gb}}=S/\Omega\approx 1/d$, $S$ is the area of boundaries in the volume $\Omega$ [6, 4]. The contribution to the specific electrical resistance of thermal vibrations of Cu atoms in the lattice sites was taken equal to $\rho_{\mathrm{TCu}}=1.724\cdot 10^{-8}$ Ω·m similar to the case of pure coarse-grained copper [5].

## 3. Results and discussion

The solubility of Cr in Cu at 1000 °C is 0.4 wt.% (0.488 at.%), the solubility of Zr in Cu is 0.15wt.% (0.104 at.%) [7]. Since the content of the alloying elements exceeds their solubility in Cu at the quenching temperature, spherical particles of undissolved phases appear in the alloy structure. After ECAP, fine particles with a size of 7–11 nm were detected [3]. During the deformation process, a partial dissolution of large particles occurred [3], and a supersaturated solid solution was formed. The solution contained Cr and Zr atoms, the atomic fractions of which were $C_{\mathrm{sCr}}$=0.3840 at.% and $C_{\mathrm{sZr}}$=0.0821 at.%, respectively. The atomic fraction $C_{\mathrm{sCr}}$ of Cr atoms dissolved in the matrix in the UFG alloy state was calculated taking into the assumed linear dependence of the variation in the lattice parameter on the concentration of Cr in solid solution. The lattice parameter in the IS was equal to $a$=3.622±0.005Å. In pure Cu, the lattice parameter was assumed to be $a_0$=0.3615 nm. The atomic fraction of Zr atoms $C_{\mathrm{sZr}}$ was determined similarly. For Cr atoms, the size of the electron scattering region was taken equal to $l_{\mathrm{Cr}}=1.35b_{\mathrm{B}}$ [4]. For Zr atoms, $l_{\mathrm{Zr}}=1.73b_{\mathrm{B}}$ according to the known experimental value of the electrical conductivity of the alloy in the initial state is $\delta$ =30.0±0.6% IACS. The increase in the size of the interaction region is due to the presence of a difference between the valences of Zr and Cu [8].

The contribution of dislocations to the resistivity (figure 1a) was calculated at $l=1.55b_{\mathrm{B}}$ [4]. The dislocation density $q$ (table 2) was calculated according to the formula $q=2\sqrt{3}\langle\varepsilon^2\rangle^{1/2}/D_{\mathrm{XRD}}b_{\mathrm{B}}$ [9] taking into account the value of the Burgers vector $b_{\mathrm{B}}$ and the known experimental data (table 1).

The volume fractions of large and small particles were calculated according to the formula $f_{\mathrm{seca}}=V_{\mathrm{pa}}N_{\mathrm{p}}/V_{\mathrm{s}}$, where $N_{\mathrm{p}}$ - the number of particles in the sample, $V_{\mathrm{s}}$ - the sample volume, equal to

$f_{secCr} \approx 6.731\cdot10^{-3}$ and $f'_{secCr} \approx 1.156\cdot10^{-2}$ respectively. The volume fraction of large and small $Cu_5Zr$ particles after ECAP took the values $f_{sec\,Zr} \approx 1.219\cdot10^{-2}$ and $f'_{sec\,Zr} \approx 2.067\cdot10^{-2}$ respectively.

The grain boundaries in the alloy after ECAP were assumed to have high-angle misorientations.

**Table 1.** Microstructure parameters of Cu-1.8 wt.%Cr-0.8 wt.%Zr alloy in 4ECAP and 4ECAP+A states.

| State | $D_{XRD}$ (*nm*) | $\langle\varepsilon^2\rangle^{1/2}$, (%) | *a*, (Å) | $d_{av}$ (nm) | *d* (nm) | $L_{av}$ (nm) |
|---|---|---|---|---|---|---|
| 4ECAP | 34±4 | 0.20±0.02 | 3.620±0.002 | $(1.9\pm0.1)\cdot10^3$ | 155±10 | $21.0\cdot10^3$ |
| 4ECAP+A | 35±2 | 0.18±0.01 | 3.615±0.002 | $(1.8\pm0.1)\cdot10^3$ | 175±10 | $18.2\cdot10^3$ |

It was assumed that the contribution of the specific resistance $\rho_T$ of the thermal vibrations of atoms in the lattice sites in the 4ECAP state is equal to $\rho_T = (1 - f_{secCr} - f'_{secCr})\rho_{TCu} + (f_{secCr} + f'_{secCr})\rho_{TCr}$ where $\rho_{TCr}$=2.7 $10^{-8}$ Ω·m - the specific electrical resistance of Cr at room temperature. It was assumed that the specific resistance of the $Cu_5Zr$ compound is of the order of the specific electrical resistance of the matrix. The contributions of all factors that increase the specific electrical resistance $\rho$ of the alloy in the 4ECAP state are presented in figure 1 and the resulting value of $\rho$ and electrical conductivity $\delta$ are summarized in table 2.

The solubility of Cr in Cu at 450 °C is 0.03 wt.% ($3.661\cdot10^{-2}$ at.%), Zr - 0.005 wt.% ($3.477\cdot10^{-3}$ at.%) [7]. It is known that as a result of primary aging of the CG Cu-0.78% Cr-0.13% Zr alloy at a temperature of 460 °C for 3 h, fine particles precipitated, containing Cr and Zr atoms, the average size of which is ≈3 nm. Additional aging contributes to the enlargement of small particles precipitated during primary aging [10]. As a result, the concentration of such particles in the sample decreases *g* times. Let us take the following model of the alloy aged after ECAP. Particles precipitated as a result of dynamic aging (primary aging) are enlarged after aging at a temperature of 450 °C for 1 h. Moreover, their volume fraction remains almost unchanged. The number of particles $N_{sec}$ is proportional to their volume fraction $f'_{sec}$: $N_{sec} = 3V_s f'_{sec} / 4\pi R^3_{sec}$, where $f'_{sec}$ and $R_{sec}$ are the volume fraction and radius of particles precipitated as a result of dynamic aging in the ECAP process correspondingly. It follows that the radius of the particles $R_{secA}$ after aging will be equal $R_{secA} = R_{sec}\sqrt[3]{g f_{secA} / f'_{sec}}$, where $f_{secA}$ is the volume fraction of particles after aging, $f_{sec\,A} \approx f'_{sec}$. The concentration of particles having an average size after ECAP, equal to 9 nm, decreases approximately by *g*=5 times [10]. Consequently, the average particle size becomes $2R_{secA}\approx16$ nm. The sizes and density of large particles after aging practically did not change and remained the same as after ECAP (table 1). Accordingly, their volume fractions did not change. During aging, decomposition of the supersaturated solid solution occurred. The volume fractions of the precipitated small particles of the secondary phases Cr and $Cu_5Zr$ were, respectively, $f''_{AsecCr} \approx 3.510\cdot10^{-3}$, $f''_{AsecZr} \approx 5.444\cdot10^{-3}$. In NS materials, diffusion processes are accelerated. Therefore, it can be expected that the average diameter of particles containing Cr and Zr will be higher than 3 nm. Let us take the value of their radius *r* close to the maximum radius $r_{max}$, which is cut by dislocations. $r_{max}=\alpha b_B/2\xi$, where $\xi$ is the factor of size mismatch, $\xi=(r_s-r_m)/r_m$, $r_s$ is the radius of the atoms of the solute, $r_m$ is the radius of the atoms of the matrix. The parameter $\alpha$ was taken equal to 0.39, based on the known value of the true yield strength $\sigma_{Yexpi}$ in the initial quenched state, $\sigma_{Yexpi}\approx77.9$ MPa. It was assumed that the yield stress of an alloy taken in the IS is determined by the contribution of four factors: $\sigma_{Yexpi}=\sigma_P+\sigma_c+\sigma_{disl}+\sigma_{sec}$. $\sigma_c$ is the solid-solution hardening ($\sigma_c \approx MG\varepsilon_s^{3/2}C_{sa} / 760$) [5]), which was calculated with the parameter $\varepsilon_s$ taken to be 2.3 for Zr and 0.6 for Cr. $\sigma_{sec}$ - hardening by incoherent particles of the secondary phases, calculated in the framework of the Orowan model $\sigma_{sec}=QMGb_B\Phi\{\ln[(L_{av}-d_{av})/4b_B]\}/[2\pi(L_{av}-d_{av})]$ taking into account the values of the parameters presented in table 1, as well as the given values of the parameters $Q$=0.83, $\Phi$=1. $\sigma_P$ is the Peierls stress, which was calculated according to the simplified formula $\sigma_P=MG\exp(2\pi/(1-\nu))/(1-\nu)$ [5], where the Poisson's ratio is $\nu$=0.343. $\sigma_{disl}$ - dislocation hardening, $\sigma_{disl}=M\alpha Gb_B(q)^{1/2}$ [5]. For Cr atoms in the Cu matrix, the size mismatch factor is $\xi$=0.0156. Accordingly, $r_{max}$≈3.2 nm. $Cu_5Zr$ particles are not cut by dislocations. The average value of the contribution of GBs into the specific electrical resistance was estimated as $\rho_{gb} = 0.705\cdot10^{-16} N_{gb}$ ($\Theta$<16°) [6, 4]. In assessing the contribution of thermal lattice vibrations $\rho_T$ to the specific electrical resistance of the alloy, the areas occupied by small Cr particles precipitated during aging were taken into account. The contributions of all factors that increase the specific electrical resistance of the alloy in ECAP+A state are summarized in table 2 and shown in figure 1.

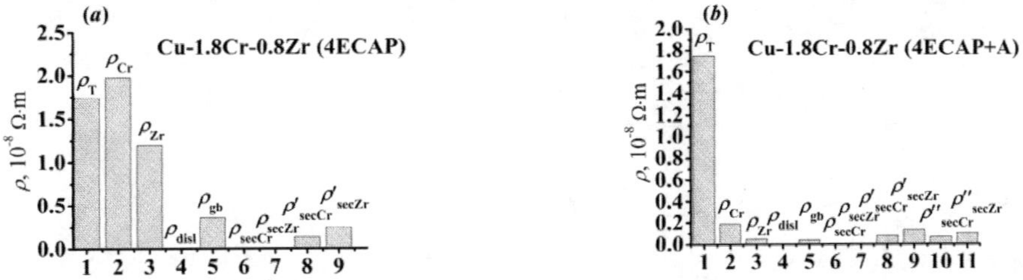


**Figure 1.** Contribution to the specific electrical resistance of alloying atoms ($\rho_{Cr}$, $\rho_{Zr}$), thermal vibrations of the crystal lattice ($\rho_T$), dislocations ($\rho_{disl}$), GBs ($\rho_{gb}$), large particles ($\rho_{secCr}$, $\rho_{secZr}$) precipitated during quenching and small particles ($\rho'_{secCr}$ $\rho'_{secZr}$) of secondary phases precipitated during the ECAP process, as well as small particles precipitated as a result of aging ($\rho''_{secCr}$, $\rho''_{secZr}$) in 4ECAP (a) and 4ECAP+A (b) states.

**Table 2.** The model values of dislocation density $q$ and specific electrical resistance $\rho$, the model $\delta$ and the experimental $\delta_{exp}$ value of the electrical conductivity.

| State | $q$ (m$^{-2}$) | $\rho$ (Ω·m) | $\delta$ (IACS, %) | $\delta_{exp}$ (IACS, %) |
|---|---|---|---|---|
| 4ECAP | $7.96\cdot10^{14}$ | $5.662\cdot10^{-8}$ | 30.5 | 29.0±0.6 |
| 4ECAP+A | $6.96\cdot10^{14}$ | $2.398\cdot10^{-8}$ | 71.9 | 70.0±0.6 |

## 4. Conclusions

According to the results of the modeling, a significant contribution to the resistivity of the alloy in the 4ECAP state is made by the dissolved alloying atoms and the thermal oscillations of the crystal lattice. An order of magnitude smaller contribution is made by the GBs, as well as particles of secondary phases of Cr and $Cu_5Zr$ precipitated during the ECAP process. The contribution of the large Cr and $Cu_5Zr$ particles precipitated during the quenching process is negligible. Aging after 4 ECAP passes contributes to the formation of a specific structure, which allows increasing the electrical conductivity as a result of separating alloying atoms from the matrix into small particles of secondary phases, coarsening the fine particles precipitated after ECAP and reducing the misorientation angle between the grains.

## Acknowledgments

Authors are grateful to be supported by the financial support of the Ministry of education and science within the frame work of the project No. 16.1969.2017/4.6.

## References

[1] Nikolaev A K and Rosenberg V M 1978 *Alloys for Resistance Welding Electrodes* (Moscow: Metallurgiya) [in Russian]
[2] Valiev R Z and Alexandrov I V 2007 *Nanostructured Metallic Materials* (Moscow: Academkniga) [in Russian]
[3] Sarkeeva E A, Sitdikov V D, Raab G I, Wei W and Alexandrov I V 2018 *IOP Conf. Ser.: Mater. Sci. Eng.* **447** 012065
[4] Chembarisova R G 2018 *IOP Conf. Ser.: Mater. Sci. Eng.* **447** 012085

IOP Conf. Series: Materials Science and Engineering **672** (2019) 012023 doi:10.1088/1757-899X/672/1/012023

[5] Chembarisova R G, Aleksandrov I V and Yamileva A M 2019 *Tech. Phys.* **64** 162
[6] Karolik A S and Luhvich A A 1994 *J. Phys.: Condens. Matter* **6** 873
[7] Osintsev O E and Fedorov V N 2004 *Copper and Copper Alloys* (Moscow: Mashinostroyeniye) [in Russian]
[8] Ziman J M 1962 *Electrons and Phonons* (Oxford: Clarendon)
[9] Zhao Y H, Liao X Z, Jin Z, Valiev R Z and Zhu Y T 2004 *Acta Mater.* **52** 4589
[10] Hatakeyama M, Toyama T, Nagai Y, Hasegawal M and Singh B N 2008 *Mater. Trans.* **49** 518

# Contribution of structural factors to the deformation behaviour of nanostructured Cu-Zn alloys

**R G Chembarisova**[1], **I V Alexandrov**[1,2] **and L I Zaynullina**[1,2]

[1] Ufa State Aviation Technical University, Ufa, 450008, Russia
[2] Sino-Russian Joint Laboratory of Functional Nanostructured Materials, Changzhou University, Changzhou 213164, Peoples Republic of China

E-mail: chroza@yandex.ru

**Abstract**. The results of the analysis of the deformation behavior of nanostructured Cu-Zn alloys, obtained as a result of severe plastic deformation, implemented by the equal-channel angular pressing method, are presented. The role of structural factors responsible for increasing the strength characteristics and uniform elongation of the Cu-Zn alloys has been revealed.

## 1. Introduction

In [1] it was shown that in low-carbon steels, as well as in metals with a FCC lattice at room temperature and α-Ti with an HCP lattice, the true uniform elongation $\varepsilon_u$ in the region of large grain sizes $d$ ($d$>10 μm) does not depend on the grain size $d$, while in the region of small grains the value $\varepsilon_u$ decreases with a decrease in the value $d$. Moreover, $\varepsilon_u$ in alloys with a FCC lattice is higher than in alloys with a BCC lattice due to high rates of deformation hardening.

In metals in which the flow stress $\sigma$ is sensitive to the strain rate, the strain is localized subject to the Hart criterion:

$$\frac{1}{\sigma}\frac{\partial \sigma}{\partial \varepsilon} \le 1 - m, \quad (1)$$

$m$ – the rate sensitivity, $\varepsilon$ - the strain. An increase in the value of $m$ contributes to maintaining the plasticity of the sample. Only nanostructured (NS) FCC and HCP metals have a higher rate sensitivity as compared to their coarse-grained (CG) analogues. NS BCC metals have even lower rate sensitivity $m$ than BCC metals in the CG state [1]. Therefore, it is more essential to maintain high values of the strain hardening rate. It can be assumed that the nature of the dependence of $\varepsilon_u$ on the grain size $d$ is not due to its value, but to the dislocation microstructure formed during the deformation process. A similar relationship exists between the flow stress $\sigma$ and $d$. The grain size is a function of the dislocation density $\rho_{tot}$ [2]. Therefore, it can be expected that there is a definite relationship between the quantities $\varepsilon_u$ and $\sigma$. The dependence of the value of $\varepsilon_u$ on the true yield strength of $Y_i$ in steels was obtained. It turned out that the value of uniform elongation is a function of the strength of the alloy: the value of true uniform elongation $\varepsilon_u$ decreases with increasing $Y_i$ [1]. Thus, the value of $\varepsilon_u$ is determined by the state of the formed microstructure corresponding to a given grain size $d$.

The observed differences in the deformation behavior of alloys with BCC and FCC lattices and the different strain hardening of alloys with FCC lattice can be explained by the difference in their stacking fault energy (SFE) [3]. It is known that alloying leads to a decrease in the SFE, an increase in the activity of deformation twinning with a decrease in the activity of dislocation slip [4, 5]. In NS Cu and alloys of the Cu-Al system obtained by the method of equal-channel angular pressing (ECAP) [6, 7, 4], a monotonous increase in the yield stress and uniform elongation with decreasing SFE was revealed.

The purpose of this work is to analyze the role of structural factors responsible for increasing the strength and the uniform elongation of the Cu-10 wt.% Zn alloy subjected to 1 and 8 ECAP passes, as well as flat rolling (FR) after 8 ECAP passes. As a result of these alloy treatments, ultrafine-grained (UFG) states 1 ECAP, 8 ECAP and 8 ECAP+FR were formed, respectively.

## 2. Experimental

A detailed description of the experiments is presented in [8]. Samples of the Cu-10 wt.% Zn alloy, were kindly provided by prof. Wei Wei, Changzhou University, P.R.C., in the initial CG state and were a subject for annealing at a temperature of 1073 K for 2 h and subsequent quenching in water from a temperature of 773 K. As a result, a homogeneous structure with an average grain size of 117±50 μm was obtained (initial state - IS). The SFE of pure Cu is ~78 mJ/m$^2$, the Cu-10% Zn alloy is ~35 mJ/m$^2$. Then, the samples in the IS were subjected to 8 ECAP passes (route Bc) at a temperature of 150 °C and subsequent FR at a temperature of 150 °C to reduction degree 95%. Mechanical tensile testing of the samples was carried out on an Instron 8801 tensile testing machine at room temperature. Microstructure studies were carried out using a transmission electron microscope. Investigations of the microstructure of alloy samples subjected to ECAP revealed a significant refinement of grains, the average size of which after 1 pass was 85±5 μm, after 8 passes - 121±50 nm. Consequently, a structure with finer grains is formed in the studied alloy during the ECAP process. The average grain size after FR slightly increased and amounted to 210±60 nm. Inside the grains of the alloy in UFG states, nano-twins were observed, the volume fraction and thickness of which increased with increasing number of ECAP passes. As a result of FR after 8 ECAP passes, their volume fraction and thickness decreased. The dimensions of coherent domains $D_{XRD}$, the lattice parameter $a$, and the elastic microdistortion of the crystal lattice $\langle\varepsilon^2\rangle^{1/2}$ obtained by the X-ray diffraction method [9] are summarized in table 1. The atomic fraction $C_{sZn}$ of Zn atoms dissolved in the matrix in the UFG alloy states was calculated taking into the assumed linear dependence of lattice parameter on the concentration of Zn in solid solution. Due to the absence of the experimental lattice parameter of pure Cu, a reference value of $a_0$=0.3615 nm was used.

**Table 1.** Microstructure parameters of Cu-10 wt.% Zn alloy taken in different structural states.

| The parameters of the microstructure | IS | 1ECAP | 8ECAP | 8ECAP+FR |
|---|---|---|---|---|
| $D_{XRD}$ (nm) | - | 47±7 | 35±3 | 29±2 |
| $\langle\varepsilon^2\rangle^{1/2}$ (%) | - | 0.20±0.01 | 0.31±0.02 | 0.34±0.01 |
| $a$ (Å) | 3.6379±0.0001 | 3.6377±0.0005 | 3.6363±0.0007 | 3.6364±0.0005 |

## 3. Results and discussion

The value of dislocation density (table 2) in various structural states of the alloy was estimated taking into account the experimental data presented in table 1 and the values of the Burgers vector equal to $b=a\sqrt{2}/2$ nm according to the formula $\rho_{tot}=2\sqrt{3}\langle\varepsilon^2\rangle^{1/2}/D_{XRD}b$ [10]. It was assumed that the yield stress $\sigma_Y$ of an alloy taken in the initial state is determined by the contribution of three active factors: $\sigma_Y=\sigma_P+\sigma_c+\sigma_{disl}$ (table 2). The Peierls stress $\sigma_P$ was calculated according to the simplified formula $\sigma_P=[MG/1-\nu]\exp[2\pi/1-\nu]$, where the Poisson coefficient is $\nu$=0.343, the Taylor factor is $M$=3.05, the shear modulus is $G$=42.1 GPa [11]. Solid-solution hardening $\sigma_c$ was calculated according to the dependence presented in [12]: $\sigma_c \approx MG\varepsilon_s^{3/2}C_{sZn}/760$, where the parameter $\varepsilon_s$ is 0.55. In this case, the atomic fraction $C_{sZn}$ of Zn atoms dissolved in the matrix was calculated according to the equation

$$C_{sZn} = W_{sZn}\frac{\mu_{Cu}}{\mu_{Zn}}\left(\frac{W_{Zn}}{1-W_{Zn}}+1\right)\left(\frac{W_{Zn}}{1-W_{Zn}}\frac{\mu_{Cu}}{\mu_{Zn}}+1\right)^{-1}, \qquad (2)$$

$W_{Zn}$ is the mass fraction of Zn atoms in the alloy, $\mu_{Zn}$ and $\mu_{Cu}$ are the relative atomic masses of Zn and Cu, respectively. In evaluating the dislocation hardening $\sigma_{disl} = M\alpha Gb\sqrt{\rho_{tot}}$ [13], the parameter $\alpha$ was taken equal to 0.37 [11]. The experimental value of the true yield stress $Y_i$, calculated according to the formula $Y_i=Y(1+\varepsilon_Y)$, where $\varepsilon_Y$ - experimental value of deformation corresponding to stress $Y$ (table 3).

With an increase in the number of ECAP passes, the dislocation density increases. This is caused by the limitation of the dislocation annihilation at their double cross-slip and climb as a result of a low SFE. At the same time, a low SFE facilitates deformation by twinning. The presence of twin boundaries contributes to the accumulation of dislocations in the inner regions of the grains.

**Table 2.** Model values of dislocation density, strengthening factors and model value of yield strength $\sigma_Y$ of Cu-10 wt.% Zn alloy.

| State | $\rho_{tot}$ (m$^{-2}$) | $\sigma_{disl}$ (MPa) | $\sigma_t$ (MPa) | $\sigma_c$ (MPa) | $\sigma_P$ (MPa) | $\sigma_Y$ (MPa) |
|---|---|---|---|---|---|---|
| IS | $1.50\cdot10^{13}$ | 47.3 | - | 6.73 | 13.8 | 67.8 |
| 1ECAP | $5.76\cdot10^{14}$ | 292.8 | 63.3 | 6.67 | 13.8 | 376.6 |
| 8ECAP | $1.20\cdot10^{15}$ | 422.7 | 65.8 | 6.26 | 13.8 | 508.6 |
| 8ECAP+FR | $1.59\cdot10^{15}$ | 486.6 | 31.0 | 6.28 | 13.8 | 537.7 |

**Table 3.** The values of experimental $Y$ and true $Y_i$ yield stress, experimental strain value $\varepsilon_Y$, volume fraction $f_t$ and thickness $t$ of twins, atomic fraction $C_{sZn}$ of Zn atoms dissolved in the matrix and the number of dislocations $N_t$ blocked at the twin boundaries.

| State | $Y$ (MPa) | $\varepsilon_Y$ (%) | $Y_i$ (MPa) | $f_t$ (%) | $t$ (nm) | $C_{sZn}$ (%) | $N_t$ |
|---|---|---|---|---|---|---|---|
| IS | 120.0 | 1.0 | 121.2 | - | - | 9.74 | - |
| 1ECAP | 368.0 | 3.8 | 382.0 | 2.4 | 16.0 | 9.65 | 5 |
| 8ECAP | 500.0 | 4.5 | 522.5 | 5.3 | 21.0 | 9.06 | 3 |
| 8ECAP+FR | 520.0 | 4.0 | 540.8 | 0.5 | 8.0 | 9.09 | 6 |

In the case of the alloy subjected to ECAP, in addition to the above three strengthening factors, it is necessary to take into account strengthening along the boundaries of the twins. The backpressure created by the boundaries of the twins was calculated according to the formula $\sigma_t = K_t z^{-1}$, where $K_t=MGbN_t/2$, $z$ is the distance between neighboring twins, $z^{-1} = f_t/\left[2t\ 1-f_t\ \right]$. The number of dislocations blocked at the boundaries of the twins $N_t$ [14] (table 3) was calculated taking into account the known experimental value of the true yield stress $Y_i$ and known values of the contributions of the active hardening factors $\sigma_{disl}$, $\sigma_c$ and $\sigma_P$ (table 2). With an increase in the number of ECAP passes, no saturation of the dislocation density is observed. The experimental value of the uniform elongation increases from 4% to 6% after 1 and 8 ECAP passes, respectively.

The subsequent FR of the samples after 8 ECAP passes leads to a further increase in the dislocation density (table 2), which was calculated taking into account the experimental data presented in table 1. The experimental value of uniform elongation took the value of 8.6%. The yield stress also increased (table 3). The number of dislocations $N_t$ blocked at the twin boundaries, calculated taking into account the experimental value of the true yield strength $Y_i$ and the calculated values of the strengthening factors $\sigma_{disl}$, $\sigma_c$, and $\sigma_P$, turned out to be larger than in the states after ECAP (table 3). At the same time, the contribution of hardening by the boundaries of the twins was smaller due to a decrease in the volume fraction of twins (table 2). A decrease in the thickness and volume fraction of twins gave an opportunity for further propagation of dislocations. There is an opportunity for further propagation and annihilation of dislocations. In this state, dislocation glide is an even more active deformation mechanism than in 8ECAP state.

## 4. Conclusions

The assumption was confirmed that the nature of the dependence of the uniform elongation of the Cu-10% Zn alloy on the grain size *d* is due to the microstructure formed during the deformation process. According to the results of the analysis, the boundaries of the twins, which appear in the process of ECAP in the Cu-10 wt% Zn alloy, increase their strength. An increase in the strain hardening of the samples of the alloy is realized as a result of the accumulation of dislocations in the inner regions of the grains in front of the twin boundaries, the volume fraction of which increases with decreasing grain size *d*. At the same time, a low SFE limits the annihilation of dislocations during their double cross-slip, as well as the annihilation of dislocations during their non-conservative motion. An increase in strain hardening is accompanied by an increase in the uniform elongation. Additional flat rolling with the reduction degree 95% after 8 passes ECAP leads to the decrease in the volume fraction of the twins, an increase in the grain size. It can be assumed that this occurs as a result of the injection of sufficient energy into the system during the flat rolling process, which leads to a change in the morphology of the grain. Their appearance resembles pancakes elongated in rolling direction. At the same time, the dislocation mechanism of deformation is activated, the dislocation density increases, and the strain hardening increases, accompanied by an increase in the degree of uniform elongation at room temperature.

**Acknowledgments**
The work was carried out with the financial support of the Ministry of Education and Science of Russia within the framework of the project № 16.1969.2017/4.6.

## References

[1] Morrison V B and Miller R L 1973 Plasticity of alloys with ultrafine grains *Ultrafine-Grain Metals* ed L K Gordienko (Moscow: Metallurgiya) pp 181–205 [in Russian]
[2] Edington J W and Smallman R E 1964 *Acta Metall.* **12** 1313
[3] Takahashi M and Suzuki A 1968 *Trans. Japan Inst. Met.* **9** 547
[4] An X H, Lin Q Y, Wu S. D, Zhang Z F, Figueiredo R B, Gao N and Langdon T G 2011 *Scripta Mater.* **64** 954
[5] Stremel M A 1997 *Strength of Alloys. Part II. Deformation* (Moscow: MISIS) [in Russian]
[6] An X H, Han W Z, Huang C X, Zhang P, Yang G, Wu S D and Zhang Z F 2008 *Appl. Phys. Lett.* **92** 201915
[7] Qu S, An X H, Yang H J, Huang C X, Yang G, Zang Q S, Wang Z G, Wu S D and Zhang Z F 2009 *Acta Mater.* **57** 1586
[8] Zaynullina L I, Sitdikov V D, Alexandrov I V and Wei W 2018 *IOP Conf. Ser.: Mater. Sci. and Eng.* **447** 012041
[9] Ungar T and Borbely A 1996 *Appl. Phys. Lett.* **69** 3173
[10] Zhao Y H, Liao X Z, Jin Z, Valiev R Z and Zhu Y T 2004 *Acta Mater.* **52** 4589
[11] Chembarisova R G, Aleksandrov I V and Yamileva A M 2019 *Tech. Phys.* **64** 162
[12] Fleischer R and Hibbard W 1967 Hardening during the formation of a solid solution *The Relation between the Structure and Mechanical Properties of Metals* ed V A Alekseev (Moscow: Metallurgiya) pp 85–111 [in Russian]
[13] Konrad X 1973 Model of strain hardening to explain the effect of grain size on the flow stress of metals *Ultrafine-grain metals* ed L K Gordienko (Moscow: Metallurgiya) pp 206–219 [in Russian]
[14] Remy L 1978 *Acta Metall.* **26** 443

IOP Conf. Series: Materials Science and Engineering **672** (2019) 012025 doi:10.1088/1757-899X/672/1/012025

# Microstructure and mechanical properties of the ultrafine-grained TiNi alloy after multiple martensitic transformations and subsequent aging

**A A Churakova[1,2] and D V Gunderov[1,2]**

[1] Institute of Molecule and Crystal Physics - Subdivision of the Ufa Federal Research Center of the Russian Academy of Sciences, 151 pr. Oktyabrya, Ufa, 450075, Russia
[2] Ufa State Aviation Technical University, 12 K. Marx str., Ufa, 450008, Russia

E-mail: churakovaa_a@mail.ru

**Abstract.** In this paper, we investigate the effect of preliminary thermal cycling in the temperature range of martensitic transformation on the aging processes in the ultrafine-grained $Ti_{49.0}Ni_{51.0}$ alloy, using as an example the examination of the microstructure, mechanical properties and fractography. Our studies show that preliminary thermal cycling leads to a significant increase in the density of dislocations that promote the intensification of the aging process and thereby enhance the strength and functional characteristics of the TiNi alloy in the ultrafine-grained state.

## 1. Introduction

TiNi alloys with the shape-memory effect (SME) possess high functional characteristics, enhanced strength, ductility, corrosion resistance, biocompatibility, etc. [1-2]. It is known that a cycle of martensitic transformations (MTs) under cooling and heating leads to the generation of dislocations in the crystalline lattice. Understanding the effect of multiple "cooling and heating" cycles below and above the martensitic transformation points – thermal cycling (TC) – on the structure and properties of materials is of great importance for TiNi alloys and products made thereof [3-4]. Thermal cycling brings about changes in the structural and stress states, therefore it can be used to control the functional properties of shape-memory alloys (SMAs) [5-10]. For TiNi alloys, an important role is played by the aging processes – precipitation of the aging particles $Ti_3Ni_4$, $Ti_2Ni_3$ during the solid solution decomposition [11]. In [12], the authors proposed a method for improving the functional stability of NiTi alloys having a high content of Ni. The method consisted in combining two methods – multiple phase transformations and low-temperature aging. Grain size is of an important decisive significance for improving functional characteristics. The new approach stipulates that first the generation of dislocation occurs in the interior part of a grain through cyclic phase transformations, then the low-temperature treatment by aging is performed to introduce nanoprecipitates, and dislocations can act as heterogeneous centers for the nucleation of precipitates. As a result, the homogeneous distribution and improvement of functional characteristics due to particle precipitation [12].

IOP Conf. Series: Materials Science and Engineering **672** (2019) 012025 doi:10.1088/1757-899X/672/1/012025

However, in that paper only the material in the coarse-grained state was considered. Scientifically interesting is the combination of an ultrafine-grained structure and thermal cycling during subsequent aging and their role in producing a material with high functional and strength characteristics.

## 2. Material and methods

In this work, as the material for investigation we used a TiNi intermetallic – $Ti_{49.0}Ni_{51.0}$ (manufactured by MATEX, Russia), having a B2 crystalline lattice in the initial state after quenching from a temperature of 800 °C in water. To produce an ultrafine-grained structure, ECAP processing was conducted with the number of passes n=6 at a temperature of T=450 °C with an angle of 120°. Thermal cycling of the samples was performed via a successive immersion of the samples into liquid nitrogen (-196 °C) followed by heating to a temperature of 140±5 °C, the temperatures which are known to be lower and higher than the temperature of the direct martensitic transformation, $M_s$, and that of the reverse martensitic transformation, $A_f$. The number of "heating – cooling" thermal cycles was from 0 to 100 (n=20, n=100). The exposure time was t=8 min to ensure the complete heating/cooling of the samples. Heat treatment in the aging regime was performed in a Nabertherm furnace in a temperature range of 250-400 °C with exposure for 1 hour at each temperature. Tensile mechanical tests of small-sized flat samples with a gauge portion of 1*0.25*4 mm were conducted at room temperature with a strain rate of $1*10^{-3}$ $s^{-1}$ on a Shimadzu AG-50kNXD machine.

## 3. Results and discussions

The structure of the $Ti_{49.0}Ni_{51.0}$ alloy produced by ECAP consists of grains-subgrains with a size of about 300±10 nm (Figure 1, a). The subsequent annealings at a temperature of T=400 °C do not result in a change in the grain size of the ECAP-sample. The structure after TC with n=20 in the UFG is martensitic; martensite packets with an average size of 300±15 nm can be observed (Figure 1, b). After n=100 cycles, martensite laths are also observed in the structure and the boundaries of the parent austenite phase are still preserved. The annealing at T=400 °C of the samples after n=20 and n=100 has produced similar structures with a size of the structural element of 250±10 nm (Figure 1, c).

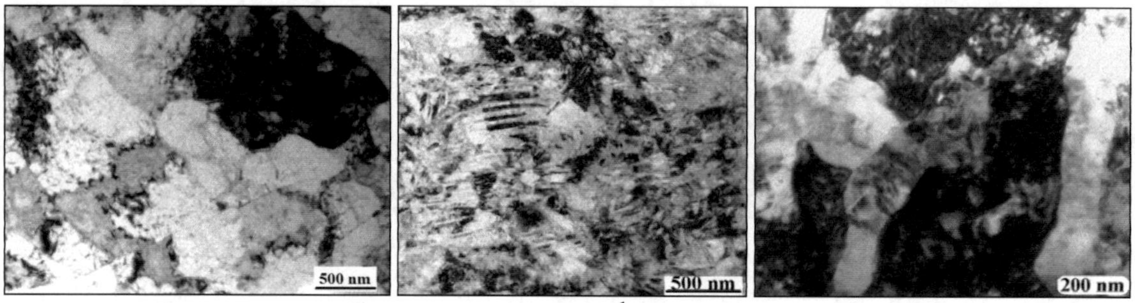


a b c

**Figure 1.** Structure of the $Ti_{49.0}Ni_{51.0}$ alloy: a) after ECAP, b) ECAP + TC (n = 20), c) ECAP + TC (n = 100) + annealing 400 °C.

The mechanical tests were performed on the $Ti_{49}Ni_{51}$ alloy samples in the UFG state before and after TC, with subsequent aging in a temperature range of 250-400 °C (Table 1). The ECAP-processed samples have the ultimate tensile strength (UTS) of 1125 MPa; aging at T=400 °C raises the strength to 1185 MPa, which is related to the grain refinement accompanied by the precipitation hardening, the yield stress (YS) also increases from 960 to 1040 MPa, and the ductility slightly declines from 42 to 36.

In the condition after TC with n=20 and subsequent aging at T=400 °C, YS grows to 1220 MPa as compared to the condition before aging, which can be attributed to the homogeneous precipitation and distribution of aging particles on dislocation networks formed in the process of thermal cycling; UTS also increases considerable, which is apparently related to the large quantity of strengthening $Ti_3Ni_4$ aging particles. The substantial increase in YS to 1300 MPa after TC with n=100 and aging at T=400 °C can be related to the large density of defects accumulated during multiple martensitic transformations. Thus, preliminary thermal cycling before aging leads to a large increase in UTS and YS, as compared to the conditions after aging without thermal cycling.

A fractographic analysis was performed to evaluate fracture viscosity and to examine aging particles. Figure 2 (a) shows the fracture surface of the ECAP-processed alloy, the fracture is ductile, the dimple size is 20±2 μm. After aging at 400 °C, the fracture is still ductile, and the dimple size is unchanged (figure 2, b).

**Table 1.** Results of mechanical tensile tests.

| State | UTS, MPa | $\sigma_m$, MPa | YS, MPa | $\sigma_r$, MPa | δ, % |
|---|---|---|---|---|---|
| CG initial | 840±30 | 255±15 | 460±20 | 205±15 | 33±5 |
| ECAPed | 1125±20 | 275±15 | 960±20 | 685±10 | 42±5 |
| ECAP + annealing 250 °C | 1120±15 | 260±15 | 980±20 | 720±15 | 35±5 |
| ECAP + annealing 400 °C | 1185±20 | 265±15 | 1040±20 | 775±10 | 36±5 |
| ECAP + TC (n=20) | 1180±25 | 342±15 | 950±20 | 608±15 | 20±5 |
| ECAP + TC (n=100) | 1365±25 | 420±15 | 1170±20 | 750±15 | 20±5 |
| ECAP + TC (n=20) + annealing 250 °C | 1242±25 | 365±15 | 1036±20 | 671±10 | 22±5 |
| ECAP + TC (n=100) + anneal. 250 °C | 1318±25 | 411±15 | 1160±25 | 749±10 | 24±5 |
| ECAP + TC (n=20) + anneal.400 °C | 1220±25 | 400±15 | 1040±25 | 640±10 | 20±5 |
| ECAP + TC (n=100) + anneal. 400 °C | 1300±25 | 421±15 | 1165±25 | 742±15 | 20±5 |

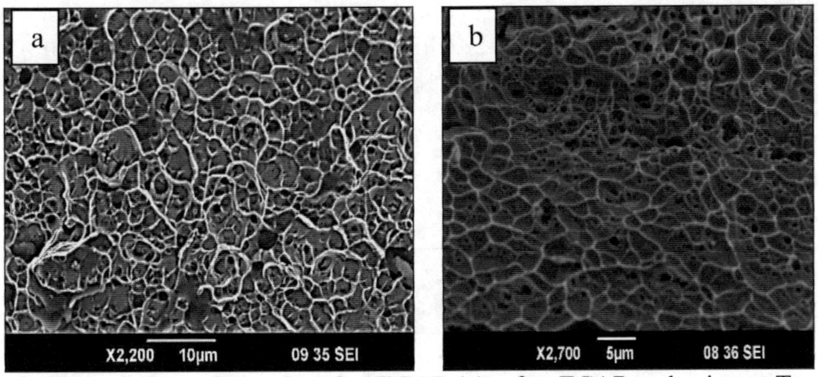


**Figure 2.** The fracture surface of sample: after ECAP (a); after ECAP and aging at T= 400 °C (b).

Figure 3, a shows the fracture surface of the alloy after TC with n=20 at T=400 °C; the fracture is ductile, the dimple size is 20±3 μm. The fracture surface of the alloy after TC with n=100 at T=400 °C is also ductile, the dimple size is 22±3 μm, but the dimples appear to be more elongated, presumably (Fig. 3, b) $Ti_3Ni_4$ aging particles are observed inside the dimples [13].

It can be noted that the dimple size (20 μm) on the fracture surface of the sample after ECAP is significantly larger than the grain / subgrain size, which indicates the intercrystalline propagation of the crack upon fracture.

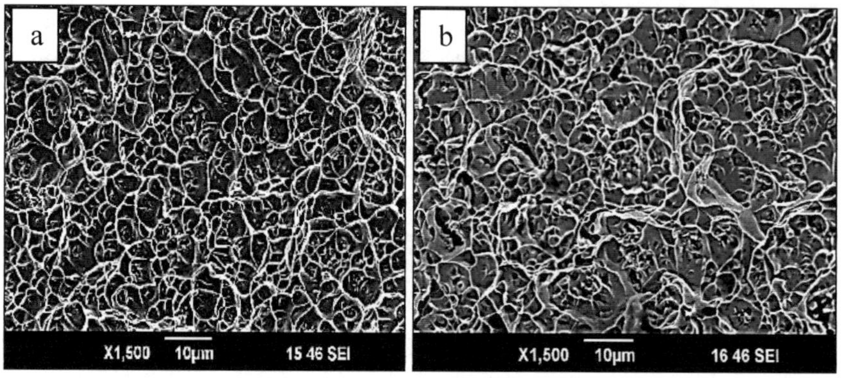


**Figure 3.** Fractography of the $Ti_{49}Ni_{51}$ alloy sample in the UFG+TC condition: a) n=20 T=400 °C, b) n=100 T=400 °C (SEM).

## 4. Conclusions

The structure of the $Ti_{49}Ni_{51}$ alloy produced by ECAP consists of martensite laths with the size of the austenite grains-subgrains with a size of about 300 nm. TC and subsequent annealings at a temperature of T=400 °C do not result in a change of the grain size. According to TEM, the size of martensite packets is 300±15 nm. The results of tensile mechanical tests show that preliminary TC (n=20 and n=100) during subsequent aging promotes to an increase in the ultimate tensile strength and yield stress in the UFG state. Fractographic analysis reveals that in the $Ti_{49.0}Ni_{51.0}$ alloy in the UFG state there is a pronounced ductile fracture, the dimple size is on average 20±5 μm, and the dimples have inclusions in individual regions. The fracture behavior does not change as a result of thermal cycling and subsequent aging, the dimple size varies within the measurement error, but in separate regions in the condition TC n=100 T=400 °C aging presumably $Ti_3Ni_4$ particles are observed inside the dimples.

## Acknowledgments

This work was supported by a Grant of the Bashkortostan Republic of the Russian Federation for young scientists (No. 28 GR of 07.03.2019).

## References

[1] Khachin V N 1992 *Titanium Nickelide: Structure and Properties* (Moscow: Nauka) [in Russian]
[2] Brailovski V, Prokoshkin S, Terriaultet P and Trochu F 2003 *Shape Memory Alloys: Fundamentals, Modeling, Applications* (Montreal: ETS Publ.)
[3] Otsuka K 1999 *Shape Memory Materials* (Cambridge: Cambridge University Press)
[4] Gunther V E *et al.* 1998 *Medical Materials and Implants with Shape Memory*, ed V E Gunther (Tomsk: TSU) [in Russian]
[5] Miyazaki M, Igo Y and Otsuka K 1986 *Acta Metall.* **34** 2045
[6] Erofeev V Y, Monasevich L A, Pavskaya V A, Paskal Y I 1982 *Fiz. Met. Metalloved.* **53** 963 [in Russian]
[7] Mironov Yu P, Erokhin P G and Kul'kov S N 1997 *Russian Physics Journal* **40** 210
[8] Furuya Y, Park Y C 1992 *Nondestructive Testing and Evaluation* **8–9** 541
[9] Lin G M, Lai J K L, Chung C Y 1995 *Scripta Metall. Mater.* **32** 1865
[10] Belyaev S, Resnina N 2013 *Int. J. Mater. Res.* **104** 11
[11] Kustov S, Mas B, Salas D, Cesari E, Raufov S, Nikolaev V, Van Humbeeck J 2015 *Scripta Mater.* **103** 10
[12] Wang X, Pu Z, Yang Q, Huang S, Wang Z, Kustov S, Humbeeck J V 2019 *Scripta Mater.* **163** 57
[13] Xu G X, Zheng L J, Zhang F X, Zhang H 2019 *Journal of Alloys and Compounds* **775** 698

IOP Conf. Series: Materials Science and Engineering **672** (2019) 012026 doi:10.1088/1757-899X/672/1/012026

# Effects of grain refinement by HPT processing in carbon steel with various cementite morphology

**A V Ganeev**[1], **Y Ivanisenko**[2] **and R Z Valiev**[1]

[1] Institute of Physics of Advanced Materials, Ufa State Aviation Technical University, 12 K. Marx str., Ufa 450008 Russia

[2] Institute of Nanotechnology, KIT, Main Building 640 Hermann-von-Helmholtz-Platz 176344 Eggenstein-Leopoldshafen, Germany

E-mail: artur_ganeev@mail.ru

**Abstract**. This article reports on the effect of the initial phase composition (martensitic, patented, spheroidized) of the middle carbon steel C45 (0.45 wt.% C) on grain refinement during high-pressure torsion (HPT) at elevated temperature. It is established that the cementite dissolution and the final grain size after HPT strongly depend on the processing regime, as well as on the cementite morphology in the initial state of the steel. As result of this study, new approaches to obtain a very high-strength state in carbon steels have been proposed.

## 1. Introduction

Studies of the influence of the structural phase state of the starting metal materials on the formation of a nanostructure in them during processing by the methods of severe plastic deformation (SPD) are of great interest [1]. In particular, it was established that the formation of the initial martensitic structure leads to a stronger refinement of the structure in low-carbon steels using SPD [2]. In [3], in industrial pearlite steel (~ 0.7% C), the formation of a nanostructure with a grain size of about 10 nm was observed, as well as the complete dissolution of cementite plates during HPT at room temperature. Obviously, the initial structure has a significant effect not only on the formation of the nanostructured state and their mechanical properties, but also on the choice of modes of severe plastic deformation in steels. This topical issue was the subject of study in this paper.

## 2. Material and methods

To assess the influence of the initial structure on the processes of nanostructuring, we used the high pressure torsion (HPT) technique. The work was carried out on high-quality carbon steel C 45 (C - 0.44, Mn - 0.65, Si - 0.18, P0.01, S - 0.0006, Cu - 0.05 wt.%). Several variants of the initial structure states were proposed: stable, metastable and unstable. The stable ferritic-pearlite structure of steel C45 was obtained by austenization at 800 °C, quenching in water and subsequent high tempering for 1 hour at 650 °C (spheroidized state). A metastable bainitic structure was obtained by heating at 800 °C for 1 hour followed by isothermal conversion to salt melting at 450 °C and cooling with water (patented state). An unstable martensitic structure was obtained by austenization at 800 °C for 1 hour and quenching in water (martensitic state).

HPT was performed at 350 °C for 10 revolutions under a pressure of 6GPa and speed of 0.2 rpm. Specimens after HPT were disk-shaped with a diameter of 20 mm diameter and a thickness of1.1 mm.

Microstructural, mechanical and electrical data were collected from the specific location in the middle of the radius of the sample (r – 10 mm). Tensile tests were carried out on an Instron 5982 machine at room temperature with a strain rate of $10^{-3}$ $s^{-1}$. Yield stress (YS), ultimate tensile stress (UTS) and ductility (el.%), measured as elongation-to-failure, were obtained on the small plane samples with a gauge dimensions of 1.0 x 1.0 x 4.0 mm prepared by the spark wire-cutting technique. The microstructure was characterized using a JEM-2100 transmission electron microscope. Thin foils for TEM investigations were punched from the slices of the samples. Then they were mechanically ground to a thickness of 0.12 mm, and finally double-jet electropolished to perforation using an electrolyte based on n-butyl alcohol. To study the relief surface of the fractured samples, as well as to further study the structure, a JSM-6390 scanning electron microscope with an accelerating voltage of 30 kV was employed.

## 3. Results and discussion

An analysis of the structure by SEM showed that after HPT in steel C45 with a different initial state, the grain size is refined in all cases. So, for a spheroidized state, the average grain size decreased from 13 μm to 1 μm. The size of the globule carbide decreased from 2 μm to 0.5 μm. The fraction of carbides observed in SEM increased from 10 to 26%. This value significantly exceeds the equilibrium value of 7%, which, apparently, is associated with the etching effect. In the patented state, the initial structure of bainite after HPT changes not so noticeably at low magnification. In SEM images, some dissolution of the cementite lamellas is observed. The distance between the lamellas after HPT changed from 0.3 μm to 0.2 μm. Finally, in steel C45, after the deformation of the martensitic state, a dispersed ferrite-cementite mixture is observed in the structure. At a small magnification in SEM, a metallographic texture is visible, which is characteristic of the microstructure of steels obtained by deformation according to the simple shear pattern.

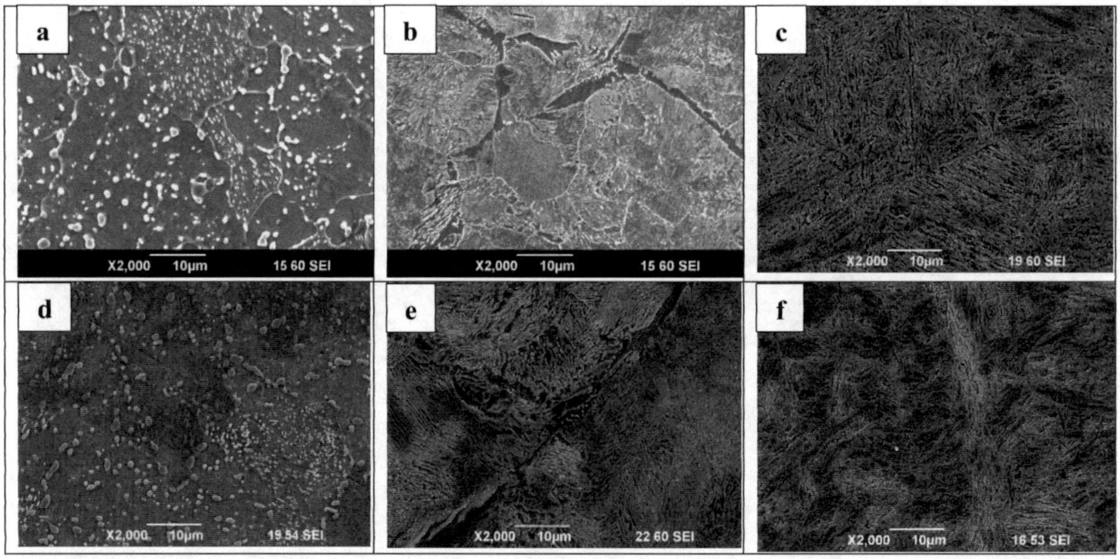


**Figure 1.** The evolution of the structure of steel C45 during HPT at 350 °C, SEM: initial heat treatment (a) spheroidized state; (b) - patenting state; (c) martensitic state; after HPT: (d) spheroidized state, (e) - patenting state; (f) - martensitic state.

Analysis of the microstructure of steel C45 in the spheroidized state showed that, despite a significant degree of deformation, the dissolution of coarse carbides practically does not occur. If the size of the ferrite grains decreases by 13 times, then the size of carbides changes only by 4 times. Apparently, this is due to the excessively high hardness of cementite (more than 800 HB).

The evolution of the structure of the patented state of steel C45 in the process of HPT at elevated temperature was considered in detail in a prelimenary study [4]. It was shown by TEM that the initial cementite plates fragment during processing. The particles in this case are in most cases along the grain boundaries. The microstructure is formed by grains with an axial axis and elongated along the shear direction with grains with an average size of about 40±10 nm in the cross section (figure 2a).

The initial martensitic structure with lamellar morphology of the matrix phase accelerates the fragmentation of the structure due to the large extent of the boundaries of martensitic origin [5]. Since the HPT processing occurred at an elevated temperature of 350 °C, a partial decomposition of martensite already has time to pass in the structure of steel C45. The absence of large cementite particles in the initial structure ensures the release of highly dispersed cementite particles uniformly distributed over the grain boundaries as a result of deformation). The average grain size calculated from dark-field images was 120 ± 20 nm. Dispersed carbides were also observed, which are located mainly along the grain boundaries of the α-phase. The average particle size of carbides is 15 ± 5 nm. The distance between cementite particles is about 100 nm [5]. An analysis of the microstructure of steel C45 in the cross section of the disk after HPT obtained by STEM showed that grains elongated along the shear axis are formed with an average length of 130 nm and a width of 25 nm (average grain size is 50±10 nm) (figure 2b).

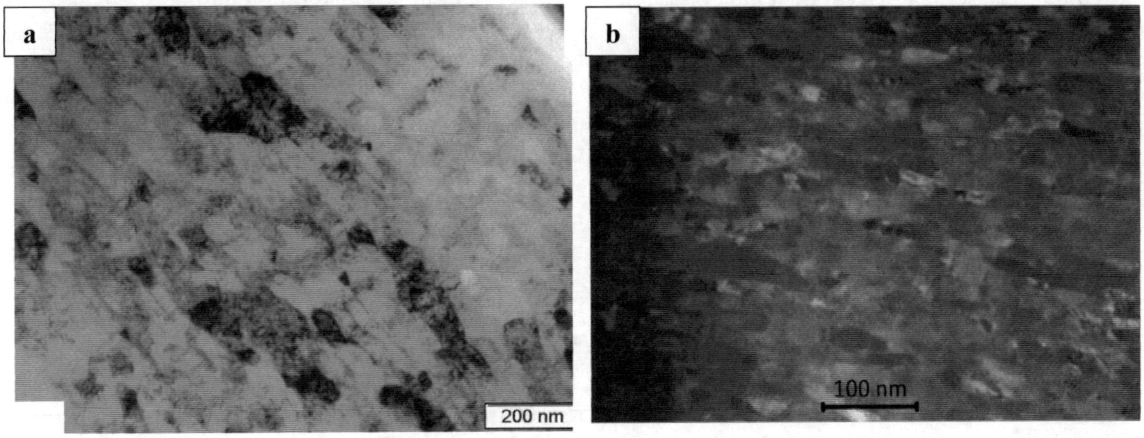


**Figure 2.** The structure of steel C45 during HPT at 350 °C in cross section: (a) patenting state [4]; (b) martensitic state, STEM.

The difference in the structure of steel C45 after HPT from the initial state can be explained by the difference in the share of the ferrite/cementite interfaces. Smart analysis shows that the length of the interphase boundaries in the patented state increases by 11 times, compared with the spheroidized state. In addition, a thin lamellar structure actively dissolves during HPT [3,4] and dispersed carbides stabilize the ferrite structure, which leads to the formation of a metallographic texture along the shear direction (figure 2). An analysis of the structure in the cross section for the patented and martensitic states shows that the grain shape and carbide size have a significant coincidence. This is explained by the fact that at 350 °C the decomposition of martensite already begins and thin lamellas of cementite are formed [5].

However, despite the similarity, these states after the HPT have differences in mechanical behavior. The results of the tensile tests are presented in table 1.

**Table 1.** Structure parameter & mechanical properties of steel C45 after heat treatment and HPT.

| State | Grain size/ plate width, nm | Cementite size/ plate width, nm | YS, MPa | UTS, MPa | El%, % | Ref. |
|---|---|---|---|---|---|---|
| Quenching | 120 | - | 2000 | 2190 | 1 | [5] |
| Patenting | 300±20 | 150±50 | 350 | 800 | 16 | [4] |
| Spheroidized | 13000±800 | 2000 | 320 | 620 | 22 | |
| Quenching + HPT 350 °C | 50±10 | 10±5 | 2390 | 2650 | 3 | [5] |
| Patenting + HPT 350 °C | 40±10 | 10±5 | 1390 | 2180 | 4.5 | [5] |
| Spheroidized + HPT 350 °C | 1000 | 500±50 | 800 | 1200 | 12 | |

In order to understand the nature of the high-strength state of the UFG steel under study, the contribution of different structural parameters to the yield strength following our previous investigations was estimated using a superposition of the contributions of structural components [5]. Where $\Delta\sigma_0$ is the friction stress of the α-iron lattice; $\Delta\sigma_{hp}$ - grain boundary; $\Delta\sigma_{ss}$ - solid solution; $\Delta\sigma_{disp}$ - dispersion; $\Delta\sigma_d$ - dislocation hardening. The obtained values are listed in table 2. It can be noted that grain boundary hardening in all cases makes the main contribution to the strength. The maximum hardening in steel 45 is achieved due to the formation of the UFG structure and the incomplete decomposition of the supersaturated solid solution.

**Table 2.** Contributions of various hardening mechanisms to the tensile strength of UFG C45 steel.

| State | $\Delta\sigma_0 + \Delta\sigma_{ss}$, MPa | $\Delta\sigma_{hp}$, MPa | $\Delta\sigma_{disp}$, MPa | $\Delta\sigma_d$, MPa | YScalc, MPa | $YS_{exp}$, MPa |
|---|---|---|---|---|---|---|
| Quenching + HPT 350 °C | **560** | 1350 | 190 | 110 | 2200 | **2390** |
| Patenting + HPT 350 °C | 102 | 1373 | 190 | 102 | 1760 | 1400 |
| Spheroidized + HPT 350 °C | 100 | 475 | 332 | 90 | 994 | 800 |

## 4. Conclusions

The influence of the initial state (martensitic, patented, spheroidized) of carbon steel C45 on the formation of the UFG structure during high-pressure torsion at elevated temperature is determined. It is shown that coarse carbides practically do not deform during HPT; on the contrary, the thin lamellar form of cementite or its partial dissolution in a solid solution contributes to a significant refinement of the grain structure of the steel and made it possible to obtain a high-strength state in carbon steels. Thus, for the first time in steel 45, record values of the yield stress - 2390 MPa, and the ultimate tensile strength - 2650 MPa were achieved.

**Acknowledgements**

A.V. Ganeev is grateful for the support for the fabrication of the UFG samples and their studies by TEM from RFBR under project No. 18-08-00340.

**References**

[1] Valiev R, Islamgaliev R and Alexandrov I 2000 *Prog. Mater. Sci.* **45** 103
[2] Maier G G, Astafurova E G, Koshovkina V S, Chomyakova G V, Naydenkin E V, Odessky P D and Dobatkin S V 2014 *IOP Conf. Ser.: Mater. Sci. Eng.* **63** 012133
[3] Ivanisenko Yu, Lojkowski W, Valiev R Z *et al.* 2003 *Acta Mater.* **51** 5555
[4] Haddad M, Ivanisenko Y and Fecht H-J 2015 *International Journal of Applied Science and Technology* **5(1)** 144
[5] Karavaeva M, Kiseleva S, Ganeev A *et al.* 2015 *J. Mater. Sci.* **50** 6730

IOP Conf. Series: Materials Science and Engineering **672** (2019) 012027 doi:10.1088/1757-899X/672/1/012027

# Comparative analysis of the microstructure and mechanical properties of an Al-Cu-Mg-Ag alloy peak-aged at relatively low and high temperatures

**M Gazizov[1], R Holmestad[2] and R Kaibyshev[1]**

[1] Belgorod State University, Pobeda 85, Belgorod, Russia
[2] Department of Physics, Norwegian University of Science and Technology (NTNU), Trondheim, Norway

E-mail: gazizov@bsu.edu.ru

**Abstract.** The mechanical properties and microstructure have been investigated in the Al-4.35Cu-0.46Mg-0.63Ag-0.36Mn-0.12Ti-0.04Fe alloy (in wt. %) peak-aged at 150 °C and 190 °C. The yield stresses (YS) and the ultimate tensile strength (UTS) were insignificantly higher by 15 MPa and 5 MPa, respectively, after peak-aging at 150 °C for 24 h compared with 190 °C for 1.5 h. Elongation-to-failure (δ) was the same (~6%) in both states. Microstructure analysis showed great differences in the morphology of precipitates within grain/subgrain interiors, as well as in the structure of regions adjacent to grain boundaries (GBs). Relationships between strength/ductile properties and microstructure were discussed.

## 1. Introduction

Al-Cu-Mg-Ag alloys exhibit high specific strength, good fracture toughness and resistance to fatigue fracture, as well as superior creep resistance. This attractive combination of mechanical properties is attributed to the Ω-phase precipitated during aging as uniformly distributed plates along the $\{111\}_{Al}$ planes [1–5]. The Ω-phase has a sandwich-like structure composed from a core having a nominal composition and a crystal structure similar to the equilibrium θ-phase (*I4/mcm*) observed in Al-Cu alloys [1,2,6]. Furthermore, the transition ordered layers of Mg and Ag provide a coherent structure of broad interfacial boundaries and relatively high resistance to coarsening of the plates at elevated temperatures [1–4,7]. It has been shown [3] that disperse $\{111\}_{Al}$ plates are highly efficient in strengthening aluminum alloys compared with spherical particles or other rationally oriented precipitates, such as $<100>_{Al}$ rods and $\{100\}_{Al}$ plates. One of the simplest ways to additionally enhance the strength of an age-hardenable aluminium alloy is to optimize the aging temperature, which leads to the increased contribution of the precipitate strengthening to the overall strength of the material. This aim can be reached by the formation of finer precipitates of metastable phases and/or an increase in its volume fraction occurring in accordance with thermodynamics at lower aging temperatures [2,6]. Regarding the Ω-phase, the relationships between the strength properties and morphology of the $\{111\}_{Al}$ plates of this phase at relatively low temperatures in Al-Cu-Mg-Ag alloys, have not been well studied. Thus, the aim of the present work is to characterize the precipitate microstructure in relationship with the peak mechanical properties of an Al-Cu-Mg-Ag alloy reached after aging at relatively low (150 °C) and high (190 °C) temperatures.

## 2. Experimental procedure

An aluminum alloy with the chemical composition Al-4.5Cu-0.56Mg-0.77Ag-0.42Mn-0.12Ti-0.05V-0.02Fe (in wt. %) was prepared using a direct-chill semi-continuous casting process. The alloy was homogenized at 500 °C for 24 h, followed by cooling in a furnace, then it was extruded at ~400 °C with a ratio of ~2.6 and subjected to hot rolling with a reduction of ~60%. The resulting 20 mm thick plates were used to cut rectangular samples with a size of ~20×20×3 mm and flat "dog-bone" samples with gauge sizes of 35×7×3 mm using a wire electric discharge machine. The samples were solution heat treated at 510 °C for 1 hour, followed by water-quenching and immediately aged at 150 °C and 190 °C at different times (up to 200 h).

Hardness was measured in arbitrarily selected areas of the flat surface of rectangular samples using a Wilson Wolpert 402 MVD hardness tester at a constant load of 5 N and a loading time of 15 s. Tensile properties (YS, UTS and δ) were measured using an Instron 5882 tensile testing machine at an initial strain rate of $\sim 10^{-3}$ $s^{-1}$. At least ten hardness indentations and three tensile tests were carried out for each condition to determine the standard deviation and the average tensile properties, respectively.

The methodology for preparing TEM foils was described in [3]. TEM characterization was performed using a JEOL JEM-2100F microscope operating at 200 kV. Precipitate parameters (diameter, thickness, number density and volume fraction) were analyzed in accordance with [3,4].

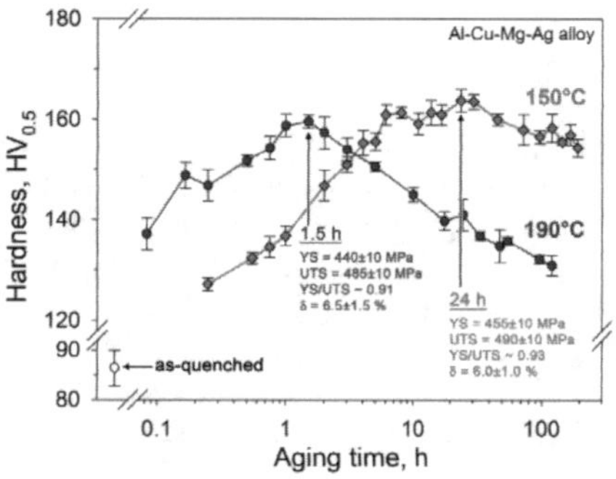


**Figure 1.** Hardness evolution during aging at 150 °C and 190 °C.

**Figure 2.** Microstructure of the alloy at peak-aging conditions after aging at 150 °C (a) and 190 °C (b).

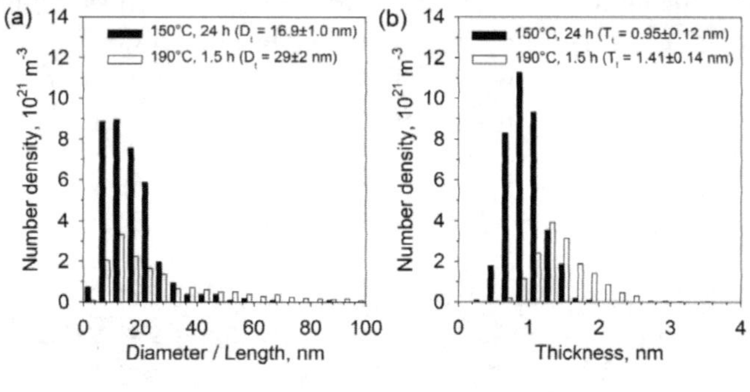


**Figure 3.** Effect of aging on the distribution of the Ω-plate diameters (a) and thicknesses (b) within the grains.

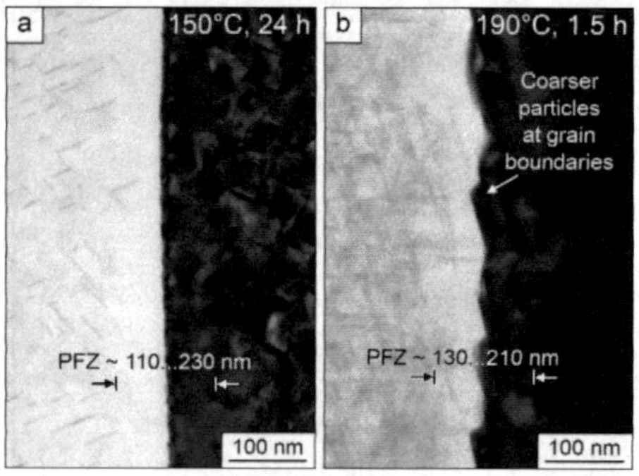


**Figure 4.** PFZs adjacent to the GBs after aging at 150°C (a) and 190°C (b).

## 3. Results and discussions

### *3.1. Mechanical properties*

Figure 1 shows the hardness evolution as a function of the aging time at 150 °C and 190 °C. It is seen that the three stages (under-, peak- and over-aging) being consistent with thermodynamics at different aging temperatures [1,2,6] can be identified at both aging temperatures. The hardness was ~86 $HV_{0.5}$ in the as-quenched state. At the initial aging stage, the hardness increased dramatically with prolonged aging. Peak hardness values of ~160 $HV_{0.5}$ and ~165 $HV_{0.5}$ were obtained in the range of 1-2 h at 190 °C and 14-30 h at 150 °C, respectively. The results of the analysis of tensile properties are put into figure 1. During further aging, the hardness decreased slowly, but preserved a relatively high level during aging at 150 °C even after ~200 h compared with aging at 190 °C, where the decrease in hardness was much more pronounced. This can indicate that the precipitation kinetics was much higher at 190 °C than at 150 °C, but the relatively close maximum strength indicates a small difference in precipitate morphology obtained in peak hardness for both temperatures [1,2,6]. Samples after 150 °C and 190 °C for 1.5 h and 24 h, respectively, were chosen for the TEM characterization.

**Table 1.** The Ω precipitate parameters within the grains/subgrains in peak-aging stage.

| Aging | $N_V$, $10^{21}$ $m^{-3}$ | $F_V$, % | Aging | $N_V$, $10^{21}$ $m^{-3}$ | $F_V$, % |
|---|---|---|---|---|---|
| **150°C, 24 h** | 36.4±1.3 | 0.8±0.3 | **190°C, 1.5 h** | 15.9±1.8 | 1.5±0.3 |

### *3.2. Microstructure*

The initial microstructure of the alloy in the as-quenched state (not shown) consists of elongated grains/subgrains having average sizes of ~27 and ~17 μm in the longitudinal and transverse directions, respectively. Uniformly distributed dispersoids were observed in the matrix. The average diameter size and volume fraction of them were ~50 nm and 0.5%, respectively.

Representative TEM images showing precipitate microstructures at the peak hardness stages, are given in figure 2. Particles in the form of platelets with the $\{111\}_{Al}$ habit planes were identified as Ω-phases. The Ω-phase plates nucleated homogeneously in the grain/subgrain interiors. The distributions of the apparent diameters and thicknesses of the $\{111\}_{Al}$ plates (in figures 3a and b, respectively) show that the size of the Ω-phase is different in the two states. It seems that a bimodal distribution of the plate diameters takes place after peak-aging at 190 °C. As a result, the diameter distribution peak is narrower at 150 °C compared with aging at 190 °C, where a competitive growth of relatively long and short Ω plates is observed. It should be noted that, firstly, for the short and long plates at 190 °C there is no obvious difference in thickness between them, but they are thicker at 190 °C than at 150 °C; secondly, the diameters of the short ones at 190 °C can be commensurable with the diameters of the Ω plates at 150 °C. Number density ($N_V$) and volume fraction ($F_V$) of the $\{111\}_{Al}$ plates have also been estimated (table 1). It is seen that the low temperature aging provides the formation of smaller Ω plates having a higher number density but lower volume fraction compared with aging at 190 °C. The precipitation strengthening contribution to the overall strength of the alloy due to shearing the $\{111\}_{Al}$ plates has been roughly estimated using equations from work [3] and an interfacial energy of 0.165 J×$m^{-2}$ equal to ~214 MPa and ~225 MPa after aging at 150 °C and 190 °C, respectively.

A TEM analysis of structures of the GBs has been performed due to their importance for the mechanical behavior of the aged aluminum alloys [8]. Examples of TEM images of the alloy in peak-aging states at 150 °C and 190 °C are shown in figure 4. The widths of the precipitate free zones (PFZ) adjacent to GBs were measured to be in a similar range in both states. This is inconsistent with the hypothesis of precipitation in the age-hardenable alloys – wider PFZs can be formed at higher aging temperatures due to the lower driving force for precipitation and the higher critical supersaturation of vacancies needed for nucleation to occur, in comparison with low temperatures [2,6]. It has also been shown [5,7] that Mg-Ag clusters play an important role in the heterogeneous mechanism of nucleation of the Ω-phase. It seems that the effect of the vacancy concentration on this type of cluster formation is less pronounced at the studied aging temperatures. Further, the most obvious difference in the studied peak-aged alloys was in the particle dispersion at the GBs. The PFZ widths varied in the same range for 150 °C and 190 °C, meanwhile coarser particles with a higher volume fraction were observed along the GBs at 190 °C. This is inconsistent with the amount of solute elements, which are available for precipitation close to the GBs, and is probably the result of a more complicated diffusion process. For the age-hardenable Al alloys, the narrower PFZ and fine precipitations on the GBs are favorable for the plasticity of these alloys. For this reason, the regions adjacent to the GBs occupied by coarser particles should exhibit a lower plasticity in the alloy aged at 190 °C compared with 150 °C. However, the δ values (figure 1) are the same after aging at both temperatures. For this reason, the two factors contrary affecting the plasticity can act together. An example of the latter one can be grain/subgrain interiors, demonstrating higher plasticity due to the different precipitate microstructure [1,2].

Precipitates elongated along the $\langle 001\rangle_{Al}$ directions in the $\{110\}_{Al}$ projection were identified as plates of the θ'-phase. At both aging temperatures, the θ' plates were non-uniformly distributed in the matrix. They can be found as nucleated along dislocation lines. It should be noted that the fraction of the θ'-phase was lower at 190 °C compared with aging at 150 °C, where its number density was less than 5% of the total number density of all precipitates ($\{111\}_{Al}$ and $\{001\}_{Al}$ plates) found. Thus, the Ω-phase plates are the main strengthening phase because of their dominant fraction compared to other secondary precipitates, such as the θ′-phase, as well as dispersoids in the alloy after 150 °C and 190 °C peak-aging. Thus, aging at 150 °C and 190 °C provides a large difference in the precipitate morphology, meanwhile similar strength and ductile properties were obtained. To better understand this phenomenon, further comprehensive investigations are required.

## 4. Conclusions

The microstructure and precipitation in relationship with the mechanical properties were analyzed in the Al-4.35Cu-0.46Mg-0.63Ag-0.36Mn-0.12Ti-0.04Fe (wt. %) alloy aged at 150 °C for 24 h and 190 °C for 1.5 h. Small differences in the tensile strength properties were observed. Analysis of the precipitate strengthening contributions showed that similar increments to the overall strength of the alloy took place despite a significant difference between the precipitate morphology in the grain/subgrain interiors after aging at 150 °C and 190 °C. PFZ widths varied in similar ranges in both states, but coarser particles at the GBs were found at 190 °C. Ductility attributed to the plastic grain interiors and PFZ, expecting to show a less plastic or even brittle fracture mode after aging at 190 °C, were similar to the samples aged at 150 °C, where finer particles appear at the GBs.

## Acknowledgments

This work is financially supported by Council for Grants of the President of the Russian Federation (No. 075-02-2018-164) and the Faculty of Natural Sciences at NTNU, Norway (Project No. 81617879). The TEM work was carried out on the NORTEM infrastructure (Grant 197405) at NTNU. The authors are grateful to the staff of the TEM Gemini Center at NTNU and the Joint Research Center at BelSU for their assistance with the structural and mechanical characterizations.

## References

[1] Polmear I J 2006 *Light alloys: from Traditional Alloys to Nanocrystals* (Oxford: Butterworth-Heinemann/Elsevier)
[2] Nie J F 2014 Physical Metallurgy of Light Alloys *Physical Metallurgy: Fifth Edition* vol 3 (Elsevier) pp 2009–156
[3] Gazizov M and Kaibyshev R 2017 *Mater. Sci. Eng.* A **702** 29
[4] Gazizov M and Kaibyshev R 2017 *Mater. Sci. Technol.* **33** 688
[5] Gazizov M and Kaibyshev R 2015 *Mater. Sci. Eng.* A **625** 119
[6] Porter D A, Easterling K E and Sherif M Y 2014 *Phase Transformations in Metals and Alloys* (New York: CRC press)
[7] Bai S, Liu Z, Zhou X, Xia P and Zeng S 2014 *J. Alloys Compd.* **602** 193
[8] Christiansen E, Marioara C D, Marthinsen K, Hopperstad O S and Holmestad R 2018 *Mater. Charact.* **144** 522

IOP Conf. Series: Materials Science and Engineering **672** (2019) 012028 doi:10.1088/1757-899X/672/1/012028

# Effect of high-pressure torsion on the mechanical behavior of a Zr-based BMG

**D Gunderov[1,3], E Boltynjuk[2], A Churakova[1,3], E Batirshin[4], A Mullayanov[4], V Titov[1] and J Ivanisenko[5]**

[1] Ufa State Aviation Technical University, 12 K. Marx str., Ufa 450008 Russia
[2] Saint Petersburg State University, 7-9 Universitetskaya nab., Saint Petersburg 119034 Russia
[3] Institute of Molecule and Crystal Physics UFRC RAS, 71 pr. Oktyabrya, Ufa 450054 Russia
[4] Bashkir State University, 32 Zaki Validi st., Ufa 450076 Russia
[5] Institute of Nanotechnology, Karlsruhe Institute of Technology, 1 Hermann-von-Helmholtz-Platz, 76344 Eggenstein-Leopoldshafen, Germany

E-mail: dimagun@mail.ru

**Abstract.** The effect of high-pressure torsion (HPT) on the structure and mechanical properties of a Zr-based bulk metallic glass (BMG) was investigated. The structure of the BMG changes significantly after HPT processing. In tensile tests, the BMG samples before and after HPT processing are brittle under a stress of about 1700 MPa. For the first time, three-point bending tests of the HPT-processed BMG were performed. In the three-point bending tests of the initial BMG and that after HPT processing, the strength exceeds 3000 MPa and plastic flow is observed.

## 1. Introduction

Bulk metallic glasses (BMGs) arouse a keen interest due to their specific structure and unique properties [1,2]. However, BMGs exhibit a limited tensile ductility [1,2]. In previous years, studies of the effect of various treatments on the structure and properties of BMGs were conducted [3-5]. High-pressure torsion (HPT) is of great interest for the processing of metallic materials. The HPT processing of amorphous alloys leads to considerable changes in their structure and properties [6-13]. HPT processing may lead to nanocrystallization in the amorphous phase [8,9] or the formation of a cluster-type amorphous structure [10,11]. During tensile tests at room temperature, initial BMGs undergo brittle fracture [1,2]. This makes it impossible to trace the evolution of ductility in BMGs under various treatments. Amorphous alloys exhibit ductility during upsetting tests and three-point bending tests [14]. Three-point bending is a much «milder» test procedure than tension. However, samples of amorphous materials produced by HPT have the shape of disks with a diameter of 10 mm and a thickness of 0.2-0.7 mm. Conducting the three-point bending tests of microspecimens that can be cut out from disk-shaped samples of such a size is a complicated task. To perform such tests, we used a unique facility for the three-point bending testing of small-sized specimens, designed in Ufa (Russia). This paper reports the first results from the three-point bending tests of the HPT-processed BMG.

IOP Conf. Series: Materials Science and Engineering **672** (2019) 012028 doi:10.1088/1757-899X/672/1/012028

## 2. Results and discussion

As the initial material, we used plates with a size of 75x24x0.75 mm from the Vit105 BMG $Zr_{52.5}Cu_{17.9}Ni_{14.6}Al_{10}Ti_5$ (at.%), produced by LIQUIDMETAL Ltd, USA. HPT processing was carried out under an applied pressure of 6 GPa for 5 and 30 complete revolutions with a rotation speed of 1 rpm at a temperature of 20 °C. As a result, samples with a thickness of 0.2-0.3 mm and a diameter of 10 mm were produced.

The amorphous structure was investigated by X-ray diffraction (XRD) under Cu radiation employing a Rigaku Ultima IV device. The diffraction patterns were taken with a step size of 0.05 degree and 10-second exposure per point. According to XRD, the structure of the as-cast Vit105 BMG and that after HPT processing is amorphous [14].

The principle of the three-point bending test is shown in figure 1. A specimen is subjected to bending with a constant rate of $10^{-3}$ $s^{-1}$ in the middle between supports until it fractures. During the testing, the load applied to the specimen measured, the deflection values were recorded by a videocamera.

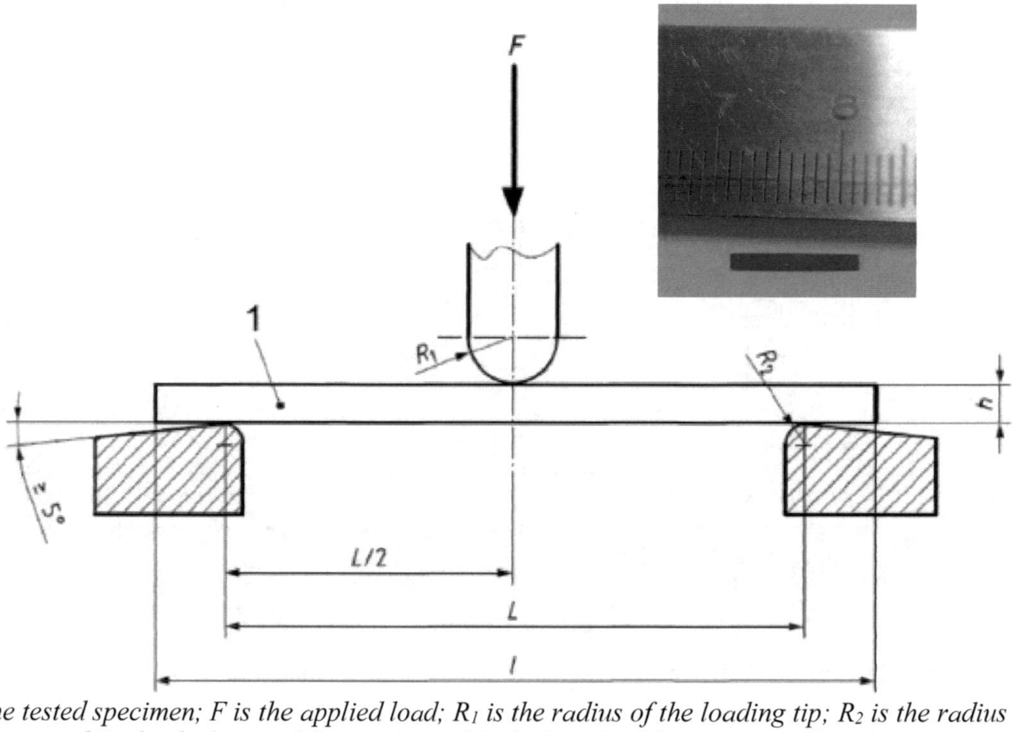


*1 is the tested specimen; F is the applied load; $R_1$ is the radius of the loading tip; $R_2$ is the radius of the supports; h is the thickness of the specimen; l is the length of the specimen; L is the support span*

**Figure 1.** Principle of the three-point bending test and a photo of a microspecimen for testing.

The flexural stress $\sigma_f$, MPa, is calculated according to the formula:

$$\sigma_f = \frac{3FL}{2bh^2}$$

where $F$ is the applied load, N; $L$ is the support span, mm; $b$ is the width of the specimen, mm; $h$ is the thickness of the specimen, mm. The relative flexural strain $\varepsilon_f$, a dimensionless quantity or %, is calculated according to the formula:

$$\varepsilon_f = \frac{6sh}{L^2}$$

where s is the deflection, mm.

During tensile tests, the Vit105 BMG before and after HPT processing undergoes brittle fracture under a fracture stress of about 1700 MPa. At the same time, during three-point bending tests the fracture stress of the Vit105 BMG exceeds 3000 MPa and plastic flow is observed (figure 2). It is worth noting that the strength values of BMG depend on the testing technique. Much higher strength values of the BMG were observed during three-point bending tests, as compared to the strength values observed during tensile tests [15]. The plastic deformation to failure during the three-point bending tests of the initial Vit 105 BMG is 2.5 % (for the specimen thickness 0.37 mm) (table 1).

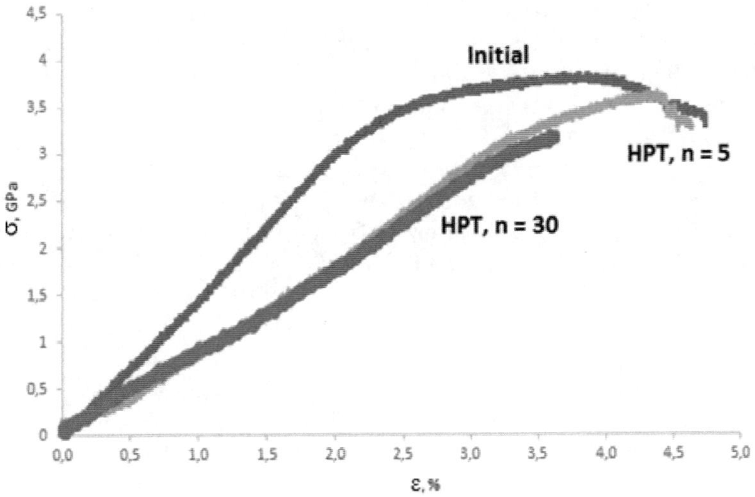


**Figure 2**. Three-point bending test results of the Vit105 BMG in the initial state and after HPT for n=5, HPT for n=30.

**Table 1.** Three-point bending test results of the Vit105 BMG in the initial state and after HPT.

| Vit105 BMG (manufactured in the USA) | Thickness h, mm | YS, MPa | Maximum stress, MPa | ε, % | Plastic deformation to failure, % |
|---|---|---|---|---|---|
| initial | 0.37 | 3100 | 3840 | 4.6 | 2.5 |
| HPT n = 5, 20 °C | 0.37 | 3250 | 3660 | 4.3 | 1.4 |
| HPT n = 30, 20 °C | 0.39 | 2870 | 3200 | 3.4 | 0.2 |

As it was expected, the plastic deformation to failure increases with decreasing specimen thickness. After HPT processing for n=5, some ductility is still observed, although HPT does not increase ductility relative to the initial state. After HPT n=30, the brittle fracture of the specimens observed. We should note that this paper reports only on the first results, and further studies with more statistical data are required for come to definite conclusions. However, the used procedure of the three-point bending test enables evaluating the variation of ductility in the BMG after HPT processing and opens a possibility of varying further the HPT processing regimes (temperature, strain, groove shape) in order to identify the HPT regimes providing an increase in ductility.

## 3. Conclusions

The procedure of the three-point bending testing was used for measuring the mechanical properties of the Vit105 BMG after HPT in microspecimens with sizes of 10x2x0.5 mm. The studies have shown that at tensile tests, the Vit105 BMG before and after HPT undergoes brittle fracture under a fracture stress of about 1700 MPa, whereas during three-point bending tests the fracture stress exceeds 3000

MPa and plastic flow is observed. This procedure opens a possibility for finding the HPT processing regimes, which provide an increase in the ductility of the BMG.

**Acknowledgements**
The authors are grateful to Saint-Petersburg State University for research grant No. 6.65.43.2017, for the research grant from the DFG under contract No. HA1344/30-1. The authors acknowledge the support of RFBR grant No. 17-08-00974.

**References**
[1] Greer A L 1995 *Science* **267** 1947
[2] Louzguine-Luzgin D V and Inoue A 2013 Bulk metallic glasses: formation, structure, properties, and applications *Handbook of Magnetic Materials* (vol 21) ed K H J Buschow (Amsterdam: Elsevier) chapter 3 pp 131–171
[3] Ma E and Ding J 2016 *Materials Today* **19** 568
[4] Abrosimova G E 2011 *Physics Uspekhi* **181** 1265
[5] Lee M H, Lee K S, Das J, Thomas J, Kühn U and Eckert J 2010 *Scripta Mater.* **62** 678
[6] Wang X D, Cao Q P, Jiang J Z, Franz H, Schroers J, Valiev R Z, Ivanisenko Y, Gleiter H and Fecht H-J 2011 *Scripta Mater.* **64** 81
[7] Meng F, Tsuchiya K, Seiichiro II and Yokoyama Y 2012 *Applied Physics Letters* **101** 121914
[8] Valiev R, Gunderov D, Zhilyaev A P, Popov A G and Pushin V 2004 *Journal of Metastable and Nanocrystalline Materials* **22** 21
[9] Glezer A M, Sundeev R V and Shalimova A V 2011 *Doklady Physics* **56** 476
[10] Gunderov D, Slesarenko V, Lukyanov A, Churakova A, Boltynjuk E, Pushin V, Ubyivovk E, Shelyakov A and Valiev R 2015 *Adv. Eng. Mater.* **17** 1728
[11] Ubyivovk E V, Boltynjuk E V, Gunderov D V, Churakova A A, Kilmametov A R and Valiev R Z 2017*Mater. Lett.* **209** 327
[12] Boltynjuk E V *et al.* 2018 *Journal of Alloys and Compounds* **747** 595
[13] Boltynjuk E, Ubyivovk E, Gunderov D, Mikhalovskii V and Valiev R 2018 *Defect and Diffusion Forum* **385** 319
[14] Gunderov D V, Churakova A A, Boltynjuk E V, Ubyivovk E V, Astanin V, Asfandiyarov R N, Valiev R Z, Xioang W and Wang J T 2019 *Journal of Alloys and Compounds* **800** 58
[15] Conner R D, Johnson W L,. Paton N E and Nix W D 2003 Journal of Applied Physics **94** 904

IOP Conf. Series: Materials Science and Engineering **672** (2019) 012029 doi:10.1088/1757-899X/672/1/012029

# Combining methods of severe plastic deformation as an effective way to increase the strength of metallic materials

**A M Ivanov**

V.P. Larionov Institute of Physical-Technical Problems of the North of Siberian Branch of the Russian Academy of Sciences, 1, Oktyabrskaya St., Yakutsk, 677980, Russia

E-mail: a.m.ivanov@iptpn.ysn.ru

**Abstract**. The article reviews the combined methods of severe plastic deformation (SPD) for hardening metallic materials. The combined deformation processing, extrusion with the equal-channel angular pressing (ECAP) is presented. The comparison is made of the mechanical characteristics and the fracture mode of the Fe360 low carbon steel strengthened by the combined methods of extrusion with ECAP and ECAP with extrusion.

## 1. Introduction

The application of combined methods in the deformation processing of metallic materials, compared to single methods, provides greater opportunities in solving the problems of hardening, the selection of processing modes and, ultimately, achieving the desired physical and mechanical properties.

At the present stage of the development of deformation processing, SPD methods provide unique properties of metallic materials [1–12, etc.]. The SPD can result in changes in structure, phase state, strength, plasticity, microhardness, and other stress-related properties.

By the combined methods of processing materials, we understand the application of procedures of either one physical nature (e.g. mechanical) implemented in one technological process (even, possibly, in a single device), or of a different nature, or in a certain sequence at different time intervals [13]. Since each of the separately applied processing methods has certain limitations in achieving cumulative deformation and changing the structural-phase state in the material, the combination of methods has certain opportunities in solving the problem of optimal combination of high strength and ductility values of metallic materials [14, 15, etc.].

The processing methods can be combined in various sequences. The question is, if the sequence of stages in the execution of certain processing types can affect the physical and mechanical properties of materials.

A significant increase in strength and a decrease in ductility during ECAP with the subsequent extrusion of the Fe360 low carbon steel was shown in [13]. An increase in the ductility of the material was also noted in tensile tests at 213 K compared with tests at 293 K. The fracture behavior of the Fe360 steel specimens in the initial state and the ECAP and extrusion-hardened state is qualitatively identical, viscous with the formation of dimples by merging micropores [13].

The purpose of the research is to study and compare the mechanical properties and the fracture behavior of the low carbon Fe360 steel subjected to extrusion and ECAP with those after ECAP and extrusion.

IOP Conf. Series: Materials Science and Engineering **672** (2019) 012029 doi:10.1088/1757-899X/672/1/012029

## 2. Material and methods

The material under study is the Fe360 low carbon structural steel (0.18% C, 0.23% Si, 0.60% Mn, 0.01% Cr, 0.003% Ni, 0.01% P, 0.01% S, 0.05% Cu, 0.05% Al, remaining Fe). Chemical analysis was performed on a Foundry-Master atomic emission spectrometer (Worldwide Analytical Systems AG, Germany).

The combined processing was carried out as follows. First, a Fe360 steel bulk billet with a diameter of 11.8 mm and a length of 40 mm was extruded in a single pass at 673 K. During extrusion, the bulk billet was inserted into the input channel with a diameter of 12 mm, then it passed through a narrowing segment with a cone angle of 34°, and it was subsequently pressed into the output channel with a square cross section of 8×8 mm. Another billet with a diameter of 6 mm and a length of 50 mm was made from an extruded billet, which was pressed into a 6/19.8×50 mm bushing made from the same material. Then, ECAP of a composite billet having a diameter of 19.8 mm and a length of 50 mm was performed in a single pass at the same temperature as extrusion. The angle of the channels intersection during ECAP was 120°. Extrusion and ECAP were made on a PSU 125 3IM hydraulic press at 1250 kN with special equipment. After extrusion, the equipment was changed for ECAP. Prior to the deformation processing, the billet was preheated and kept in a muffle furnace to 673 K, then placed in processing equipment heated to 673 K with a rotary furnace. To reduce the friction of the surface of the billet against the walls of the die mouth, a ROSOIL-ANGELINA lubricant with flake graphite additives was applied. The equipment provided a deformation degree of 0.4 during extrusion and 0.64 during ECAP.

Samples for mechanical uniaxial tensile testing were made from the strengthened billets. Cylindrical specimens with a working area of 2 mm in diameter and 10 mm in length were tested for uniaxial tension using a UTS-20k machine (UTS TestSysteme GmbH, Lammerweg 29.D-89079 Ulm, Germany) at a constant loading rate of $\approx 3.33 \cdot 10^{-5}$ $m \cdot s^{-1}$ at room temperature.

Fractures of the sample were studied by scanning electron microscopy (SEM) using a HITACHI TM3030 (Japan) scanning electron microscope at the SE mode.

## 3. Results and discussion

The strength characteristics of Fe360 steel are given in table 1. As can be seen, the yield stress and ultimate tensile strength for the steel hardened by extrusion and ECAP are ~15% lower than those for the material after ECAP and extrusion. This is probably due to the fact that the deformation processing method used at the initial stage of the combined treatment, which provides greater grain refinement with a higher deformation degree (0.64 during ECAP vs. 0.4 during extrusion), eventually leads to greater strength values.

**Table 1.** Strength characteristics of Fe360 steel.

| Steel state | Yield stress (MPa) | Ultimate tensile strength (MPa) | Relative narrowing $\psi_{um}$ (%) |
|---|---|---|---|
| As delivered | 324 | 484 | 44,9 |
| Extrusion, 673 K, n=1 + ECAP, 673 K, n=1 | 686 | 730 | 31,8 |
| [14]: ECAP, 673 K, n=1 + extrusion, 673 K, n=1 | 818 | 865 | 25,1 |

The macrorelief of the fracture surface of the Fe360 steel samples at 293 K is featured by the plastic fracture with a cup fracture formed on the neck of the stretched specimen with tightening in the deformation localization zone and the formation of shear lips. The fracture macro-surface is heterogeneous, matte. In all cases, the fracture surface includes two main zones: the central zone where the fracture originated and the fracture growth occurred, and the cut zone.

For the initial state, the fractures of the fibrous and peripheral parts of the fracture surface of the samples at 293 K, as in [16], are formed by the ductile fracture pattern (figure 1a). The fracture mode is determined by the formation of dimples by merging micropores.

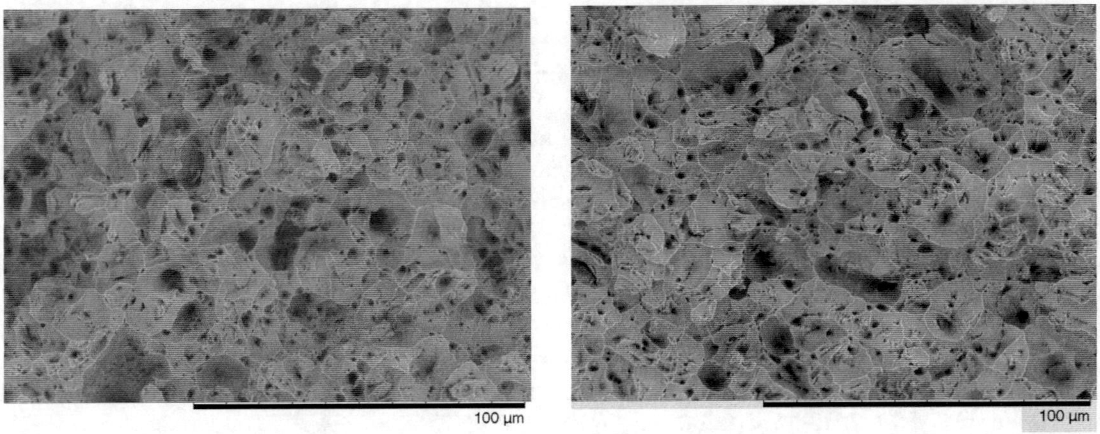


a b

**Figure 1.** Fracture microreliefs (central fibrous zone) of the Fe360 steel samples fractured at room: a – initial state; b – after extrusion and ECAP.

The fracture microrelief of the Fe360 steel specimens hardened by extrusion and ECAP, is qualitatively similar to that of the material in the initial state, as well as of the material hardened by ECAP and extrusion. The sizes of microdimples and dimples cover a wide range, which is caused by significant fluctuations in the grain and dispersed inclusions size in the examined steel. In the case of extrusion and ECAP, the micro dimples are of 0.25 μm, and the maximum size of the dimples in the fracture reaches ~20 μm, which is somewhat larger than in the case of ECAP and extrusion. The relative narrowing $\psi_{um}$ at the specimen rupture extrusion and ECAP is 31.8 %, which is 1.4 times less than that for steel in the initial state ($\psi_{um}$ = 44.9 % at 293 K) and ~1.25 times higher than that for ECAP and extruded steel ($\psi_{um}$ = 25.1 % at 293 K). The lined up micro dimples are seen on the fins of a normal fracture.

## 4. Conclusion

1. ECAP and extrusion of bulk billets of low carbon Fe360 steel increased the yield stress by 2.1 times, the ultimate tensile strength by 1.5 times and reduced ductility by 1.4 times. At this, the values of yield stress and ultimate tensile strengths during extrusion and ECAP were ~20% lower than those during ECAP and extrusion.
2. There were no qualitative differences in the fracture mode of steel before and after the combined processing by extrusion and ECAP. The fracture of the Fe360 steel specimens in its initial state and the extrusion and ECAP hardened state, as well as in the case of ECAP and extrusion, was viscous with the formation of dimples by merging micropores. The specimen fractures had a characteristic appearance with a central fibrous initiation fracture zone and slow fracture growth, as well as a peripheral cut zone. Some differences in the size and concentration of the dimples for the states after "extrusion + ECAP" and "ECAP + extrusion", are apparently explained by the different ability to resist plastic deformation, nucleation and growth of microfracture due to differences in the size and density of inclusions in grain sizes, caused by the deformation processing methods in different sequences.

**Acknowledgments**
The research was carried out as part of the research project No. III.28.1.1 under the program III.28.1 of the Siberian Branch of the Russian Academy of Sciences.

## References

[1] Segal V M, Reznikov V I, Kopylov V I, Pavlik D A and Malyshev V F 1994 *Processes of*

*Plastic Structuring of Metals* (Minsk: Nauka i Tekhnika) [in Russian]
[2] Valiev R Z and Alexandrov I V 2000 *Nanostructured Materials Obtained by Severe Plastic Deformation* (Moscow: Logos) [in Russian]
[3] Zhu Y T, Jiang H, Huang J and Lowe T C 2001 *Met. Mater. Trans.* A **32** 1559
[4] Zehetbauer M J, Stuwe H P, Vorhauer A, Schafler E and Kohout J 2004 The role of hydrostatic pressure in severe plastic deformation *Nanomaterials by Severe Plastic Deformation: Proc. 2nd Int. Conf. "Nanomaterials by Severe Plastic Deformation – NANOSPD2"* ed M J Zehetbauer, R Z Valiev (Weinheim: Wiley) pp 435–446
[5] Sestri Sh M L, Dobatkin S V and Sidorova S V 2004 *Metally* **2** 28 [in Russian]
[6] Valiev R Z and Langdon T G 2006 *Progr. Mater. Sci.* **51** 881
[7] Helmig R J, Janacek M, Hadzima B, Gendelman O V, Shapiro M, Molodova X, Springer A and Estrin Y A 2008 *Mater. Trans.* **49** 31
[8] Li L and Virta J 2011 *Mater. Sci. Technol.* **27** 845
[9] Maier G G *et al.* 2013 *Mater. Sci. Eng.* A **581** 104
[10] Donic T, Martikán M and Hadzima B 2014 *IOP Conf. Ser.: Mater. Sci. and Eng.* **63** 012047
[11] Park K T, Park L, Kim H J, Kim S B and Lee C S 2014 *IOP Conf. Ser.: Mater. Sci. and Eng.* **63** 012144
[12] Ivanisenko Yu, Kulagin R, Fedorov V, Mazilkin A, Scherer T, Baretzky B and Hahn H 2016 *Mater. Sci. Eng.* A **664** 247
[13] Ivanov A M 2018 *Prirodnye Resursy Arktiki i Subarktiki* **23(1)** 60 [in Russian]
[14] Stolyarov V V 2010 *Rev. Adv. Mater. Sci.* **25** 194
[15] Ivanov A M, Ugurchiev U H, Stolyarov V V, Petrova N D and Platonov A A 2012 *Izvestiya VUZov. Chernaya Metallurgiya* **6** 54 [in Russiam]
[16] Kovalenko N D and Ivanov A M 2016 *Aktual'nye Problemy v Mashinostroenii* **3** 424 [in Russian]

IOP Conf. Series: Materials Science and Engineering **672** (2019) 012030 doi:10.1088/1757-899X/672/1/012030

# Microstructure and mechanical properties of the Mg–Zn–Ca biodegradable alloy after severe plastic deformation

**G D Khudododova[1], O B Kulyasova[1], R K Islamgaliev[1] and R Z Valiev[1,2]**

[1] Institute of Physics of Advanced Materials, Ufa State Aviation Technical University, 12 Karl Marx St., Ufa 450008 Russia
[2] Laboratory for Mechanics of Advanced Bulk Nanomaterials for Innovative Engineering Applications, Saint Petersburg State University, 18 Universitetskiy pr., Peterhof, St. Petersburg 198504 Russia

E-mail: khudododova.gd@gmail.com

**Abstract.** Mg-Zn-Ca alloys are considered as promising biodegradable metals for implants because of their biocompatibility and proper physical properties. In this study, microstructure and strength of magnesium alloy Mg-1% Zn-0.2% Ca after a combination of the severe plastic deformation (SPD) methods have been investigated. It was shown that with an increase in the degree of deformation, an increase in the strength properties occurs. It was established that the ultimate tensile strength of the initial state was 125 MPa and after SPD a 2-fold increase in strength up to 283.3 MPa was achieved with sufficient ductility.

## 1. Introduction

Magnesium alloys have a great potential for use in medicine as implants, since they are completely biocompatible, have mechanical properties close to that of natural bones, do not cause inflammatory reactions and stimulate the growth of new bone tissue [1]. The disadvantage of magnesium-based alloys is their low strength. To increase the strength, zinc and calcium alloying elements are chosen, which are also vital for humans and non-toxic to the human body. Alloying with zinc increases the strength of the alloy [2,3], the addition of calcium in the amount of a few tenths of a percent increases the corrosion resistance of the alloy [4]. It is known also that the advanced approach to improve the mechanical properties is the grain refinement by the SPD methods [5,6].

## 2. Materials and methods

Magnesium alloy Mg-1% Zn-0.2%Ca was chosen as the initial material for the research. The initial cast samples were subjected to homogenization annealing at a temperature of 430 °C for 22 hours with cooling in water. Heat treatment of the samples was carried out in a Nabertherm muffle furnace.

The grain refinement was carried out by the method of equal-channel angular pressing (ECAP). Samples with a diameter of 20 mm were pre-aged at a pressing temperature for 20 minutes. ECAP was carried out through a die-set with the angle of intersection of the channels 120° at a speed of 6 mm/min. During the ECAP, 2 pressing passes were carried out at the temperature of 400 °C, 2 passes at a temperature of 350 °C, and 2 passes at the temperature of 300 °C (the total number of passes was 6).

The formation of the nanostructured state was carried out by the method of high pressure torsion (HPT) at a room temperature. To do this, discs with a diameter of 20 mm and a thickness of 2 mm were cut off from the ECAP bar, which were subjected to high pressure torsion (6 GPa) with a number of revolutions of 0.5 and 1.

The structure of the samples in the initial state and after ECAP was examined by an Olympus GX51 optical microscope. For this, the specimens were etched in a solution containing 2.5 g of picric acid, 2.5 ml of nitric acid, 5 ml of water, 50 ml of ethanol. The microstructure was studied using a JEM-2100 transmission electron microscope at an accelerating voltage of 200 kV. Thin foils were prepared by electro-polishing on Tenupol-5 using a solution of nitric acid - 30% and methanol - 70% electrolyte. Microhardness (HV) was measured by the Vickers method on a Micromet 5101 device with a load of 50 g and a holding time of 10 s. Mechanical tensile tests were carried out on an Instron 5982 testing machine at a room temperature with a strain rate of $10^{-3}$ $s^{-1}$ on flat specimens with the dimensions of the gauge section: 4x1x0.5mm$^3$. At least 3 samples were tested for each condition.

## 3. Results and discussion

The structure of the Mg-1% Zn-0.2% Ca alloy after homogenizing annealing was coarse-grained with an average grain size of 280 μm (figure 1a). The structure also contained particles with a size of 4μm. Using the ECAP method, a homogeneous structure with an average size of 3.7 μm was formed (figure 1b). As a result of warm deformation at the ECAP process, the particles dissolve in the solid solution. For further grain refinement, additional deformation by HPT was applied to the ECAP samples. The structure after HPT transformed into a nanoscale with an average grain size of about 250 nm at 0.5 revolution (figure 1c) and less than 200 nm after 1 revolution (figure 1d).

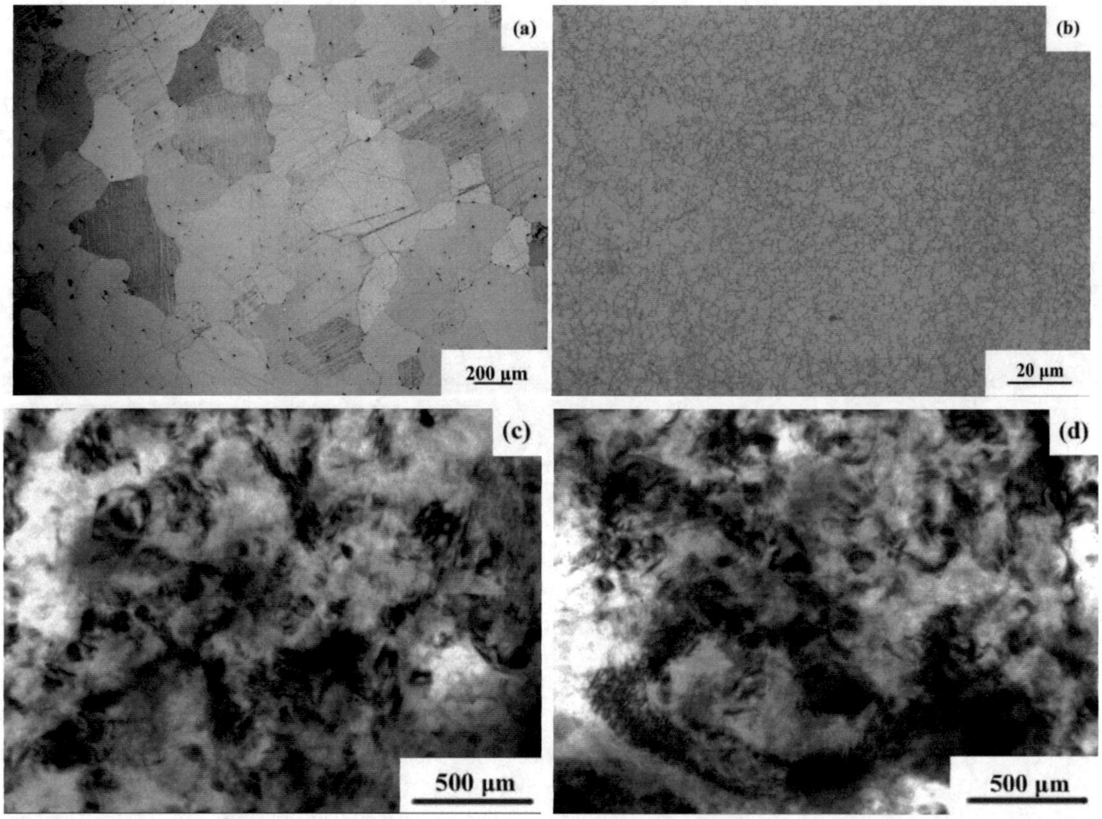


**Figure 1.** Microstructure of Mg-1% Zn-0.2%Ca at different conditions: (a) – after homogenization (initial state), (b) – after ECAP, (c) – after ECAP+HPT 0.5 revolution (d) – after ECAP+HPT 1 revolution.

In the initial state, the microhardness of the samples was 41.4 HV. As a result of deformation by ECAP, the microhardness was increased to 63.1 HV (figure 2a). Further grain refinement using the HPT method at 0.5 and 1 revolutions resulted in an increase in the microhardness to 74.7 HV and 80.1 HV, respectively (figure 2a). It should be noted that the increase in the microhardness after the deformation by ECAP and additional deformation by HPT with 1 revolution was almost twice as compared to the initial state, which occurred due to the strong grain refinement (figure 1). Tensile tests of the Mg-1% Zn-0.2%Ca alloy in the homogenized state showed that the samples demonstrated an ultimate tensile strength of 125 MPa and an elongation of 8%. Deformation by ECAP led to an increase of the ultimate tensile strength to 225 MPa and an increase in ductility to 16%. The increase of strength is likely due to the grain refinement. The increase in plasticity is probably due to the dissolution of coarse particles in the α-matrix of the alloy, which embrittle the material in the coarse-grained condition. As a result of additional deformation at 0.5 and 1 revolutions of the HPT, the ultimate tensile strength was increased and amounted to 263.7 and 283.3 MPa, respectively. It means that the greater the degree of deformation in the process of deformation by torsion, the higher the strength characteristics and less the ductility of the alloy. It should be noted that during the combined SPD, the strong grain refinement takes place similar to such simple SPD processing like ECAP and HPT. Therefore, the Zn and Ca atoms should redistribute during the combined processing between the bulk solid solution and the grain boundary adsorption layers [8], because the initial alloy contains precipitates (figure 1). Besides, the combined SPD processing could modify the composition of the matrix solid solution [9] thus changing the mechanical properties of the whole alloy.

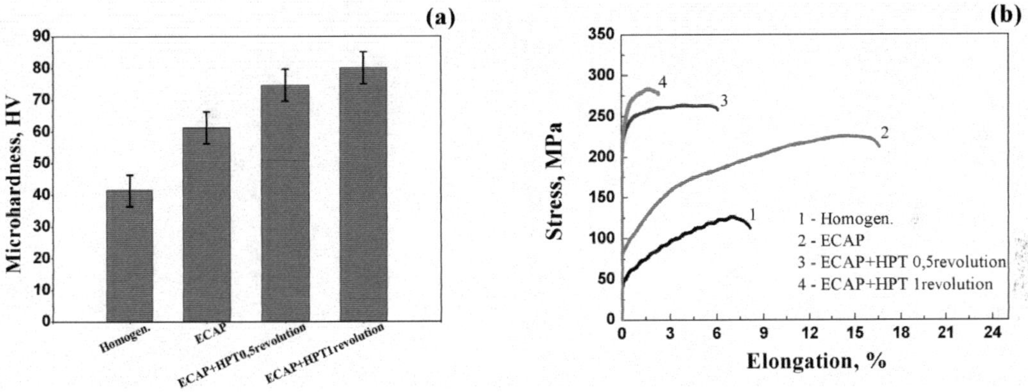


**Figure 2.** (a) Microhardness of the Mg-1% Zn-0.2% Ca alloy at various treatments; (b) tensile tests of the Mg-1% Zn-0.2% Ca alloy.

## 4. Conclusions

It is shown that the use of ECAP and additional HPT (1 revolution) leads to the strong grain refinement in the Mg-1% Zn-0.2% Ca alloy, as a result of which the microhardness of the samples increased by a factor of 2, and the UTS of 283 MPa was achieved, which is higher than the UTS of the initial state (125 MPa) more than 2 times.

## Acknowledgments

The work was suppored by the Ministry of Science and High Education of the Russian Federation through a grant in the field of basic research (No. 16.7268.2017/8.9)

## References

[1] Peng T and Xuanyong L 2015 *Regenerative Biomaterials* **2** 135
[2] Cha P R *et al.* 2013 *Scientific Reports* **3** 2367
[3] Bakhsheshi Rad H R *et al.* 2012 *Applied Mechanics and Materials* **121** 568

IOP Conf. Series: Materials Science and Engineering **672** (2019) 012030 doi:10.1088/1757-899X/672/1/012030

[4] Nowosielski R 2013 *J. Achievements in Materials and Manufacturing Engineering* **61** 367
[5] Dobatkin S V *et al.* 2017 *IOP Conf. Ser.: Mater. Sci. Eng.* **194** 012004
[6] Kulyasova O B, Islamgaliev R K, Zhao Y and Valiev R Z 2015 *Adv. Eng. Mater.* **17** 1738
[7] Vinogradov A, Vasilev E, Kopylov V I, Linderov M, Brilevesky A and Merson D 2019 *Metals* **9** 186
[8] Mazilkin A A, Straumal B B, Kilmametov A R, Boll T, Baretzky B, Kogtenkova O A, Korveva A and Zięba P 2019 *Scripta Mater.* **173** 46
[9] Straumal B B, Kilmametov A R, Korneva A, Mazilkin A A, Straumal P B, Zięba P and Baretzky B 2017 *J. All. Comp.* **707** 20

IOP Conf. Series: Materials Science and Engineering **672** (2019) 012031 doi:10.1088/1757-899X/672/1/012031

# Relationship between the stress-strain state and the imperfection of the structure of metals with SPD shear

**G I Raab, A G Raab, G N Aleshin and I S Kodirov**

Ufa State Aviation Technical University, 12 K. Marx str., Ufa 450008 Russia

E-mail: giraab@mail.ru

**Abstract.** The paper presents the results of the computer simulation and physical experiment for the free and constrained transverse shear of round specimens of steel 10 with an accumulated strain of e> 1. The stress-strain state was investigated using computer simulation, and the degree of imperfection of the emerging structures was studied by metallographic methods. The laws governing the relationship between the parameters under study and the degree of imperfection of the structure during severe plastic deformation by shear are established.

## 1. Introduction

Currently, the metallurgical industry is searching for and developing new technologies for producing high-strength metallic materials with a ductility sufficient for structural applications. As is known [1,2], such methods of severe plastic deformation (SPD) as equal-channel angular pressing (ECAP) and ECAP-Conform, have a high potential for producing a set of enhanced physical, mechanical and performance properties of metals and alloys, but they have a number of tribological problems that hinder the industrial implementation of these methods. In this connection, we consider in this paper two alternative, more technologically advanced methods that use shear deformation in free and constrained conditions, such as free torsion of workpieces and torsion during drawing in a split die, respectively, under elevated hydrostatic pressure [3-4]. These methods use combined non-monotonic deformations, which enable changing effectively the structural states of metallic materials [5-6].

## 2. Methods

The material under study was low-carbon steel 10 widely used in industry, in the form of a calibrated long rod with a diameter of 10 mm. For the experiments, a modernized laboratory-scale chain drawbench was used to ensure the rotation of the workpieces during axial drawing. The drawing dies had an eccentric profile to provide conditions for torque transfer when twisting the workpieces with respect to the longitudinal axis, the relative compression over the pass was 56%. A phosphate sub-lubricant coating and a lubricant were applied onto the workpieces to improve the tribological conditions during drawing. To implement the method of free torsion, a lathe was used. The number of revolutions relative to the longitudinal axis was 15. The stress-strain state of the produced workpieces studied by computer simulation using the DEFORM-3D software. The used processing methods are schematically presented in figure 1.

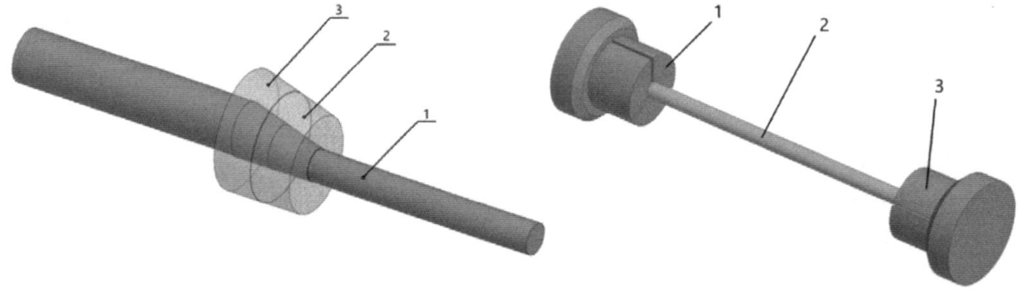


a) b)

**Figure 1.** Principles of the processing methods:
a – drawing with shear (1 – workpiece, 2 – moving die, 3 – fixed die),
b – free torsion (1 – moving holder, 2 – workpiece, 3 – fixed holder).

## 3. Results and discussion

This section compares the deformational and processing parameters of workpieces produced by two methods of plastic treatment – free torsion (FT) and drawing with shear (DS) (torsion in constrained conditions) (see figure 1). Figure 2 shows the patterns of the strain state in the cross section of the workpieces. An analysis of the patterns shows that strain does not accumulate uniformly in the workpieces. In the peripheral regions of the workpieces the maximum values of strain intensity are observed, while in the central region the minimum values of strain intensity are revealed, and the diagram of the variation in the accumulated strain has a parabolic shape. Moreover, the difference between the maximum and minimum values is more than two-fold.

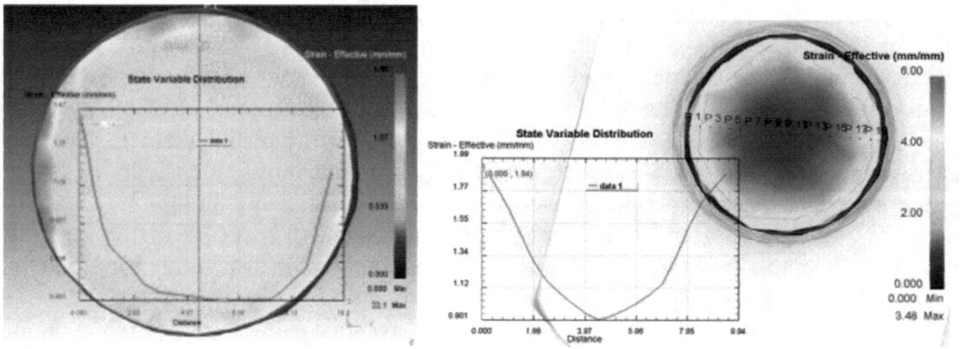


a) b)

**Figure 2.** Patterns of strain intensity in the cross section: a – drawing with shear; b – free torsion.

Study of the produced structural states revealed a number of features (see figure 3). In the case of drawing with shear, the maximum grain refinement is observed in the peripheral layers of the workpiece that are in contact with the tools. No structural defects in the form of pores are observed. In the case of free torsion at room temperature, after an accumulated strain of e = 1.5 is attained, the emergence of pores is observed, as shown in figure 3 b, and their quantity and size grow with increasing level of accumulated strain. An increase in the temperature of free torsion to 600 °C enables increasing the level of accumulated strain without pore formation to e = 2. In this case, the average size of structural fragments is 3.7 ± 1.1 μm.

IOP Conf. Series: Materials Science and Engineering **672** (2019) 012031 doi:10.1088/1757-899X/672/1/012031

It can be noted that when using the DS method, a gradient type of structure is formed, and the peripheral layers of the workpieces are refined more intensively, down to submicron sizes, at a depth of up to 0.2 mm. In the samples produced by FT at 600 °C, this phenomenon is not observed and a gradient structure is formed with a grain size of 25.3 ± 6.5 μm in the center to 3.7 ± 1.1 μm in the periphery, without any pore formation.

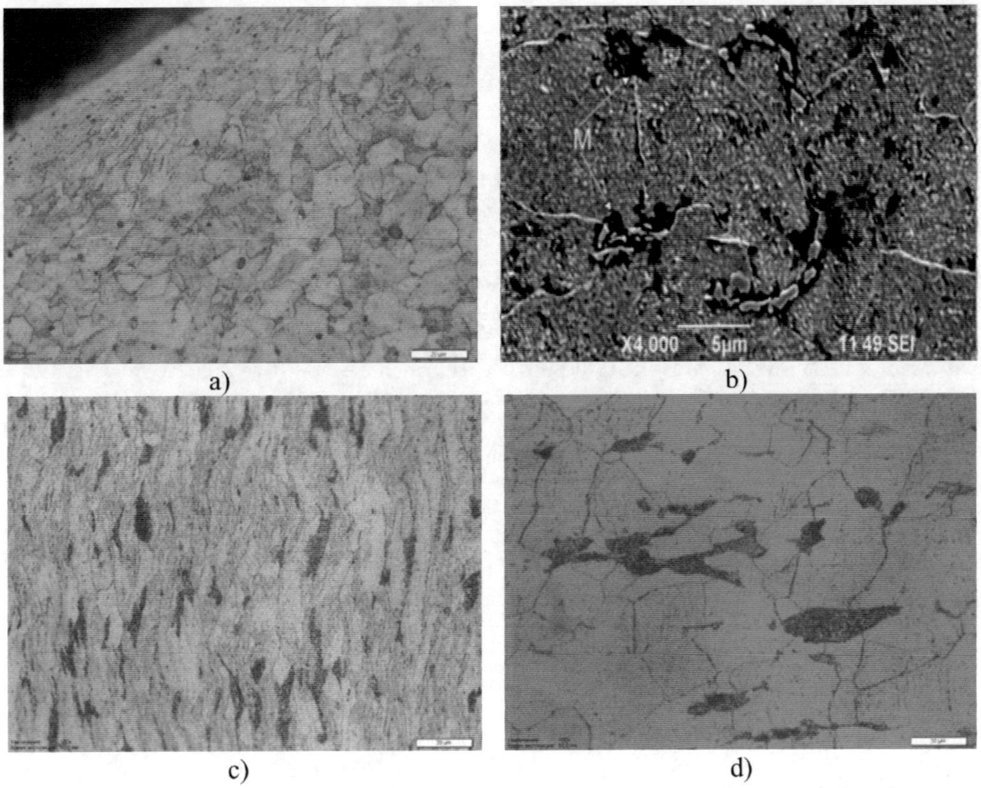


**Figure 3.** Structural states of the workpieces: a) the microstructure in the cross-section of the workpiece, DS at room temperature; b) the microstructure in the cross-section of the workpiece, FT at room temperature (upon reaching e = 3.2, the center of the section); c) the microstructure in the longitudinal section of the workpiece, FT at 600 °C (upon reaching e = 2, the periphery of the section); d) the microstructure in the longitudinal section of the workpiece, FT at a temperature of 600 °C (upon reaching e = 2, the center of the section)

## 4. Conclusions

- It has been established that in low-carbon steel 10 processed by the studied methods of DS and FT, a gradient type of structure is formed, and the peripheral layers of the workpieces are refined more intensively. At the same time, the samples produced by DS have a surface layer with a thickness of up to 0.2 mm that is very intensively refined, down to submicron sizes, and in the samples produced by FT after accumulation of shear strain over 1.5 pore formation is observed, progressing with increasing level of accumulated strain
- Computer simulation has revealed that strain in the workpieces does not accumulate uniformly. In the peripheral regions of the workpieces, the maximum values of strain intensity are observed, while in the central region the minimum values of strain intensity are revealed, and the diagram of the variation in the accumulated strain across the cross section has a parabolic shape.

Moreover, the maximum values of strain intensity exceed e = 1, and the difference between the maximum and minimum values is more than two-fold.

**Acknowledgments**
The authors gratefully acknowledge the financial support from the Russian Foundation for Basic Research under project No. 17-08-00720/19.

**References**

[1] Valiev R Z, Zhilyaev A P and Langdon T G 2014 *Bulk Nanostructured Materials: Fundamentals and Applications* (Hoboken: Wiley STM)
[2] Utyashev F Z and Raab G I 2013 *Deformation Methods for the Fabrication and Processing of Ultrafine-Grained and Nanostructured Materials* (Ufa: Gilem) [in Russian]
[3] Raab G I and Raab A G *Method for Production of Ultrafine-Grained Semi-Finished Products by Drawing with Shear*. Russian Federation patent No. 2347633, IPC B21C 1/00. Publication date 27.02.2009
[4] Raab G I, Simonova L A and Aleshin G N 2016 *Metalurgija* **55** 177
[5] Lu K and Lu J 2004 *Mater. Sci. Eng.* A **375–377** 38
[6] Raab A G, Chukin M V, Aleshin G N and Raab G I 2014 *IOP Conf. Ser.: Mater. Sci. Eng.* **63** 012008

IOP Conf. Series: Materials Science and Engineering **672** (2019) 012032 doi:10.1088/1757-899X/672/1/012032

# Effect of point defects and functionalization on structural stability and electron properties of borophene as investigated by means of density functional theory

**S Kh Khadiullin[1], A A Kistanov[2], A Y Morkina[1] and E A Korznikova[1,2]**

[1] Ufa State Aviation Technical University, 12 K. Marx st., Ufa 450008 Russia
[2] Institute for Metals Superplasticity Problems of RAS, Khalturin St. 39, 450001 Ufa, Russia

E-mail: elena.a.korznikova@gmail.com

**Abstract**. The discovery of a new two dimensional material is a new trend emerged after the discovery of graphene in 2004. Hundreds of new materials have been predicted, designed and investigated during last decade. Many of them have outstanding properties for potential application in new technologies. One of those materials is so-called borophene, being isolated atomically thin boron. This material can show different electronic properties, including semiconducting, semi-metallic, and metallic states due to its unique structure. Borophene demonstrates unusually high in-plane elasticity and strength. The crucial point for application of borophene is its low structural stability at room temperature. A possible way to solve this problem is the functionalization of its surface by external atoms like hydrogen and fluorine. This work presents a study on the effect of O dopants and the presence of vacancy defects on electronic properties of borophene by using first-principles calculations. It was revealed that unlike other 2D materials, the band structure of borophene is stable by respect to the above-mentioned factors. However, the doping by O atoms is found as an effective way to increase the stability of the borophene sheet.

## 1. Introduction

The active development of studies on new two dimensional materials emerged by the discovery of graphene in 2004 has been inspired by an outstanding set of properties, explained by their reduced dimension. Unusual features are revealed both in the mechanical behavior [1–7], physical properties [5,8–13] and in the nonlinear dynamics [14–16] of such materials. Recently, a two-dimensional boron named borophene has been obtained upon deposition of boron atoms on (111) Ag substrate. This material has attracted great interest due to its unique properties [17,18]. The possibility of the existence of various phases depending on the processing conditions, such as the deposition rate and processing temperature, indicates a large polymorphism of borophene. It was predicted that borophene has anisotropic metallic properties [7,19]. In addition, borophene exhibits extraordinary mechanical properties, for example, its Young's modulus (398 GPa·nm) exceeds that for most known two-dimensional materials [17]. All these features characterize borophene as a promising material for use in nanoelectronics.

The main obstacle in the study of borophene is its low structural stability. It was proposed to solve this problem by functionalizing the surface of borophene, which allows one to stabilize its structure and also contributes to a change in its electronic properties [17]. For example, it was predicted that the structure of borophene can be stabilized by hydrogen atoms [20]. However, detailed studies of the functionalization of the surface of a two-dimensional borophene sheet by oxygen atoms were not mentioned in literature reviews.

In addition, it is well known that the structural persistence and electronic properties of two-dimensional materials significantly depend on the presence of vacancies in them. For example, lattice defects can easily arise instability of atomically thin two-dimensional materials [6]. Although there is a fairly large number of studies on defects in various two-dimensional materials, defects in borophene have not yet been considered.

Considering the abovementioned facts, the purpose of the present work was to estimate the influence of oxygen functionalization (O-functionalization) and point defects on the structural stability and electronic properties of borophene.

## 2. Simulation details

The studies were carried out by means of the density functional theory. First-principle calculations were performed using the VASP software package [21]. The exchange – correlation interaction was specified by the generalized GGA gradient approximation using the Perdew–Burke–Ernzerhof (PBE) function [22]. Summation over the Brillouin zone using a $10\times10\times1$ wave vector lattice was performed for structure relaxation and calculation of the electronic structure of the borophene sheet. The maximum effect of functionalization while maintaining the stability of the structure, corresponded to the 50% concentration of O atoms. Periodic boundary conditions were applied in two transverse directions in the plane of the borophene nanosheet under investigation, and free boundary conditions were applied along the direction perpendicular to the plane of the 2D structure by creating a vacuum space of 15 Å.

## 3. Results

### 3.1. Structure and electronic properties of pure and O-functionalized borophene

In figure 1a, b optimized structures of pure and O-functionalized borophene sheets are shown, respectively. For a pure borophene sheet, the optimized lattice parameters are $a = 1.618$ Å and $b = 2.864$ Å, which is in good correspondence with previously published results [19]. For O-functionalized borophene, the optimized Γ lattice parameters were found to be $a$ = 3.324 Å and $b$ = 5.558 Å respectively.

Figure 1c,d shows the electronic structure of pure and O-functionalized borophene sheets, respectively. According to figure 1c, borophene bands cross the Fermi level along the Γ–X and S–Y directions, which indicate the metallic nature of pure borophene. These results are in good agreement with previous studies [19]. For O-functionalized borophene (figure 1d), there is no gap in the phonon spectrum, thus, O-functionalized borophene retains its metallic nature.

Next, we examined the structural stability of pure (equation (1)) and O-functionalized (equation (2)) borophene by calculating the average binding energy $E_b$

$$E_b = (n_B E_B - E_{tot})/n, \tag{1}$$

$$E_b = (n_B E_B + n_O E_O - E_{tot}) / n, \tag{2}$$

where $E_{tot}$ is the total energy of the system, $E_B$ is the energy of one boron atom, $E_X$ is the energy of one oxygen atom, $n_B$ is the total number of boron atoms, $n_x$ is the total number of oxygen atoms, and $n$ is the total number of atoms in the system. The calculated $Eb$ values for pure and O-functionalized borophene are 5.86 and 5.32 eV, respectively. Therefore, the functionalization of borophene by oxygen enhances its structural stability.

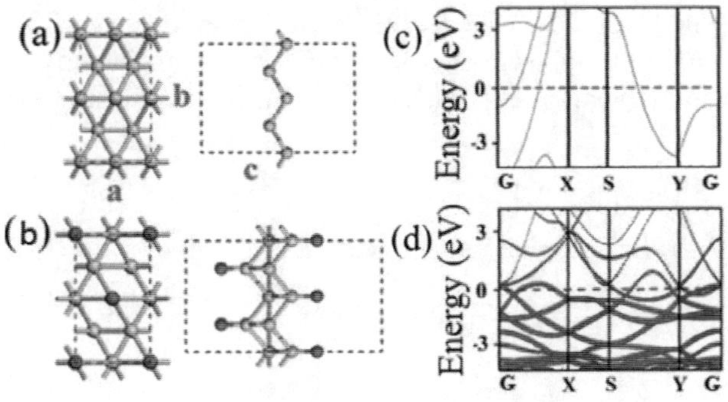


**Figure 1.** Atomic configurations of (a) pure and (b) O-functionalized borophene sheet; electronic structure of (c) pure and (d) O-functionalized borophenee sheet. The black dashed line corresponds to the Fermi level. The red lines indicate the oxygen induced bands.

The work functions of all the considered structures were calculated as the difference between the electrostatic potential and the Fermi energy [14, 15]. Figure 2e shows the work function of pure and O-functionalized borophene sheets in comparison with that of other materials.
From the presented results it follows that the work function of a pure borophene sheet (-5.31 eV) is higher than that of graphene (-4.50 eV) [16]. It is also important to note that the O-functionalization of the borophene surface contributes to a significant decrease in the work function up to -9.46 eV. Thus, the O-functionalization of borophene sheet is an effective way to change the work function of borophene allowing to control its contact resistance.

*3.2. The effect of vacancy defects on the electronic structure of borophene*
An important point is the estimation of the effect of vacancy defects on the electronic structure of two-dimensional materials and, in particular, borophene. In this paper, various vacancy defects in borophene were considered: monovacancy (figure 2b) and two possible bivacancies (figure 2c, d). An analysis of electronic structure of borophene sheets containing vacancy defects revealed a significant overlap and cross of the Fermi level, mainly due to $p_z$ orbitals of boron atoms. Thus, the electronic structure of borophene practically does not change both with the appearance of vacancy defects on its surface and with its O-functionalization, which is a hallmark of borophene compared to other two-dimensional materials [13].

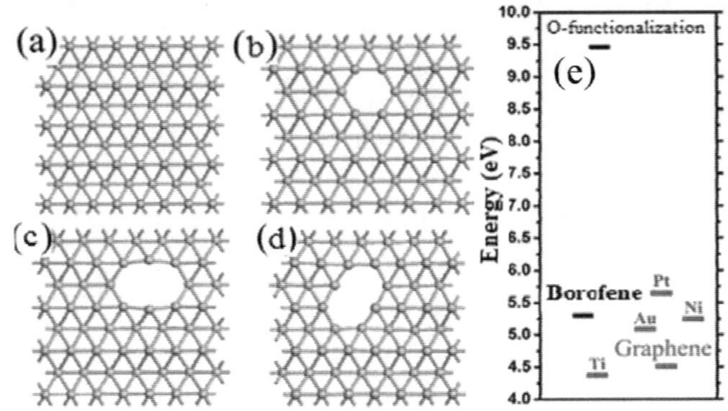


**Figure 2.** Atomic structures of (a) pure, containing (b) monovacancy, (c) bivacancy (100) and (d) bivacancy (110) of borophene sheets; graph comparing the work functions of various materials. Black markers indicate the work function of pure and O-functionalized borophene sheets, while red markers indicate the work function of other materials (values taken from [23]).

## 4. Discussion of results and conclusions

The first-principle calculations were used to study the effect of surface O-functionalization and presence of point defects of a borophene sheet on its properties, such as structural stability, electronic structure, and work function. It was found that, unlike many other two-dimensional materials, borophene is inert to the influence of impurity atoms and vacancies and retain metallic type of conduction. Nevertheless, O-functionalization of the borophene surface sheet can be used as an effective way to control the work function of borophene, which is significantly reduced from -5.31 eV (for the pure borophene sheet) to -9.46 eV (for O-functionalized borophene). Thus, the present work contributes significantly to understanding of the effect of O-functionalization and vacancies on the stability and electronic structure of borophene.

**Acknowledgments**
E. A. Korznikova and A. A. Kistanov are grateful for financial support to Russian Foundation for Basic Research, grant 18-32-20158 mol_a_ved.

**References**
[1] Savin A V, Korznikova E A and Dmitriev S V 2015 *Phys. Rev.* B **92** 035412
[2] Savin A V, Korznikova E A and Dmitriev S V 2015 *Phys. Solid State* **57** 2348
[3] Savin A V, Korznikova E A, Dmitriev S V and Soboleva E G 2017 *Computational Materials Science* **135** 99
[4] Savin A V, Korznikova E A and Dmitriev S V 2019 *Phys. Rev.* B **99** 235411
[5] Davletshin A R, Ustiuzhanina S V, Kistanov A A, Saadatmand D, Dmitriev S V, Zhou K and Korznikova E A 2018 *Physica B: Condensed Matter* **534** 63
[6] Liu B, Bai L, Korznikova E A, Dmitriev S V, Law A W-K and Zhou K 2017 *J. Phys. Chem.* C **121** 13876
[7] Liu Y, Dong Y-J, Tang Z, Wang X-F, Wang L, Hou T, Lin H and Li Y 2016 *Journal of Materials Chemistry* C **4** 6380
[8] Kistanov A A, Khadiullin S Kh, Dmitriev S V and Korznikova E A 2019 *ChemPhysChem* **20** 575
[9] Kistanov A A, Khadiullin S Kh, Dmitriev S V and Korznikova E A 2019 *Russ. J. Phys. Chem.* **93** 1088
[10] Kistanov A A, Khadiullin S Kh, Dmitriev S V and Korznikova E A 2019 *Chemical Physics Letters* **728** 53
[11] Kistanov A A, Davletshin A R, Ustiuzhanina S V, Evazzade I, Saadatmand D, Dmitriev S V and Korznikova E A 2018 *J. Mater. Sci.* **53** 15559
[12] Kistanov A A, Cai Y, Zhou K, Srikanth N, Dmitriev S V and Zhang Y-W 2018 *Nanoscale* **10** 1403
[13] Kistanov A, Khadiullin S, Zhou K, Dmitriev S and Korznikova E 2019 *Journal of Materials Chemistry* C **7** 9195
[14] Korznikova E A, Shcherbinin S A, Ryabov D S, Chechin G M, Ekomasov E G, Barani E, Zhou K and Dmitriev S V 2019 *Phys. Status Solidi* B **256** 1800061
[15] Evazzade I, Lobzenko I P, Korznikova E A, Ovid'ko I A, Roknabadi M R and Dmitriev S V 2017 *Phys. Rev.* B **95** 035423
[16] Barani E, Korznikova E A, Chetverikov A P, Zhou K and Dmitriev S V 2017 *Physics Letters* A **381** 3553
[17] Mannix A J *et al.* 2015 Science **350** 1513
[18] Feng B, Zhang J, Zhong Q, Li W, Li S, Li H, Cheng P, Meng S, Chen L and Wu K 2016 *Nature Chemistry* **8** 563
[19] Peng B, Zhang H, Shao H, Xu Y, Zhang R and Zhu H 2016 *Journal of Materials Chemistry* C **4** 3592
[20] Yang W-H, Lu W-C, Li S-D, Xue X-Y, Zang Q-J, Ho K M and Wang C Z 2019 *Physical Chemistry Chemical Physics* **21** 5466
[21] Kresse G and Furthmüller J 1996 *Physical Review* B **54** 11169
[22] Perdew J P, Burke K and Ernzerhof M 1996 *Physical Review Letters* **77** 3865
[23] Lee E J H, Balasubramanian K, Weitz R T, Burghard M and Kern K 2008 *Nature Nanotechnology* **3** 486

IOP Conf. Series: Materials Science and Engineering **672** (2019) 012033 doi:10.1088/1757-899X/672/1/012033

# Features of mass transfer by N-crowdions in fcc $Ni_3Al$ lattice

**A M Bayazitov[1], S V Dmitriev[2], P V Zakharov[3], I A Shepelev[4], S Yu Fomin[5] and E A Korznikova[2,5]**

[1] Bashkir State University, Validi St., 32, 450000, Ufa, Russia
[2] Institute for Metals Superplasticity Problems of RAS, Khalturin St. 39, 450001 Ufa, Russia
[3] The Shukshin Altai State Humanitarian and Pedagogical University, 53 Korolenko st., 659333, Biysk, Russia
[4] Saratov State University, 83 Astrakhanskaya St. 410012, Saratov, Russia
[5] Ufa State Aviation Technical University, 12 K. Marx st., 450008 Ufa, Russia

E-mail: elena.a.korznikova@gmail.com

**Abstract**. Crowdions being interstitial atoms located in closely packed atomic rows, play an important role in relaxation processes occurring in metals and alloys under severe external impacts effectively transferring mass and energy. Dynamics of crowdions motion is well studied for pure metals, while for intermetallic alloys this issue has not previously been addressed. It was earlier shown that the so-called N crowdion that involves more than one atom in its 1D motion is much more effective in frames of propagation path and excitation energy that 1-crowdion. In the present study, the motion of 1- and 2-crowdions in the $Ni_3Al$ intermetallic alloy, was studied by means of the molecular dynamics method. N-crowdion was excited by providing the same value of initial energy to N neighbouring atoms along the close-packed atomic row. It was shown that the studied crowdions can propagate along the biatomic close-packed row, but show strong instability when moving in monatomic close-packed directions. It has been established that 2-crowdions have a mean free path 2-2.5 times longer than 1-crowdions with the same initial velocity. It was also revealed that the mean free path of the nickel atom initiated crowdion than that initiated on the aluminium atom. The results suggest a higher efficiency of 2-crowdions in mass transfer in the $Ni_3Al$ intermetallic alloy. In further works, it is proposed to study the possibility of launching 2-crowdions by bombarding the crystal surface with biatomic molecules.

## 1. Introduction

Many physical processes such as plastic deformation, heat treatment, irradiation, etc., are inevitably connected with the process of the mass transfer by means of point defects motion. Interstitial atoms have a higher energy and, therefore, their concentration at thermal equilibrium is much smaller than that of vacancies, but their role significantly increases in non-equilibrium processes. Crowdion is a type of point defect where an extra atom is placed in a closely packed row of the crystal lattice being in fact a particular case of interstitial atom[1–4]. Crowdions can exist in a stationary state or move with subsonic or supersonic velocity [1,4]. In this paper, we study supersonic crowdions in the crystal structure of $Ni_3Al$. Interest in nickel aluminides is associated with their high mechanical properties, high corrosion resistance, oxidation resistance at elevated temperatures and abnormal dependence of the yield strength on the heating temperature [5,6]. For this reason, it is extremely important to study the mass transfer processes in the aluminide $Ni_3Al$ with a particular attention to the interaction of Al and Ni atoms in one close-packed row in case of dynamic collision. The atomic mass difference of Ni and Al being a factor 2.175 is a possible obstacle for crowdion initiation analyzed in this work. It is important to notice that the propagation of crowdion relates to nanoscale processes both in space and time dimensions. Taking into account this condition, one can conclude that experimental study of this phenomenon is almost an impossible task and the power of computer simulations could provide invaluable assistance in understanding the mechanisms of mass transfer in biatomic systems. The most suitable method in this case is the molecular dynamics method, which established itself as a powerful tool for investigation of nonlinear small scale phenomena in different lattices including metals [3,7], intermetallides[8] and new low dimension materials [9–21]. Experimental validation of simulation results for the case of trapping of He clusters by inert-gas impurities in tungsten is provided in [22] and demonstrates a suitable coincidence of results. In this work, the molecular dynamics method was used for analysis of the crowdion propagation scenarios in closely packed rows of the Ni3Al lattice with different initial conditions.

## 2. Simulation details

In this paper, we consider the fcc $Ni_3Al$ lattice with the lattice parameter a = 3.589 Å and the interatomic distance $d=a/\sqrt{2}$. The *x* and *y* axes are oriented along the closely packed directions <1 1 0> and <$\bar{1}$10>, respectively and the *z* axis is oriented along the <0 0 1> direction (figure 1). A small perturbation of order $10^{-5}$ degrees was introduced along the *y* axis. Atoms in the same closely packed row (red) are numbered by the index *N*.

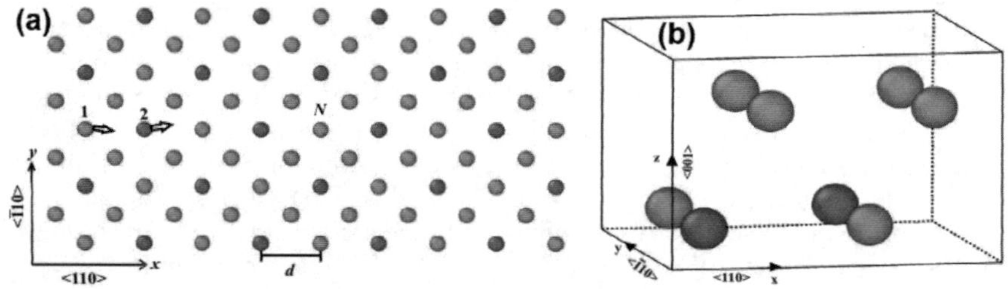


**Figure 1.** (a) Atoms of an fcc ordered $Ni_3Al$ alloy occupying lattice sites in projection onto the *xy* plane, where the axes of the Cartesian coordinate systems *x*, *y* and *z* are directed along the <110>, <$\bar{1}$10> and <001> crystallographic directions. The atoms of nickel and aluminum are shown in red and blue, respectively. (b) Calculational cell showing the coordinate axis by respect to crystal orientation.

The number of cells along the *x,y* and *z* axis was equal to 20, 6 and 8 respectively. The total number of atoms was 7,680. The aim of the work was to study solely supersonic crowdions without taking into account any corresponding disturbances. Periodic boundary conditions were used. No thermal fluctuations were introduced, i.e., the simulation was performed at a temperature of 0 K. The initial velocities of all other atoms and the initial displacements of all atoms in the computational cell were set to be zero. Thus, the total energy of the system was equal to the kinetic energy of the excited atoms at initial time was equal to $\mathbf{T_0=NM(V_x)^2/2}$ where the contribution from the small transverse component of the initial velocity is not taken into account. It should be recalled that in our model the mass of the atom Ni = 58.6934 a. e. m., the mass of the atom Al = 26.9815386 a. e. m. The LAMMPS software package was used for simulation.

## 3. Simulation results

Starting point of the work was simulation of 1- and 2-crowdions launching in close-packed biatomic and monatomic rows with the same initial velocity. It was found that in the monoatomic rows a strong instability appears shortly after crowdion emergence and thus on cannot observe the propagation of crowdions in monoatomic rows of the $Ni_3Al$ lattice. This is due to the nonlinear nature of the interaction of the studied monoatomic row with its neighboring biatomic rows. This interaction creates an on-site potential with low degree of periodicity thus making impossible the stable motion of the crowdion. Investigation of crowdions emerged in biatomic rows revealed their successful propagation that was, however, accompanied by certain peculiarities related to the difference in the masses of neighboring atoms. In figure 2, one can see the dependence of the propagation distance of the 1- and 2-crowdion on the initial energy of atoms. First of all, it is clearly seen that 2 crowdion is much more efficient in frames of mass transfer. Secondly, 1-crowdion initiated on Ni atom has almost twice longer propagation path which is probably due to larger atomic mass of Ni. The same refers to the 2-crowdion complex where the case of Ni –Al system has the propagation length about 80 interatomic distances while that for Al – Ni case is about 47 units.

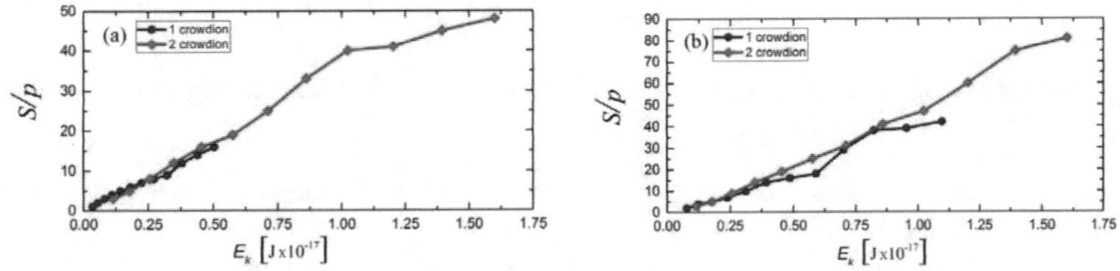


**Figure 2.** Dependence of the propagation path of the 1- and 2-crowdion on the initial energy of atoms (1). On (a) the initial kick was applied to the Al atom in the case of a 1-crowdion, and in the case of a 2-crowdion the initial impact was applied to Al-Ni 2 atom series. (b) The initial kick in the case of the 1-crowdion was attributed to the Ni atom, and in the case of the 2-crowdion, the 2 atom series Ni-Al has been excited.

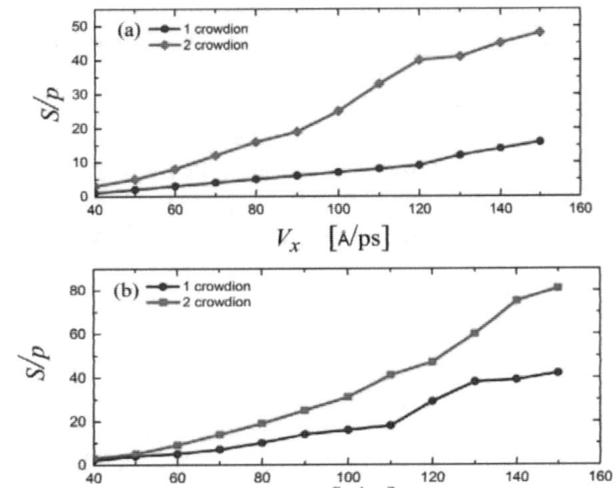


**Figure 3.** Dependence of the propagation path of 1- and 2-crowdion on the *x*-component of the initial velocity of atoms. Plots (a) and (b) correspond to initial scenarios described in figure 2.

Plots presented in figure 3 show the increase of the propagation efficiency of the 2-crowdion by comparison to that of the 1-crowdion. As it was earlier noted, both 1- and 2-crowdions with starting Ni atom propagate further than similar crowdions with an Al atom in the beginning. For instance for the case of initial velocity of 130 Å / ps, a 1-crowdion with a left Al atom travels 12 interatomic distances, and a 2-crowdion decays after passing through 41 interatomic distance. In the case of the left Ni atom with the same initial velocity, the 1-crowdion travels 38 interatomic distances, and the 2-crowdion travels 60 interatomic distances before it decays. Consequently, the mass of the starting atom plays an important role in the range of the crowdion. At an initial velocity of 120 Å/ps and higher, the propagation length of the 1-crowdion sharply increases, which is probably due to the appearance of an additional vacancy (see figure 3b) in the studied close-packed row. This is due to the emergence of the mass transfer process in dynamics similar to the movement of a 1-crowdion.

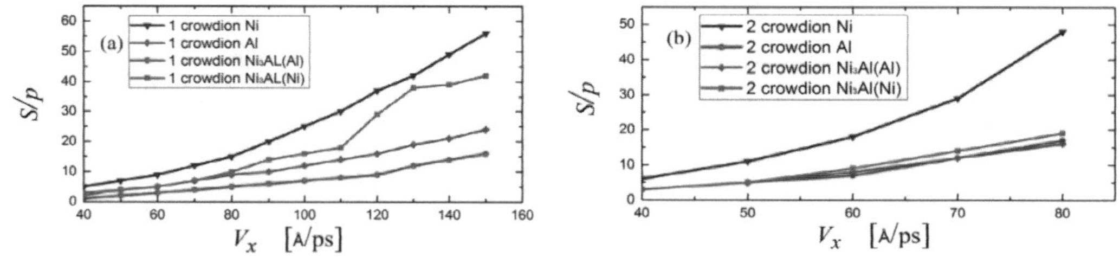


**Figure 4.** Dependence of the range of 1- and 2- crowdion on the initial kinetic energy. a) presents the results for 1-crowdion, b) for 2-crowdion.

The mean free paths of the 1- and 2-crowdions excited in $Ni_3Al$ were compared with the 1- and 2-crowdions that were triggered in the pure Ni and Al lattices (see figure 4). One can note that the 1-crowdion in pure Al and Ni propagates further than the 1-crowdion in the $Ni_3Al$ intermetallic compound with the left Al or Ni atom, respectively. For the 2-crowdion in Al and $Ni_3Al$, the mean free paths differ slightly in the studied velocity range. The latter can be explained by greater stability of the dynamics of the 2-crowdion in Al and to the faster attenuation of the 2-crowdion in $Ni_3Al$, which is associated with a larger mass difference.

## 4. Conclusions

In this paper, the possibility of the movement of 1- and 2-crowdions in the biatomic close-packed row of the $Ni_3Al$ structure was shown. The results suggest that launching of 2-crowdions is energetically more efficient comparing to 1-crowdion. It was shown that the mass of the initial atom plays an important role in the process of mass and energy transfer. Attempts to launch 1- and 2-crowdions in a monoatomic close-packed series have not been successful due to the low-symmetry arrangement of atoms of various kinds in the close-packed rows surrounding the monoatomic rows under consideration. Present study being the first attempt of analysis of the crowdion propagation in biatomic systems allowed to reveal a number of features contributing considerably to the efficiency of mass and energy transfer in intermetallic alloys.

## Acknowledgements

BAM and KEA thank for the financial support the Council of the President of the Russian Federation for state support of young Russian scientists, grant No MD-3639.2019.2 (obtaining of results, writing the manuscript) ZPV is grateful for the financial support of the Russian Foundation for Basic Research, grant No 18-42-220002 p_a (discussion of results), I A Sh acknowledges the financial support of Russian science foundation, grant No 19-72-00109 (analysis of simulation results).

## References

[1] Dmitriev S V, Korznikova E A and Chetverikov A P 2018 *J. Exp. Theor. Phys.* **126** 347

[2] Moradi Marjaneh A, Saadatmand D, Evazzade I, Babicheva R I, Soboleva E G, Srikanth N, Zhou K, Korznikova E A and Dmitriev S V 2018 *Phys. Rev.* E **98** 023003

[3] Babicheva R I, Evazzade I, Korznikova E A, Shepelev I A, Zhou K and Dmitriev S V 2019 *Computational Materials Science* **163** 248

[4] Chetverikov A P, Shepelev I A, Korznikova E A, Kistanov A A, Dmitriev S V and Velarde M G 2017 *Computational Condensed Matter* **13** 59

[5] Hu Z-Y, Cheng X-W, Zhang Z-H, Wang H, Li S-L, Korznikova G F, Gunderov D V and Wang F-C 2017 *Mater. Sci. Eng.* A **684** 1
[6] Cinca N, Lima C R C and Guilemany J M 2013 *Journal of Materials Research and Technology* **2** 75
[7] Murzaev R T, Babicheva R I, Zhou K, Korznikova E A, Fomin S Yu, Dubinko V I and Dmitriev S V 2016 *Eur. Phys. J.* B **89** 168
[8] Zakharov P V, Korznikova E A, Dmitriev S V, Ekomasov E G and Zhou K 2019 *Surface Science* **679** 1
[9] Kistanov A A, Khadiullin S Kh, Dmitriev S V and Korznikova E A 2019 *ChemPhysChem* 20 575
[10] Savin A V, Korznikova E A and Dmitriev S V 2019 *Phys. Rev.* B **99** 235411
[11] Kistanov A A, Khadiullin S Kh, Dmitriev S V and Korznikova E A 2019 *Chemical Physics Letters* **728** 53
[12] Davletshin A R, Ustiuzhanina S V, Kistanov A A, Saadatmand D, Dmitriev S V, Zhou K and Korznikova E A 2018 *Physica B: Condensed Matter* **534** 63
[13] Evazzade I, Lobzenko I P, Korznikova E A, Ovid'ko I A, Roknabadi M R and Dmitriev S V 2017 *Phys. Rev.* B **95** 035423
[14] Barani E, Korznikova E A, Chetverikov A P, Zhou K and Dmitriev S V 2017 *Physics Letters* A **381** 3553
[15] Evazzade I, Lobzenko I P, Saadatmand D, Korznikova E A, Zhou K, Liu B and Dmitriev S V 2018 *Nanotechnology* **29** 215704
[16] Korznikova E A, Fomin S Yu, Soboleva E G and Dmitriev S V 2016 *JETP Lett.* **103** 277
[17] Murzaev R T, Bachurin D V, Korznikova E A and Dmitriev S V 2017 *Physics Letters* A **381** 1003
[18] Savin A V, Korznikova E A and Dmitriev S V 2015 *Phys. Rev.* B **92** 035412
[19] Savin A V, Korznikova E A and Dmitriev S V 2015 *Phys. Solid State* **57** 2348
[20] Liu B, Bai L, Korznikova E A, Dmitriev S V, Law A W-K and Zhou K 2017 *J. Phys. Chem. C* **121** 13876
[21] Barani E, Lobzenko I P, Korznikova E A, Soboleva E G, Dmitriev S V, Zhou K and Marjaneh A M 2017 *Eur. Phys. J.* B **90** 38
[22] Nguyen-Manh D and Dudarev S L 2015 *Nuclear Instruments and Methods in Physics Research Section B: Beam Interactions with Materials and Atoms* **352** 86

IOP Conf. Series: Materials Science and Engineering **672** (2019) 012034 doi:10.1088/1757-899X/672/1/012034

# Evolution of microstructure and mechanical properties of Ti-based metal-matrix composites during high-pressure torsion

**M S Ozerov[1], M V Klimova[1], E A Shigoleva[1], T N Vershinina[2], Yu V Ivanisenko[3] N D Stepanov[1] and S V Zherebtsov[1]**

[1] Laboratory of Bulk Nanostructured Materials, Belgorod National Research University, Belgorod 308015, Russia
[2] Centre of Nanostructured Materials and Nanotechnology, Belgorod State University, Belgorod 308015, Russia
[3] Karlsruhe Institute of Technology, Institute of Nanotechnology, Karlsruhe 76021, Germany

E-mail: Zherebtsov@bsu.edu.ru

**Abstract**. The microstructure and microhardness evolution of a Ti/TiB and Ti-15(wt.%)Mo/TiB metal-matrix composites (MMC) during high-pressure torsion (HPT) at 400 °C was studied. The composites were fabricated by spark plasma sintering of either hcp α-Ti with 10 wt.% of $TiB_2$ or Ti with 13.5 wt.% of Mo (resulted in a bcc β-Ti matrix) and of 10 wt.% $TiB_2$ powders mixtures at 1000 or 1200 °C, respectively. An increase in the dislocation density, a considerable decrease in the grain size in both hcp or bcc Ti matrix, and shortening of TiB whiskers were observed in the microstructures of the composites during HPT. After five revolutions, a nanostructure with a (sub) grain size of ~30 or ~55 nm was produced in Ti or Ti-15Mo matrix, respectively. The microhardness increased with strain from 452 HV in the initial state to 520 HV after 5 revolutions for Ti/TiB MMC and from 575 to 730 HV for Ti-15Mo/TiB MMC. The contributions of various hardening mechanisms to the composites were evaluated.

## 1. Introduction

One of the well-known ways to increase strength of metallic materials is associated with a significant microstructure refinement through severe plastic deformation (SPD) [1, 2]. Due to a considerable decrease in grain size, commercially pure titanium, for example, becomes strong enough for production of surgical implants [3]. Another promising approach to increase strength of titanium without losing its biocompatibility or high corrosion properties can be associated with embedding of ceramic reinforcements, like TiB, in titanium matrix [4, 5]. SPD of such metal-matrix composites (MMCs) induces therefore the simultaneous operation of several hardening mechanisms (substructure hardening, Hall–Petch strengthening, precipitation hardening by TiB fibers) [6].

Some additional hardening mechanisms (solid solution hardening or strengthening due to the formation of ω-phase particles or twins) may also be involved as a result of the transition of the titanium matrix from the hcp α-phase to the bcc β-phase. This transition can be attained by alloying of the titanium matrix with Mo, which is a strong beta-stabilizing element.

In the present work, Ti/TiB and Ti-15(wt.%)Mo/TiB MMCs were fabricated through a spark plasma sintering process (SPS). Then, the composites were deformed using high-pressure torsion

IOP Conf. Series: Materials Science and Engineering **672** (2019) 012034 doi:10.1088/1757-899X/672/1/012034

(HPT) at 400 °C to different strain levels [7]. The microstructure evolution of the composites was comprehensively studied using X-ray diffraction (XRD), scanning (SEM) and transmission electron microscopy (TEM); microhardness was measured to evaluate the influence of HPT on their mechanical properties.

## 2. Materials and methods

Commercial purity powders of Ti (99.1% purity), Mo (99.95% purity) and $TiB_2$ (99.9% purity) were used for the sintering of either Ti/TiB (with hcp Ti matrix) or Ti-15wt.% Mo/TiB (with bcc Ti matrix) composites. The average sizes of the Ti, Mo and $TiB_2$ particles in the powders were 25, 20 and 4 μm, respectively. The powders mixture containing 90 wt.% Ti and 10 wt. % $TiB_2$ (to obtain the Ti with 17 vol.% of TiB ) or 80.75 wt.% Ti, 14.25 wt.% of Mo and 5 wt.% of $TiB_2$ (to obtain a Ti-15wt.% Mo alloy with 8.5 vol.% of TiB) were produced using a Retsch RS200 vibrating cup mill in ethanol for 1h; the milling rotation speed was 700rpm.

Specimens of Ti/TiB and Ti-15Mo/TiB MMCs, measured 19 mm diameter and 20 mm height, were produced through the SPS process using a Thermal Technology SPS10-3 machine under vacuum at 40 MPa for 5 min at 1000 °C (Ti/TiB) or 1200 °C (Ti-15Mo/TiB).

Discs measured 0.7 mm thickness and 10 mm diameter were deformed by HPT at 400 °C and 6 GPa with a speed of 1 rpm. The maximum shear strain was $\gamma$=157 (5 rotations). The corresponding shear strain level $\gamma$ can be calculated as [7]: $\gamma = \frac{2\pi N r}{h}$, where $N$ is the number of revolutions, $r$ is the radius and $h$ is the thickness of the specimen.

X-ray diffraction (XRD) analysis was done for the shear plane of the specimens using an ARL-Xtra diffractometer with $CuK_\alpha$ radiation. The structure and mechanical properties were studied in the mid-thickness of the specimens in the axial cross-section (SEM and microhardness) or in the shear plane (TEM). Vickers microhardness was determined as the average of at least 10 measurements under a load of 1 kg for 10 s.

## 3. Results and discussion

### *3.1. Initial microstructure*

XRD showed that Ti/TiB MMC (figure 1a) consisted of hcp α-Ti (78.6 vol.%), $TiB_2$ with a hexagonal lattice (2.4%), and TiB with an orthorhombic lattice (19%). Ti-15Mo/TiB MMC (figure 1b) contained a β-Ti solid solution with a bcc lattice (84 vol.%), α" martensite with an orthorhombic lattice (4%), TiB with the orthorhombic lattice (10%) and retained $TiB_2$ with the hexagonal lattice (2%).

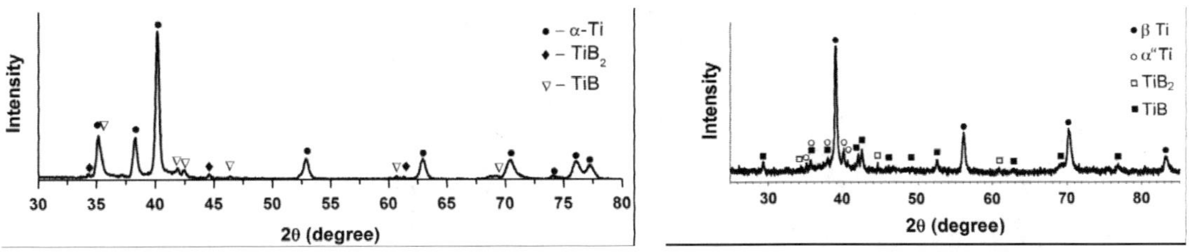


a b

**Figure 1.** XRD patterns of the Ti/TiB (a) and Ti-15Mo/TiB (b) MMCs.

The microstructure of the Ti/TiB composite was also presented by three structural components: the Ti matrix, TiB whiskers and retained $TiB_2$ particles (figure 2a). The average diameter of the TiB whiskers was 36±15 nm. TEM showed TiB non-uniformly distributed whiskers in the matrix (figure 1b). The microstructure of the Ti-15%Mo/TiB composite qualitatively similar to that obtained in Ti/TiB, however the average diameter of TiB fibers was founded to be noticeably higher (~130 nm).

The dislocation density was quite high, especially in the Ti/TiB composite with a higher content of TiB; however, in some places single TiB whiskers can be recognized (figure 2 b, c). Grain boundaries were not seen in the microstructure of both MMCs. Meanwhile, the inter-TiB-whiskers-spacing, which determines the free dislocation path was ~1.5 μm. The volume fraction of pores in the structure of both composites did not exceed 0.5%. It should be noted also that some variations in the Mo percentage were detected in the bcc matrix of the Ti-15%Mo/TiB composite (±3% around the nominal Mo content of 13.5 wt.%).

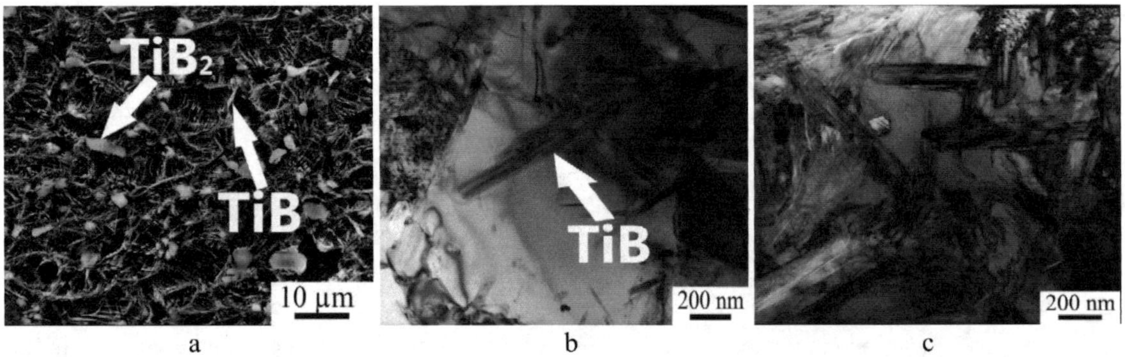


**Figure 2.** Initial microstructure of Ti/TiB (a,b) and Ti-15%Mo/TiB (c) composites (a) - SEM (etched surface) and (b, d) - TEM images.

TEM images of the severely deformed composites are shown in figure 3. In both composites, the matrix microstructure was refined considerably after HPT to 5 revolutions with the formation of dense dislocation arrays, cells and (cub)grains. The size of the strain-induced cells and (sub)grain was evaluated to be ~30 nm in Ti/TiB or ~50 nm in Ti-15%Mo/TiB. In addition, the Mo distribution throughout the Ti-15%Mo/TiB composite microstructure was found to be quite uniform after HPT.

The TiB whiskers were also refined considerably during HPT. Debris measured of 150-300 nm in length were found in both composites suggesting the formation of modestly elongated (the $l/d$ aspect ratio was ~10 in the case of Ti/TiB) or nearly equiaxial ($l/d \approx 1.5$ in Ti-15%Mo/TiB) particles.

**Figure 3.** Bright-field TEM images of Ti/TiB (a) and Ti-15%Mo/TiB (b) after five HPT revolutions ($\gamma \approx 157$).

The microhardness analysis shows a considerable hardening of both composites during HPT (figure 4). The increase in hardness at the edge (figure 4b) was expectably more intensive than that in the center (figure 4a) in both composites. On the other hand, the response of Ti-15%Mo/TiB hardness on strain was more pronounced both in the center and at the edge in comparison with that of Ti/TiB. In addition, the initial level of the microhardness was noticeably higher in Ti-15%Mo/TiB irrespective to two times lower fraction of TiB (8.5 vol. % vs. 17%).

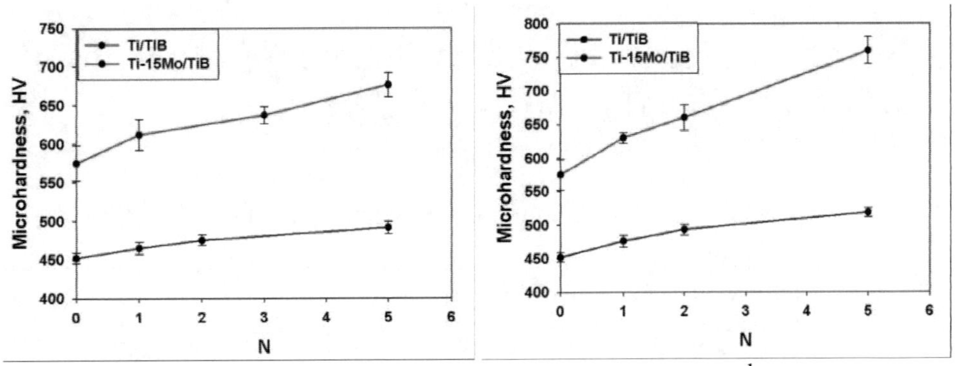


**Figure 4.** Microhardness evolution of the Ti/TiB and Ti-15%Mo/TiB MMCs during HPT at 400°C in the center (a) and at the edge (b) of the discs.

The latter was most probably associated with a higher contribution of solid-solution strengthening. Other main factors which contributed to the overall strength of the severely deformed composites are Hall-Petch strengthening and precipitation hardening (considering dislocation density similar in both composites). A rough estimate showed that after 5 revolutions of HPT, the main contribution into the strength was associated with precipitation hardening via the Orowan mechanism (~60% in both composites) although the absolute value was higher in Ti-15%Mo/TiB. Hall-Petch strengthening and substructure (dislocation) hardening were noticeably low, giving together ~40% contribution.

## 4. Conclusions

The microstructure and microhardness evolution of the Ti/TiB and Ti-15Mo/TiB composites were studied during HPT at 400 °C to a strain of $\gamma$=157. The following conclusions were drawn:

i) Five revolutions of HPT resulted in the microstructure with a (sub)grain size of ~30 or 50 nm in Ti/TiB or Ti-15%Mo/TiB, respectively. The length of the TiB whiskers decreased to 150-300 nm.

ii) The microhardness increased from 445 HV in the initial state to 510 HV after 5 revolutions and from 575 HV to 730 HV in the Ti/TiB and Ti-15Mo/TiB composites, respectively.

## Acknowledgments

The work was supported by the Russian Science Foundation under Grant № 15-19-00165. The authors are also grateful to Joint Research Center “Materials and Technologies”, Belgorod State University, for the assistance with instrumental analysis.

## References

[1] Valiev R Z, Islamgaliev R K and Alexandrov I V 2000 *Progr. Mater. Sci.* **45** 102
[2] Ovid'ko I A, Valiev R Z and Zhu Y T 2018 *Progr. Mater. Sci.* **94** 462
[3] Khorasani A M, Goldberg M, Doeven E H and Littlefair G 2015 *Journal of Biomaterials an Tissue Engineering* **5** 593
[4] Morsi K 2019 *J. Mater. Sci.* **54** 6753–6771
[5] Ozerov M, Klimova M, Kolesnikov A, Stepanov N and Zherebtsov S 2016 *Mater. Des.* **112** 17
[6] Ozerov M, Klimova M, Sokolovsky V, Stepanov N, Popov A, Boldin M and Zherebtsov S 2019 *J. Alloys Compd.* **770** 840
[7] Zhilyaev A P and Langdon T G 2008 *Prog. Mater. Sci.* **53** 893

IOP Conf. Series: Materials Science and Engineering **672** (2019) 012035 doi:10.1088/1757-899X/672/1/012035

# Evolution of the structure and mechanical properties of the Mg-1%Zn-0.2%Ca alloy during ECAP

**O B Kulyasova, I A Evdokimova and R K Islamgaliev**

Institute of Physics of Advanced Materials, Ufa State Aviation Technical University, 12 K. Marx Str., Ufa, 450008, Russia

E-mail: elokbox@mail.ru

**Abstract**. Microstructure of the Mg-1%Zn-0.2%Ca magnesium alloy after equal-channel angular pressing (ECAP) has been investigated. It was shown that an increase of the shear strain leads to significant grain refinement, as well as to the improved uniformity of microhardness. As a result of grain refinement by ECAP processing, the microhardness was increased to a value of 45.5, which is 28% higher than the initial state. Tensile tests demonstrated the enhanced ultimate tensile strength of 248.1 MPa, yield stress of 153.2 MPa, ductility of 19.4 % whereas in the initial coarse-grained state these parameters were 124 MPa, 45.3 MPa and 8.7 %, accordingly.

## 1. Introduction

Recently, the treatment of bone fractures has become widespread in traumatology [1]. For their treatment in traumatology various implants are used, mainly from bioinert metals and alloys [2]. The requirements for materials for the manufacture of implants are constantly being tightened, in particular, this concerns the increase in biochemical and biomechanical compatibility with body tissues, reduction of toxicity [3]. Bioresorbable Mg alloys are of great interest, since implants from the completely disappear when the bone has healed sufficiently, and the device becomes no longer useful [4-7]. Magnesium is a unique material for medical use also because it can gradually resorb [8]. Magnesium alloys are non-toxic, non-carcinogenic, their mechanical properties are more similar to the structure of the cortical bone [9]. But the analysis of the experiments allowed us to identify the main problems, namely, the problems of increasing strength [10]. It is known that a promising method for increasing the strength of magnesium alloys is grain refinement by the equal-channel angular pressing (ECAP) method [11-16]. Therefore, there is a special interest to study the influence of the structure evolution in the process of ECAP on the mechanical properties of the Mg-1% Zn-0.2% Ca alloy.

## 2. Experimental

The cast Mg-1%Zn-0.2%Ca alloy was obtained as an experimental material for the studies. The initial cast samples with a length of 100 mm and a diameter of 20 mm were subjected to homogenization annealing at a temperature of 723 K for 24 hours with cooling into water using a Nabertherm muffle furnace. Equal-channel angular pressing was carried out through a die-set with an angle of channels intersection of 120° at a speed of 6 mm/min. Samples were pressed by 1, 2 and 4 passes through route $B_C$ at a temperature of 673 K. The equivalent strain in the case of the number of passes 1, 2, and 4 was 0.9, 1.8 and 3.6, respectively.

The microstructure was examined after etching in 2.5 mg. picric acid; 2.5 ml. acetic acid; 5 ml. distilled water; 50 ml. ethyl alcohol using an Olympus GX51 optical microscope. The microhardness ($H_V$) was measured by the Vickers method on a Micromet 5101 measurement tool with a load of 50 g and a dwell time of 10 s. Tensile tests were performed on an Instron 5982 testing machine at room temperature and an initial strain rate of $10^{-3}c^{-1}$ using specimens with a gage dimension of 0.6×1×4 $mm^3$. For each condition, a minimum of 5 samples was tested.

## 3. Result and discussion

### *3.1. Microstructure*

Metallographic analysis showed that the structure of the initial material consists of equiaxed grains with a mean grain size of about 280 μm (figure 1a). The heterogeneous structure was observed in the sample after 1 pass ECAP (figure 1b). Undeformed grains up to 280 μm in size occupied 65% of the scanned area.

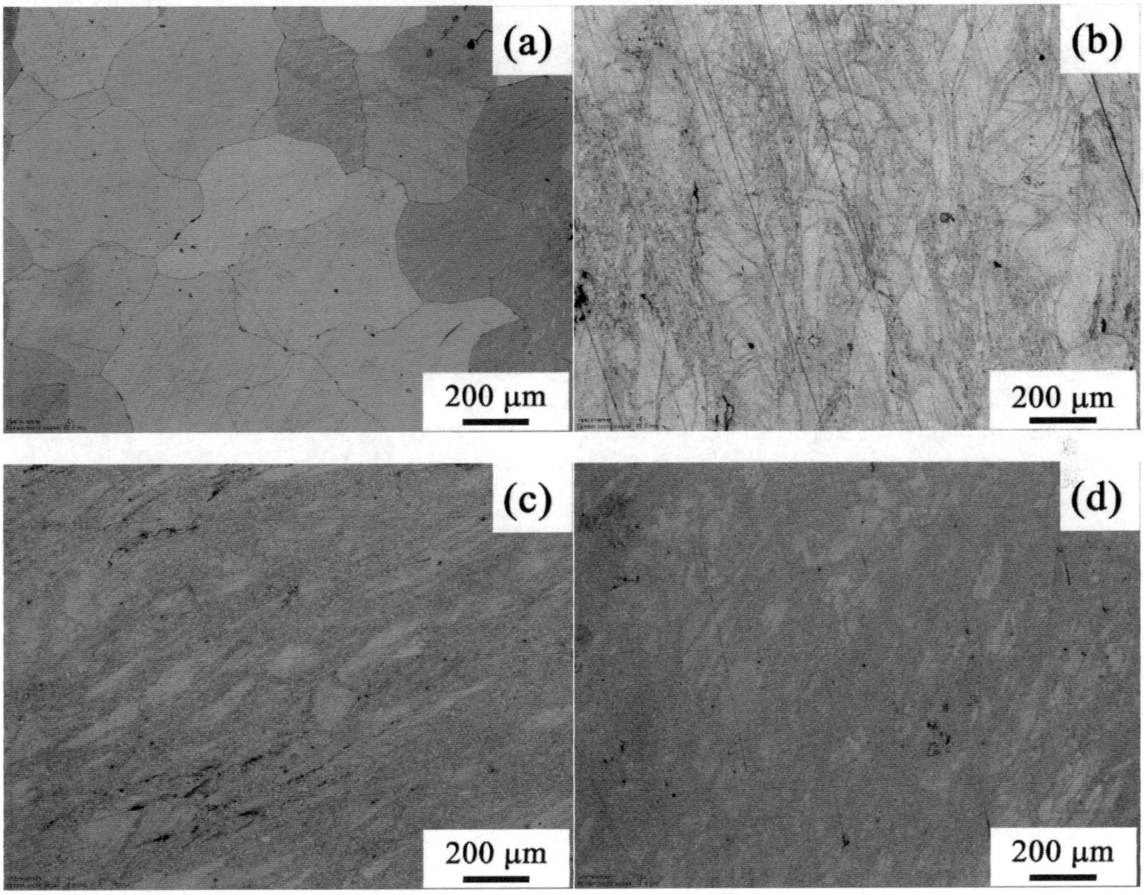


**Figure 1.** Microstructure of the Mg-1%Zn-0.2%Ca alloy: (a) initial state; (b-d) after ECAP: (b) 1, (c) 2, (d) 4 passes.

The remaining surface area consists of fine equiaxed grains with an average size of 4.5 μm. After 2 passes of deformation, slight changes in the microstructure were found. The size of large grains decreased to 230 μm and their fraction was only 31%, so the rest of the area was occupied by small grains of 4.3 μm in size (figure 1c). Further deformation up to 4 passes of ECAP leads to the most uniform structure with a mean grain size of 2.7 μm. However, large grains with a size of 120 μm existed in about 17% of the investigated area (figure 1d). It should be noted that the formation of an inhomogeneous structure after deformation by ECAP on the similar magnesium alloy Mg-1% Zn-0.3% Ca was observed also in [15].

### *3.2. Mechanical properties*

The microhardness of the investigated alloy increased with an increase in the number of passes of ECAP due to the grain refinement. Thus, the microhardness of the initial samples was 35.5 $H_V$ and it increased to 40, 42 and 45.5 $H_V$ after 1, 2 and 4 passes of ECAP, respectively (figure 2a, table 1).

The alloy in the initial coarse-grained state has rather poor mechanical properties, namely: yield stress (YS) 45.3 MPa, ultimate tensile strength (UTS) 124 MPa and ductility 8.7 % (figure 2b, table 1). An increase in the mechanical properties was found in all samples after deformation. As a result of deformation of the Mg-1% Zn-0.2% Ca alloy by 1 pass of ECAP, YS and UTS increased up to 60.7 and 181.2 MPa, respectively. An increase in strength occurred due to the grain refinement in the structure and growth of the dislocation density during severe plastic deformation. The ductility remained at the same level. After 2 passes of deformation, YS and UTS also slightly increased up to 68.7 and 195.3 MPa, respectively, in accordance with growth of the area, which was occupied by small grains (figure 1c).

The most fine-grained structure provides the highest level of YS and UTS, which were 153.2 and 248.1 MPa, respectively (figure 2b, table 1). The elongation to failure changed a lot after deformation by 4 passes and amounted to 19.4%. This result can be caused by both special texture formation and additional twinning during the ECAP processing. An increase in the strength of Mg-Zn-Ca alloys as a result of grain refinement during severe plastic deformation by ECAP was already observed in [15, 16]. In the case of the Mg-4%Zn-0.56%Ca alloy [15], the ultimate tensile strength (UTS) as a result of ECAP increased by factor 1.1 up to 215 MPa. In [16], due to the grain refinement by ECAP, UTS of the Mg-4%Zn-0.56%Ca alloy also increased up to 276 MPa that was 1.6 times higher in comparison with the initial undeformed sample. Both studies also revealed an increase in ductility as a result of the formation of an inhomogeneous (bi-modal) structure. It is noted that the increase in strength occurred due to the formation of small grains, which led to the formation of a greater length of grain boundaries. Large grains are responsible for an increase in ductility [17, 18].

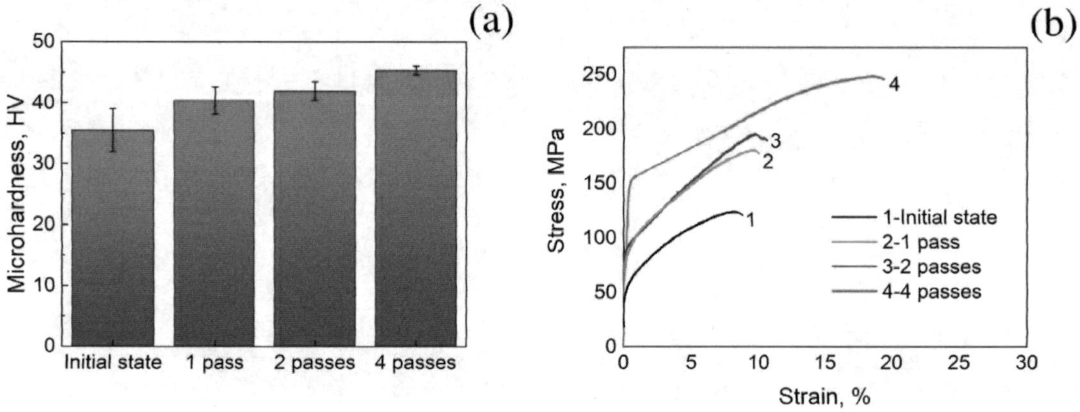


**Figure 2.** (a) Microhardness, (b) tensile tests of the magnesium alloy after ECAP processing.

**Table 1.** Mechanical characteristics of the samples in different conditions.

| Samples | Microhardness/$H_v$ | YS/MPa | UTS/MPa | Elongation (δ), % |
|---|---|---|---|---|
| Initial state | 35.5±3.5 | 45.3±2.3 | 124.0±6.2 | 8.7±1.2 |
| ECAP 1 pass | 40.0±2.2 | 60.7±3.1 | 181.2±7.1 | 10.0±1.1 |
| ECAP 2 passes | 42.0±1.5 | 68.7±3.4 | 195.3±7.7 | 10.6±1.2 |
| ECAP 4 passes | 45.5±0.7 | 153.2±5.7 | 248.1±7.4 | 19.4±1.3 |

## 4. Conclusions

An increase in the equivalent strain during the ECAP processing leads both to the grain refinement and to an increase in the homogeneity of microstructure in the Mg-1%Zn-0.2%Ca alloy. The increase in microhardness and strength occurring during ECAP can be explained by the grain refinement and enhanced dislocation density. The finest grain structure provides the highest level of $H_V$, YS and UTS.

**Acknowledgments**

This work was supported by the Ministry of Education and Science of the Russian Federation through a platform grant in the field of basic research (No. 16.7268.2017/8.9).

**References**

[1] Rahim M I, Ullah S and Mueller P P 2018 *Metals* **8** 532
[2] Patel N R and Gohil P P 2012 *International Journal of Emerging Technology and Advanced Engineering* **2** 4
[3] Zheng Y F, Gu X N and Witte F 2014 *Materials Sience and Engineering* R **77** 1
[4] Chagnon M, Guy L-G and Jackson N 2019 *Toxicologic Pathology* **1** 11
[5] Pogorielov M *et al.* 2017 *Interventional Medicine & Applied Science* **9** 1
[6] Zhao D *et al.* 2017 *Biomaterials* **112** 287
[7] Willbold E *et al.* 2016 *J. Biomed Mater Res: Part A* **105** 1
[8] Staiger M P, Pietak A M, Huadmai J and Dias G J 2006 *Biomaterials* **27** 9
[9] Castiglioni S, Cazzaniga A, Albisetti W and Maier J 2013 *Nutrients* **5** 3022
[10] Valiev R Z, Islamgaliev R K and Alexandrov I V 2000 *Progress Materials Science* **45** 2
[11] Dobatkin S V *et al.* 2017 *IOP Conf. Ser.: Mater. Sci. Eng.* **194** 012004
[12] Kulyasova O B and Islamgaliev R K 2018 *IOP Conf.Ser.: Mater. Sci. Eng.* **380** 012013
[13] Mostaed E, Vedani M, Hashempour M and Bestetti M 2014 *Biomatter* **4** e28283
[14] Tong L B *et al.* 2010 *Materials Science and Engineering* A **527**
[15] Martynenko N *et al.* 2019 *Materials Letters* **238**
[16] Vinogradov Yu *et al.* 2019 *Metals* **9** 2
[17] Ma E 2006 *JOM* **58(4)** 49
[18] Lapovok R *et al.* 2005 *J. Mater. Res.* **20** 1375

# Anodic plasma electrolytic saturation of the Ti6Al4V alloy with nitrogen and oxygen

**S A Kusmanov, I V Tambovskiy, S S Korableva and P N Belkin**

Department of Physical, Mathematical and Natural Science, Kostroma State University, 17 Dzerzhinskogo Street, Kostroma 156005, Russia

E-mail: sakusmanov@yandex.ru

**Abstract**. The structure of the Ti6Al4V alloy, its microhardness, surface roughness, and wear and corrosion resistance after anodic plasma electrolytic saturation with nitrogen and oxidation of the surface were investigated. It was shown that an electrolyte containing ammonia and ammonium chloride provided the saturation of the alloy with nitrogen and oxygen and the formation of $TiO_2$ with rutile structure and a nitrogen solid solution in the alloy. The treatment at 750 °C during 5 min results in an increase in microhardness up to 790 HV and a decrease in surface roughness by 1.4-fold owing to the anode dissolution of titanium. The dry friction coefficient of the samples after treatment at 700 °C can be 3.5-fold reduced through sliding with a speed of 0.144 and a load of 105 N. These regimes allow diminishing the wear rate of samples of titanium alloy after their wear testing by 2 orders of magnitude. Corrosion resistance in Ringer's solution does not deteriorate after treatment at a low temperature, when an oxide layer forms on the surface.

## 1. Introduction

Alpha- and beta-titanium alloy Ti6Al4V and its Russian analogue VT6 are most widely used in modern industry due to a number of advantages [1]. These include high ductility and machinability, corrosion resistance, low thermal conductivity, etc. The poor wear resistance of Ti6Al4V, like that in other titanium alloys, can be improved by conventional thermal chemical treatment, particularly, nitriding. Long gas nitriding (up to 100 hours) can be significantly reduced to 3 hours by carrying out the process in pure nitrogen at a temperature of 1400 °C [2] or by plasma nitriding in an anomalous glow [3] as well as non-self-maintained discharge with the emission of $N^+$, $O^+$, and $C_mH_n^+$ ions [4]. Some coatings obtained by plasma electrolytic oxidation are characteristic by good biocompatibility and bactericidal properties [5]. Plasma electrolytic saturation (PES) of titanium alloys with atoms of light elements is a promising area, which permits to reduce the processing time to several minutes.

Cathodic nitrocarburising of commercial pure titanium in an electrolyte containing formamide and triethanolamine allows obtaining a thin carbonitride layer with a microhardness up to 2080 HV [6]. Saturation of commercial pure titanium with boron in a borax solution leads to the formation of nanocrystalline borides in the surface layer with an increase in the corrosion resistance in Ringer's solution [7]. Borocarburising of the Ti48Al2Cr2Nb alloy in a solution containing glycerol, borax and boric acid also results in the formation of nanocrystals in the surface layer [8]. Research has established an increase in the wear resistance of Ti6Al4V alloy after its cathodic carbonitriding in a carbamide-based solution [9]. The wear rate decreases by 3 orders of magnitude under lubricant friction conditions against steel hardened to 50 HRC.

The anodic processes of PES are characterized by a decrease in surface roughness due to anodic dissolution and the absence of electric discharges, as well as additional saturation of the surface with oxygen. Anodic carburising of commercial pure titanium using an acetone-based solution allows increasing its surface hardness to 340 HV [10], and alpha- and beta-titanium alloy VT22 to 520 HV [11]. Anodic nitrocarburising of VT22 in a carbamide-based electrolyte shows positive results where microhardness of the layer reaches up to 780 HV [12]. Enhancement of the wear resistance of Ti6Al4V alloy is detected after its anodic carburising in an electrolyte containing ethylene glycol [13]. The weight wear rate after lubricant friction against the bearing steel as a counter body decreases by two orders of magnitude.

We could not find the results of PES of theTi6Al4V alloy with nitrogen. It is known that anodic nitriding of commercial pure titanium in an ammonia-based electrolyte provides a significant decrease in the weight wear rate (by two orders of magnitude) with a slight increase in surface hardness to 220 HV [14]. Similar results are obtained after anodic nitriding of the alloy VT22 [15]. It is of interest to study the possibility of anodic nitriding of the Ti6Al4V alloy to enhance its wear resistance. Therefore, the aim of this work is to study the effect of the temperature of the anodic nitriding of the Ti6Al4V alloy in an ammonia electrolyte on the composition of the surface layer, its microhardness and roughness, as well as on wear resistance and corrosion resistance.

## 2. Experimental

Cylindrical samples (a diameter of 10 mm, a length of 15 mm) of titanium alloy Ti6Al4V were ground with SiC abrasive paper to a roughness of Ra (1.00 ± 0.10 μm) and ultrasonically cleaned with acetone. Vertically oriented samples were completely immersed in a cylindrical electrolyzer under the condition of a longitudinal flow of a cooled aqueous solution containing ammonium chloride (10 wt.%) and ammonia (5 wt.%) with a temperature of (20±2) °C. The electrolyte volume was 4 L, the duration of saturation with nitrogen and oxygen was 5 min. The electrolyte flow rate was measured with a RMF-0.16 GUZ flowmeter and maintained at 2.6 L/min. The samples were quenched in the same electrolyte by disconnecting the voltage after saturation with nitrogen and oxygen.

The surface morphology and the elemental composition of the layer were studied using a VEGA 3 SBH electron microscope after polishing and etching samples for 1 min, using a 5 wt.% solution of hydrochloric acid in glycerol. The arbon concentration was subtracted to evaluate the layer composition under the assumption that this signal belongs to organic pollution. The structure was examined and microhardness was measured with a Falcon 503 (Innovatest) microhardness tester with a load of 50 g. The surface roughness was measured on a TR200 profilograph-profilometer.

A ball-on-disk (normal load 105 N, sliding speed 0.144 m/s, and sliding distance 100 m) against a bearing steel ball (9 mm in diameter) was applied to evaluate the wear rate and friction coefficient of the untreated and treated samples at dry and lubricated conditions (engine oil "LITOL" containing petroleum oil with a viscosity of 60–75 $mm^2/s$ at 50 °C, lithium soap of 12-hydroxy acid and an antioxidant additive). The weight loss was measured with electronic balance with an accuracy of ±0.1 mg. Corrosion tests were carried out in a three-electrode cell using an SP-150 Biologic potentiostat-galvanostat at a sweep rate of 1 mV/s. The potential was measured against a silver chloride saturated electrode. The surface of the sample was shielded, its side surface with an area of 0.32 $cm^2$ remained in contact with the solution. Ringer's solution was chosen as a corrosive medium. The current density of corrosion was calculated by extrapolating the Tafel polarization curves.

## 3. Results and discussion

Anodic saturation of titanium alloys in aqueous electrolytes is always accompanied by the formation of an oxide layer, the morphology of which is shown in figure 1. The main part of the layer is represented by a smooth surface with small pores, which facilitate the diffusion of the saturating components and the liberation of dissolved metals. According to elemental analysis, the oxygen content in the layer is (34.5 ± 0.5) wt.%, which is close to the concentration of $TiO_2$ (rutile), the formation of which was previously confirmed by X-ray analysis [13, 14]. The concentration of titanium is (59.3 ± 0.5) wt.%, which also corresponds to rutile. The content of aluminum and vanadium is likewise reduced by 38–41%. A low chlorine content (0.11 ± 0.03) wt.% indicates the participation of electrolyte anions in the anodic dissolution. Approximately the same composition, but with a higher oxygen concentration (42.5 ± 0.5) wt.% is observed in dark areas with increased roughness.

Diffusion of oxygen and nitrogen reduces the stability of the β-phase and contributes to the martensitic transformation of the alloy with the formation of a supersaturated solid solution of the alloying components in titanium. The result of this transformation is an increase in the hardness of the alloy (figure 2). The lower surface hardness after saturation at 700 °C may be attributed to the absence of martensitic transformation owing to the insufficient diffusion of saturating components. The maximal thickness of the layer of increased hardness, exceeding 0.25 mm, is achieved at a saturation temperature of 900 °C, probably because of more intensive diffusion of nitrogen and oxygen. The surface roughness with respect to Ra parameter decreases at all saturation temperatures by 5–30%. The dependence of the roughness reduction on the saturation temperature is not detected due to uncontrolled partial exfoliation of the oxide layer.

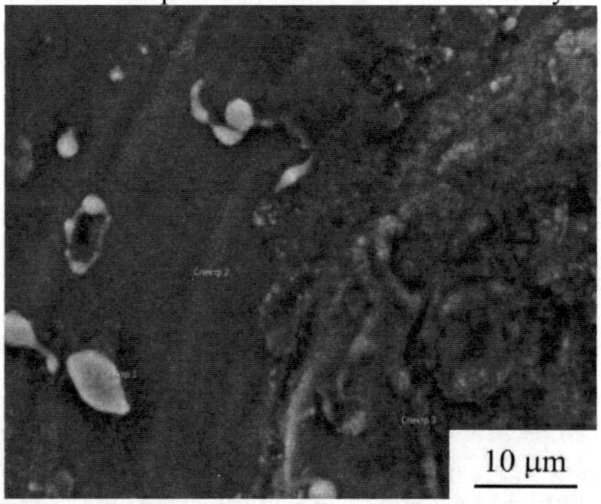


**Figure 1.** Top view of the oxide layer on the surface of the Ti6Al4V alloy after PES at 750 °C.

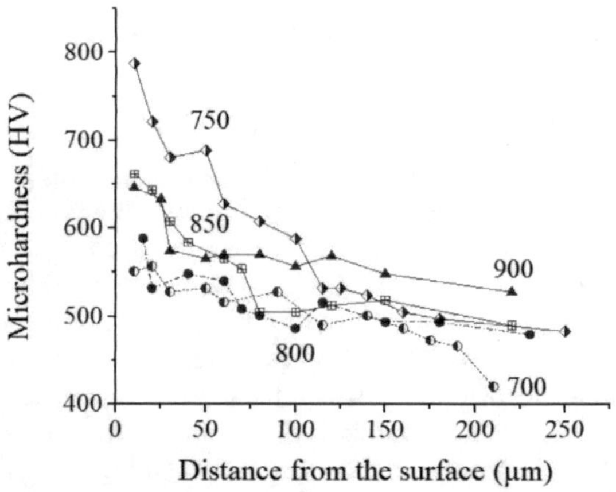


**Figure 2.** Microhardness distribution in the surface layer of the Ti6Al4V alloy after PES at different temperatures (°C).

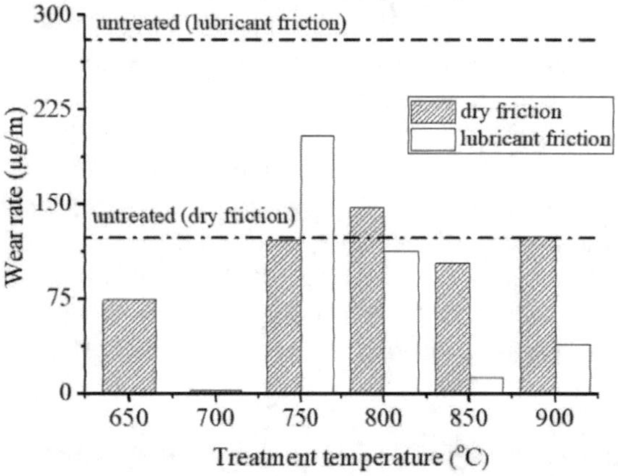


**Figure 3.** The weight wear rate of the Ti6Al4V alloy processed at different temperatures after tribological tests.

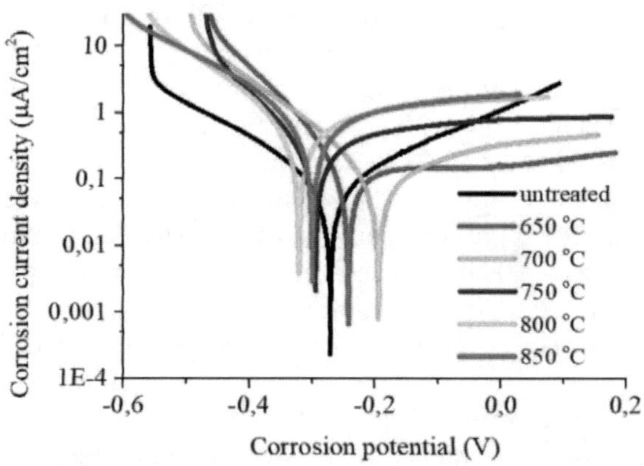


**Figure 4.** Polarization curves of corrosion tests for the Ti6Al4V alloy after PES at different temperatures. The test media is Ringer's solution.

Tribological tests show that the lubricant friction coefficient of samples treated at all temperatures is always lower than that of an untreated sample. No correlation between the friction coefficient and surface roughness Ra is found. The minimal friction coefficient, equal to 0.3, is observed after saturation at 850–900 °C, as well as the lowest wear rate (figure 3), which can result from the retention of lubricant in the pores of the oxide layer, which grows with increasing temperature. The dry friction coefficient decreases in comparison with the untreated sample only after saturation at low temperatures and reaches 0.15 at 700 °C, which is 3.5-fold less than that of the untreated sample. A minimum wear rate is also achieved under these conditions.

The results of corrosion tests show that the corrosion potential shifts in the positive direction only after treatment of the alloy at low temperatures up to 700 °C (figure 4). The minimum corrosion current density of 0.13 μA/cm$^2$ is observed after treatment at 700 °C, which is not too much higher than that of the untreated sample (0.08 μA/cm$^2$).

## 4. Conclusions

The possibility of increasing the hardness and wear resistance of alpha- and beta-titanium alloy Ti6Al4V by means of the anodic plasma electrolytic saturation with oxygen and nitrogen in an electrolyte containing ammonium chloride (10 wt.%) and ammonia (5 wt.%) followed by quenching in the same electrolyte is shown. This treatment results in an increase in surface hardness up to 790 HV, a decrease in surface roughness from Ra 1.0 μm for the untreated sample to Ra 0.7 μm, and an increase in wear resistance by two orders of magnitude under the studied friction conditions.

## Acknowledgments

The work was supported by the Council on grants of the President of the Russian Federation for state support of young Russian scientists and on state support of leading scientific schools of the Russian Federation, grant No. MD-1021.2019.8.

## References

[1] Zhecheva A, Sha W, Malinov S and Long A 2005 *Surf. Coat. Technol.* **200** 2192
[2] Gordin D M, Thibon I, Guillou A, Ansel D and Gloriant T 2007 *Ti-2007 Science and Technology* 1731
[3] Shan M S, Ikhlao U and Saleem S 2012 *Coden JNSNAC* **52** 31
[4] Timoshenko A I, Taran V S and Misiruk I O 2012 *BAHT* **82** 235
[5] Rudnev V S, Lukiyanchuk I V, Vasilyeva M S, Medkov M A, Adigamova M V and Sergienko V I 2016 *Surf. Coat. Technol.* **307** 1219
[6] Li X M and Han Y 2006 *Electroch. Commun.* **8** 267
[7] Aliev M K and Saboor A 2007 *Bull. Mater. Sci.* **30** 601
[8] Aliofkhazraei M, Hassanzadeh-Tabrizi S A, Sabour Rouhaghdam A and Heydarzadeh A 2009 *Cer. Int.* **35** 2053
[9] Li X M and Han Y 2008 *Appl. Surf. Sci.* **254** 6350
[10] Belkin P N, Kusmanov S A, Dyakov I G, Komissarova M R and Parfenyuk V I 2016 *Surf. Coat. Technol.* **307** 1303
[11] Kusmanov S A, Dyakov I G, Belkin P N and Parfenyuk V I 2017 *Surf. Eng. Appl. Electrochem.* **53** 1
[12] Tambovskiy I V, Kusmanov S A, Korableva S S, Silkin S A, Sevostyanov N V, Komissarova M R and Belkin P N 2017 *Surf. Eng. Appl. Electrochem.* **53** 407
[13] Komissarova M R, Kusmanov S A, Belkin P N, Dyakov I G, and Parfenyuk 2016 *Izv. Vyssh. Uchebn. Zaved. Khim. Khim. Tekhnol.* **59** 100
[14] Belkin P N, Zhirov A V, Belkin V S, Parfenyuk V I and Kusmanov S A 2016 *J. Mater. Sci. Technol.* **32** 1027
[15] Kusmanov S A, Smirnov A A, Silkin S A, Parfenyuk V I and Belkin P N 2016 *Surf. Coat. Technol.* **307** 1291

IOP Conf. Series: Materials Science and Engineering **672** (2019) 012037 doi:10.1088/1757-899X/672/1/012037

# Mechanical behavior of coarse- and fine-grained Al-6101 samples of different geometry under tension

**D K Magomedova[1], D V Gunderov[1,2] and M A Efimov[1]**

[1] Saint Petersburg State University, St. Petersburg, Universitetskaya emb., 7/9
[2] Ufa State Aviation Technical University, Ufa, Russia

E-mail: magmedva.dasha@mail.ru

**Abstract.** The paper presents the experimental data obtained during static tension of cylindrical samples of two structural states and two types of geometry from Al-6101 alloy at room temperature.

## 1. Introduction

Al-6101 alloy is widely used as a conductor for overhead transmission lines due to its increased strength and electrical conductivity. This alloy exhibits a tensile strength in the range from 255 to 330 MPa. This level of characteristics is achieved in it using traditional types of thermal or thermomechanical treatment [1, 2].

The approach based on the achievement of increased strength and electrical conductivity in alloys due to the formation of a regulated ultrafine-grained (UFG) structure using severe plastic deformation (SPD) has a great potential [3, 4]. The technique on the basis of equal-channel angular pressing according to the Conform scheme (ECAP-C) for producing long-length UFG semi-finished Al-6101 products (wires and wire rods) with a level of characteristics that has no world analogues was created in [5, 6].

For UFG Al alloys, it is important to study the features of their fracture, as well as the formation of pores and cracks during loading. Earlier in [7], such studies were carried out during tension of samples of coarse-grained (CG) metals, the geometry of which is shown in figure 1 (the rationale for using this geometry is given in [7]). The features of fracture and pore formation in UFG Al 6061 are investigated in our works [9], by using the method [7]. It is known that the geometry of the sample significantly affects the data obtained during tensile tests, and this influence is more significant for UFG materials [7]. In this respect, the influence of the geometry of the samples used in [7] on the behavior of CG and UFG Al-6101 alloys under static tension at room temperature was studied in this paper.

## 2. Object of study

The initial billets of Al-6101 alloy contained Si 0.3-0.7%, Mg 0.35-0.8%, Fe 0.50%, Cu 0.10%, Zn 0.10%, Cr 0.03%, Mn 0.003% in the form of bars with a diameter of 12 mm were used in this work. The initial billets were subjected to homogenization at a temperature of 550 °C for 2 hours and subsequent quenching in water; after that they were subjected to natural aging (NA) for 6 days. To obtain UFG structure, part of the billets was processed by SPD via the ECAP-C method. The regime of obtaining UFG billets and a description of their microstructure features are presented in detail in [5].

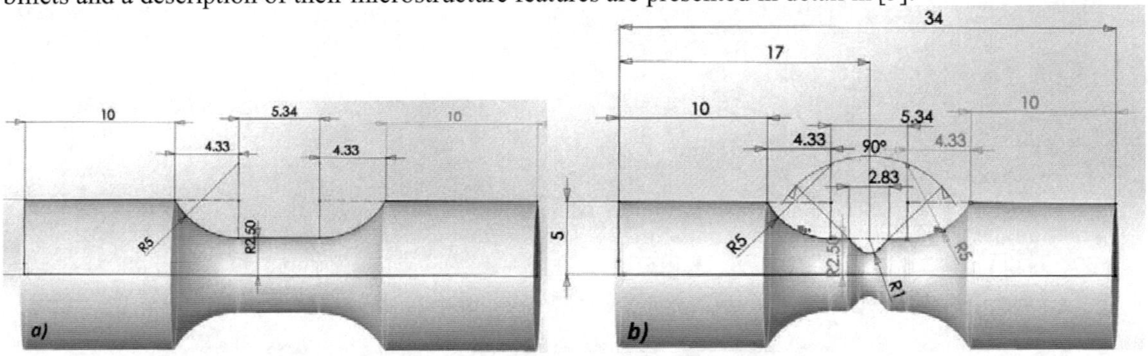


**Figure 1.** Samples with the working part diameters of 5 mm (a) and 3.2 mm (b).

Samples of a total length of 34 mm with a cylindrical working part with a diameter of 5 mm (sample No.1) and a groove ~3.2 mm in diameter (sample No. 2) were made from CG and UFG billets (figure 1).

Uniaxial tensile tests of the samples were carried out on a Shimadzu AG-50kNX testing machine at room temperature with a constant strain rate of $1.4 \cdot 10^{-4} s^{-1}$. Deformation of the working part of the samples was recorded using a Shimadzu TRViewX video extensometer. To register the deformation, horizontal marks were labelled on the surface of the working part of the samples at a distance of 3 mm from the center of the sample. At least 3 samples were tested for each microstructural state and geometry. Elongation was carried out to failure of the samples.

## 3. Results

Typical images of the CG and UFG structure of alloy samples are presented in figure 2 [9]. The grain size of the UFG sample was about 500 nm.

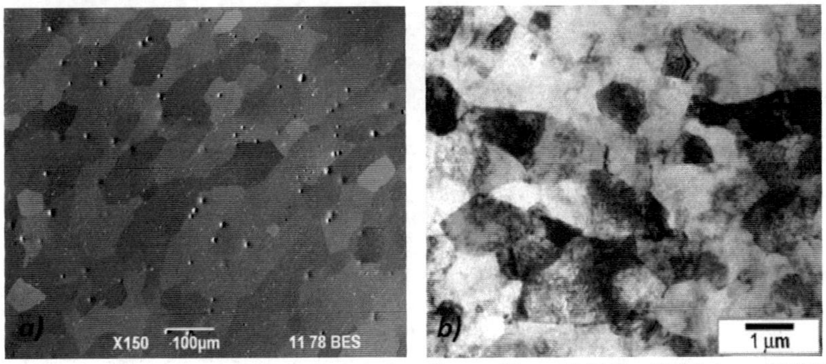


**Figure 2.** Images of a typical microstructure of samples from alloy 6101 in CG (a) and UFG (b) states.

Figures 3-4 show the stress-strain diagrams for each geometry and each structure. Table 1 presents the mechanical properties measured from the stress-strain curves (figures 3-4).

IOP Conf. Series: Materials Science and Engineering **672** (2019) 012037 doi:10.1088/1757-899X/672/1/012037

**Table 1.** The mean values of mechanical properties for CG and UFG Al-6101 samples of different geometry.

| Structure type | Sample (No.) | Yield stress (MPa) | Tensile strength (MPa) | Relative strain (%) |
|---|---|---|---|---|
| CG NA | 1 | 85 | 226 | 50.9 |
| | 2 | 92 | 237 | 18.29 |
| UFG | 1 | 304 | 351 | 32.83 |
| | 2 | 310 | 406 | 7.97 |

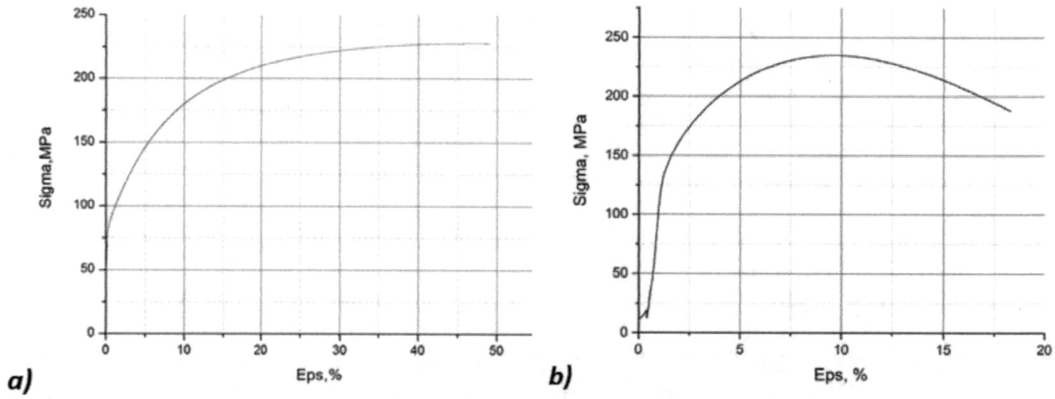


**Figure 3.** The stress-strain diagram for the sample with the working part diameter of 5 mm (a) and 3.2 mm (b), subjected to NA.

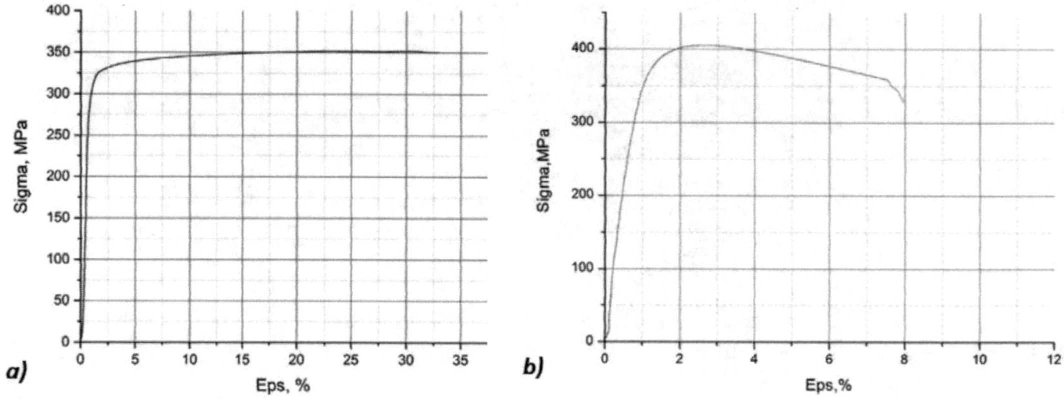


**Figure 4.** The stress-strain diagram for UFG sample with the working part diameter of 5 mm (a) and 3.2 mm (b).

## 4. Conclusions

One can draw the following conclusions during the analysis of the results:

Strength and ductility of the material are affected by the geometry of the investigated samples: Al-6101 samples with a narrower gauge length show higher strength, but they turn out to be the least ductile. The result on strength requires further analysis. The result on the ductility seems to contradict the data [8], where higher ductility was observed on the samples with a shorter gauge length. However, in our experiment, the same gauge length for both types of samples — 6 mm — was used to calculate and observe the strain via an extensometer. Sample No. 2 has a groove, respectively, it has a lower ∅ base/$L_0$ ratio ($L_0$ is the gauge length of 6 mm), and in this regard, tests on sample No. 2 show lower ductility values than on sample No.1.

The strength indexes of UFG state are noticeably higher than those of CG during testing on samples of both geometries.

The sample geometry affects the mechanical properties detected in UFG material more strongly than in CG material: the difference in strength between UFG samples of different geometries is about 50 MPa, and between CG samples – about 10 MPa.

The obtained data will be used for further study of the influence of the structure and geometry of Al-6101 samples on pore formation under static tension.

**Acknowledgements**

The authors would like to thank the Russian Science Foundation for financial support of this research in the framework of the Project 17-19-01311.

SEM investigations were performed at the Interdisciplinary Resource Center for Nanotechnology of St. Petersburg State University.

The authors thank the Resource Center for Applied Aerodynamics of St. Petersburg State University for the technical support.

## References

[1] Karabay S 2006 *Mater. Design* **27** 821
[2] Lin G, Zhang Z, Wang H, Zhou K and Wie Y 2016 *Mater. Sci. Eng.* A **650** 210
[3] Sabirov I, Murashkin M and Valiev R Z 2013 *Mater. Sci. Eng. A* **560** 1
[4] Murashkin M Yu, Sabirov I, Sauvage X, Valiev R Z 2016 *J. Mater. Sci.* **51** 33
[5] Murashkin M, Medvedev A, Kazykhanov V, Krokhin A, Raab G, Enikeev N and Valiev R Z 2015 *Metals* **5** 2148
[6] Mann V Kh, Krokhin A Yu, Matveeva I A, Raab G I, Murashkin M Yu and Valiev R Z 2014 *Light Metal Age* **December 2014** 26
[7] Argon A S, Im J and Needleman A 1975 *Metall. Trans.* ***A*** **6** 815
[8] Kulyasova O B, Valiev R Z and Islamgaliev R K 2005 *Phys. Met. Metallogr.* **100** 277
[9] Magomedova D K Efimov M A and Murashkin M Yu 2018 *AIP Conf. Proc.* **1959** 070021

IOP Conf. Series: Materials Science and Engineering **672** (2019) 012038 doi:10.1088/1757-899X/672/1/012038

# Effect of friction-stir welding on the microstructure and mechanical properties of an Al-Mg-Si alloy

**I Vysotskiy, S Malopheyev, S Mironov and R Kaibyshev**

Belgorod State University, Pobeda 85, Belgorod 308015, Russia

E-mail: malofeev@bsu.edu.ru

**Abstract**. In this work, sheets of an Al-Mg-Si alloy were joined by double-side friction stir welding and tempered by standard post-weld artificial aging. The combination of relatively high FSW tool rotation and welding speed with subsequent post-weld artificial aging allowed obtaining welds with high joint efficiency. However, a high sensibility of joint efficiency from FSW tool rotation speed at constant welding speed was found. In turns, the joint efficiency of the friction-stir welds for ultimate tensile strength was found to be 72% at 800 rpm and 92% at 1100 rpm, respectively. The reasons of difference in the mechanical properties of the welds are studied.

## 1. Introduction

Al-Mg-Si alloys are widely used in the automotive and other industries due to an attractive combination of low density, moderate strength, excellent corrosion resistance, and relatively low cost. However, the production of complex constructions from these alloys by conventional fusion welding is challenging and this essentially complicates the manufacturing of complex engineering structures. In this context, friction-stir welding (FSW), an innovative solid-state technique, appears to be very attractive for joining these materials [1–3]. The extensive research have demonstrated that FSW of Al-Mg-Si alloys involves complex precipitation phenomena, including the dissolution of particles in the stir zone and precipitation coarsening in the heat-affected zone [1–3]. Being a strong function of the weld heat input, the above effects could be efficiently controlled through welding variables, i.e. spindle rate and feed rate. In a previous study [4], for instance, the welding parameters were optimized to provide the highest cooling rate. This minimized particles coarsening in the heat-affected zone, and, thus, resulted in a relatively high joint efficiency of the produced welds. In an attempt to improve the mechanical properties of the weldments further, an alternative strategy was implemented in the present work. To suppress the precipitation phenomena at all, a relatively weld low heat input was used. To this end, the FSW trials were conducted at a relatively low spindle rate.

## 2. Material and experimental procedure

The program material employed in the present study was a commercial AA6061 alloy produced by direct continuous casting. The obtained ingot was homogenized, extruded at 380 °C to ~75% of area reduction and subjected to a standard T6 tempering, i.e., solutionized at 550 °C for 1h, water quenched and subsequently artificially aged at 160 °C for 8 h.

The tempered sheets were friction-stir butt welded along the extrusion direction using an AccuStir 1004 FSW machine. The welding tool was fabricated from a tool steel and consisted of a 12.5-mm-diameter shoulder and a M5 cylindrical probe with a length of 1.9 mm. The thickness of the welding blanks was 3 mm. To provide a full-thickness joining, a double-side FSW was applied in mutually opposite directions. To investigate the effect of the weld heat input, FSW was performed at two different spindle rates, i.e. 800 rpm and 1100 rpm, whereas a constant feed rate of 760 mm/min was employed in both cases. The principal directions of the FSW geometry were denoted as the welding direction (WD), transversal direction (TD) and normal direction (ND).

To recover the mechanical properties of the welded material, the obtained joints were artificially aged at 160 °C for 8 h. To minimize a possible effect of natural aging, the post-welding treatment was performed within ~20 min after FSW.

Microstructural observations were performed mainly by electron backscatter diffraction (EBSD) technique and optical microscopy. The final surface finish was obtained by electro-polishing in a 25 pct. solution of nitric acid in methanol. EBSD analysis was conducted by using a FEI Quanta 600 field-emission-gun scanning electron microscope (FEG-SEM) equipped with TSL OIM™ software. In the EBSD maps shown, the low-angle boundaries (LABs) ($2° < \theta < 15°$) and the high-angle boundaries (HABs) ($\geq 15°$) were depicted as red and black lines, respectively.

To evaluate broad aspects of the distribution of the microstructure within the welded material, a microhardness profile was recorded. Vickers microhardness data were obtained by applying a load of 200 g with a dwell time of 10 s by using a Wolpert 402MVD microhardness tester.

To examine mechanical properties of the weldments, transverse tensile tests were applied. The tensile specimens were centered at the weld centerline and included all characteristic microstructural zones of FSW. The tensile tests to failure were conducted at ambient temperature and a constant cross-head velocity corresponding to a nominal strain rate of $10^{-3}$ $s^{-1}$ by using an Instron 5882 universal testing machine. Further experimental details were described elsewhere [4].

## 3. Results and discussion

### 3.1. Base material

The microstructure of the *base material* was dominated by coarse elongated grains with a developed substructure. The average grain size in the direction of extrusion was measured to be ~200 μm. TEM observation revealed high density of nanoscale dispersoids evenly distributed in the grain interior. As follows from the scientific literature [5,6], these particles are a mixture of needle-shaped β" precipitates with minor fraction of lath-shaped Q' precipitates. In more detail, the microstructure of the base material was described elsewhere [4].

### 3.2. Joints

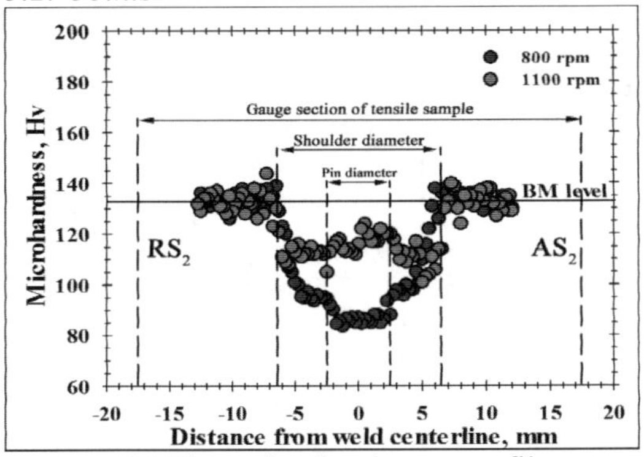


**Figure 1.** The microhardness profiles measured across the cross-section of the joints. Note: the horizontal solid line shows the average microhardness of the base material. $RS_2$ and $AS_2$ are retreating side and advancing side, respectively.

Microhardness profiles measured across mid-thickness of both heat-treated joints are shown in figure 1. For clarity, the shoulder- and probe diameters were indicated in the figure; the dimension of the gauge section of the tensile specimens was also shown. To a first approximation, the diameter of the shoulder indicates the region most influenced by the weld heat input. In both welds, softening of the material was found in this area. The effect was most pronounced in the weld produced at relatively low spindle rate of 800 rpm. Remarkably, the lowest strength in this case was found in the microstructural region outlined by the tool probe, i.e., virtually, in the stir zone. In contrast, the weld produced at a relatively high spindle rate exhibited relatively high strength level with a minimal microhardness measured in the heat-affected zone, as is normally found in friction-stir welded heat-treatable aluminum alloys [1–3].

Typical EBSD grain-boundary maps taken from the central portion of stir zone of the produced joints are shown in figures 2a and b. In both cases, the microstructure was dominated by relatively-fine grains. As expected, the weld obtained at a relatively low spindle rate (i.e., a lower heat input) was characterized by a smaller mean grain size (2.8 μm vs 7.5 μm) and larger LAB fraction (38 pct. vs 18 pct.), as shown in figures 2a and b.

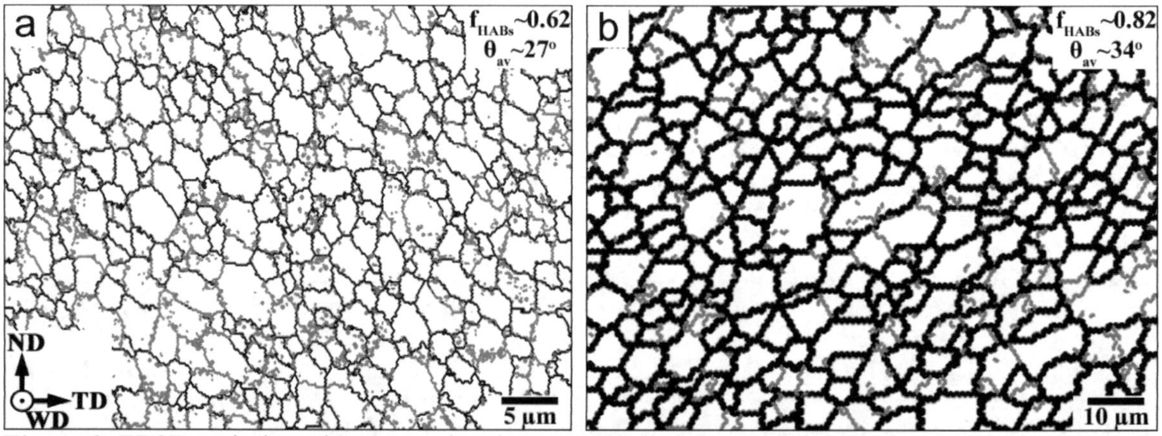


**Figure 2.** EBSD grain-boundary maps showing typical microstructures in the center of the stir zone of the joints produced at a spindle rate of 800 rpm (a) and 1100 rpm (b). The $f_{HABs}$ and $\theta_{av}$ in the inserts in the top right corner are the HABs fraction and the average misorientation, respectively.

Typical deformation diagrams recorded during the transverse tensile tests of the base material and the joints are shown in figure 3. The relevant mechanical characteristics are summarized in table 1. Both welds exhibited inferior strength and ductility than the base material. Importantly, the material softening was again most pronounced in the welds produced at relatively low spindle rate and the joint efficiency in this case was only 72 pct. (table 1). In both welds, the tensile strain was found to be localized in the softened region, i.e. either in the heat affected zone in the 1100-rpm weld or in the stir zone in the 800-rpm weld (not shown).

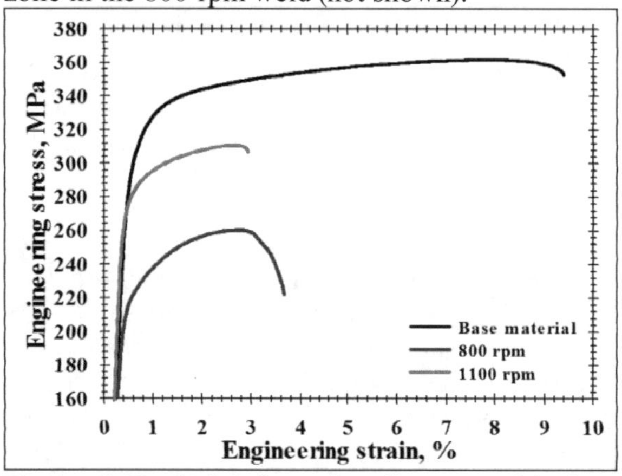


**Figure 3.** Typical deformation diagrams of the base material and the joints produced at a spindle rate of 800 rpm and 1100 rpm recorded during transverse tensile tests.

It is relatively well established that the strength characteristics of heat-treatable aluminum alloys are primarily governed by second-phase particles. Therefore, a clear difference in the tensile behavior between two welded materials was perhaps attributable to different precipitation behavior during FSW and the post-weld treatment. In the previous work [4], it was proposed that the peak temperature during FSW at 1100 rpm exceeded ~500 °C and that the strengthening precipitates in the stir zone completely dissolved. Accordingly, the post-weld aging resulted in nearly-full re-precipitation of the particles, thus essentially recovering the strength of the material.

On the other hand, the relatively low strength exhibited the 800-rpm weld suggests that the particles re-precipitation in this case was incomplete. It is likely that the peak welding temperature in this case was below the solvus temperature of the particles, and therefore the microstructure evolution in this case was governed by precipitation coarsening rather than precipitation dissolution. This effect was enhanced further by the post-welding aging, and therefore the final mechanical properties of such welds were relatively low (figure 3 and table 1).

**Table 1.** Mechanical properties of the *base material* and the *joints*.

| Material condition | Yield strength, MPa | Ultimate tensile strength, MPa | Ductility, % | Joint efficiency for UTS, % | Failure location |
|---|---|---|---|---|---|
| Base material | 305 | 360 | 9 | - | - |
| FSW at 800 rpm | 220 | 260 | 3.4 | 72 | Stir zone |
| FSW at 1100 rpm | 280 | 310 | 2.4 | 92 | Heat-affected zone |

## 4. Conclusions

In the present work, the influence of the spindle rate (800 rpm and 1100 rpm) on the microstructure and mechanical properties of friction-stir welded 6061 aluminum alloy was examined.

It was found that a reduction of the spindle rate resulted in essential degradation of the weld strength. This effect was attributed to the relatively low welding temperature, which led to precipitation coarsening rather than to precipitation dissolution in the stir zone. As a result, the post-weld aging treatment was not efficient for recovering of strength of the material.

**Acknowlegements**
This work is supported by the Ministry of Education and Science of the Russian Federation under the agreement №14.584.21.0023 (ID number RFMEFI58417X0023). The authors are grateful to the staff of the Joint Research Center, "Technology and Materials" Belgorod State National Research University for their assistance with the mechanical and structural characterizations.

## References

[1] Threadgill P L, Leonard A J, Shercliff H R and Withers P J 2009 *Int. Mater. Rev.* **54** 49
[2] Mishra R S, De P S and Kumar N 2014 *Friction Stir Welding and Processing* (Cham: Springer)
[3] Mishra R S and Ma Z Y 2005 *Mater. Sci. Eng.* R *Rep.* **50** 1
[4] Malopheyev S, Vysotskiy I, Kulitskiy V, Mironov S and Kaibyshev R 2016 *Mater. Sci. Eng.* A **662** 136
[5] Esmaeili S, Wang X, Lloyd D J and Poole W J 2003 *Metall. Mater. Trans.* A **34** 751
[6] Wang X, Esmaeili S, Lloyd D J 2006 *Metall. Mater. Trans.* A **37** 2691
[7] Mironov S, Malopheyev S, Vysotskiy I, Zhemchuzhnikova D and Kaibyshev R 2018 *Defect Diffus. Forum* **385** 359

# Abnormal grain growth in a double-sided friction-stir welded Al-Mg-Si alloy

**I Vysotskiy, S Malopheyev, S Mironov and R Kaibyshev**

Belgorod State University, Pobeda 85, Belgorod 308015, Russia

E-mail: malofeev@bsu.edu.ru

**Abstract**. In this work, a standard T6 tempering (involving solution annealing followed by artificial aging) was employed to a fine-grained Al-Mg-Si alloy produced by double-sided friction-stir welding (FSW). The material was found to exhibit abnormal grain growth, thus demonstrating a relatively low thermal stability. The kinetics of abnormal grain growth in the joint was studied.

## 1. Introduction

Friction-stir welding (FSW) is an innovative and rapidly-developing technology of solid-state joining of structural materials [1–4]. In heat-treatable aluminum alloys, FSW normally leads to the dissolution or coarsening of strengthening precipitates thus degrading the mechanical properties of the welded material. For its recovery, the produced joints typically undergo post-weld heat treatment [5-8]. Such treatment, however, often gives rise to abnormal grain growth in the stir zone [1,2,5–10], which may essentially deteriorate service characteristics of the joints. Despite the extensive recent research in this area, the mechanism of this undesirable phenomenon is still not clear.

There is some limited evidence from the scientific literature that double-sided welds demonstrate relatively good thermal stability [4]. It is believed that this variation of the FSW process promotes a relatively uniform thermal field through the weld thickness thus resulting in a reasonably homogeneous distribution of the microstructure. However, the feasibility of this approach for suppression of abnormal grain growth is still not completely clear. Therefore, the present work has been undertaken to shed more light on this issue.

## 2. Material and experimental procedure

The material for the present study was a commercial AA6061 alloy produced by direct continuous casting. The obtained ingot was appropriately homogenized and then extruded at 380 °C to ~75% of area reduction. From the extruded billet, 3-mm-thick sheets were sliced along the extrusion direction (ED) and then subjected to a standard T6 tempering, i.e., solutionized at 550 °C for 1 h, water quenched and subsequently artificially aged at 160 °C for 8 h.

The T6-tempered sheets were friction-stir butt welded using an AccuStir 1004 FSW machine equipped with a welding tool with a 12.5-mm-diameter shoulder and a M5 cylindrical probe; the length of the tool probe was 1.9 mm. Welding trials were conducted under optimized condition involving a spindle rate of 1100 rpm and a feed rate of 760 mm/min [8]. The principal directions of the FSW geometry were denoted throughout as the welding direction (WD), transversal direction (TD) and normal direction (ND). To provide a full-thickness joining, a double-sided FSW was applied in mutually opposite directions.

To recover the mechanical properties of the welded material, the obtained joints were subjected to various post-weld heat treatments. Specifically, some welds were artificially aged at 160 °C for 8 h, whereas other ones were subjected to the T6-tempering in the manner described above.

Microstructural observations were performed mainly by the electron backscatter diffraction (EBSD) technique. The final surface finish was obtained by electro-polishing in a 25 pct. solution of nitric acid in methanol. EBSD analysis was conducted by using a FEI Quanta 600 field-emission-gun scanning electron microscope (FEG-SEM) equipped with TSL OIMTM software. In the EBSD maps shown, the low-angle boundaries (LABs) (2°<θ<15°) and the high-angle boundaries (HABs) (≥15°) were depicted as white and black lines, respectively.

## 3. Results and discussion

### *3.1. Base material*
In the base material, the microstructure was dominated by relatively coarse grains, ~200 μm in length, aligned along the WD and having a developed sub-boundary structure. TEM observations revealed a high density of nanoscale dispersoids. As follows from the scientific literature [11,12], these particles are a mixture of needle-shaped β" precipitates with minor fraction of lath-shaped Q' precipitates. Detailed microstructural characterization of the base material is given elsewhere [8].

### *3.2. Aged joint*
A typical microstructure of the central portion of the stir zone of the obtained aged joint is shown figure 1a. FSW with subsequent aging led to the formation completely recrystallized fine grains with a mean size of 7.5 μm and low fraction of low-angle boundaries. From comparison with other works[1–4,8], this microstructure is typical for FSW/FSP aluminium alloys without post-weld heat treatment or subjected to low-temperature post-weld treatment.

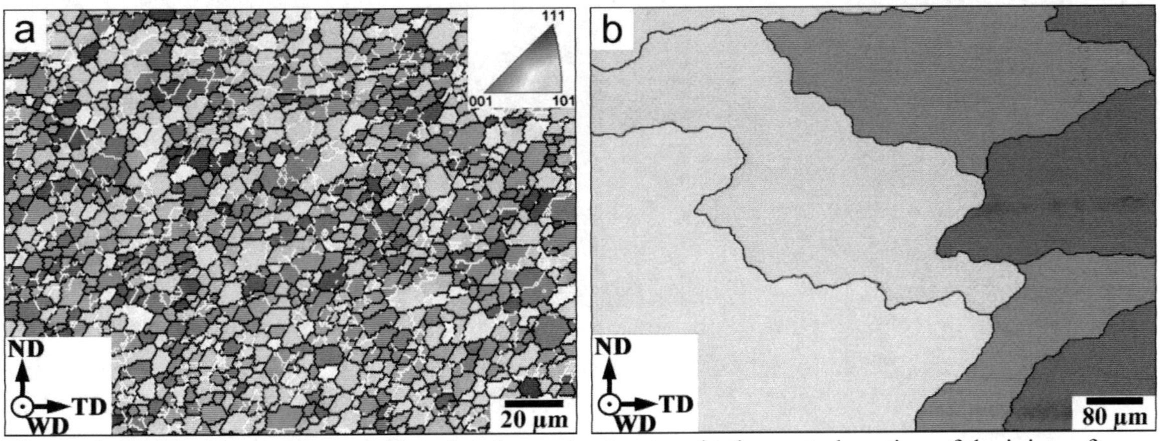


**Figure 1.** EBSD orientation maps showing the grain structure in the central portion of the joints after either post-weld aging (a) or post-weld T6 tempering (b). In the maps, individual grains are colored according to the orientation code shown in the top right corner of (a).

### *3.3. T6-tempered joint*
The post-weld T6 tempering led to abnormal grain growth in the stir zone (figure 1b). This phenomenon is often observed in the heat-treated friction-stir welded/processed aluminum alloys [1,2,8,9]. However, its origin is still unclear. Despite the catastrophic grain growth in the stir zone, the strength of T6-tempered joint was fully recovered [8,9].

IOP Conf. Series: Materials Science and Engineering **672** (2019) 012039 doi:10.1088/1757-899X/672/1/012039

### *3.4. Kinetics of abnormal grain growth at a double-sided FSW*

To examine kinetics of the abnormal grain growth in the T6-tempered welds, the as-weld joints were subjected to a series of annealing at 540 °C for 1, 4 or 10 minutes with subsequent water quenching. After each annealing step, the examined surface was slightly polished to remove the oxide layer for EBSD analysis.

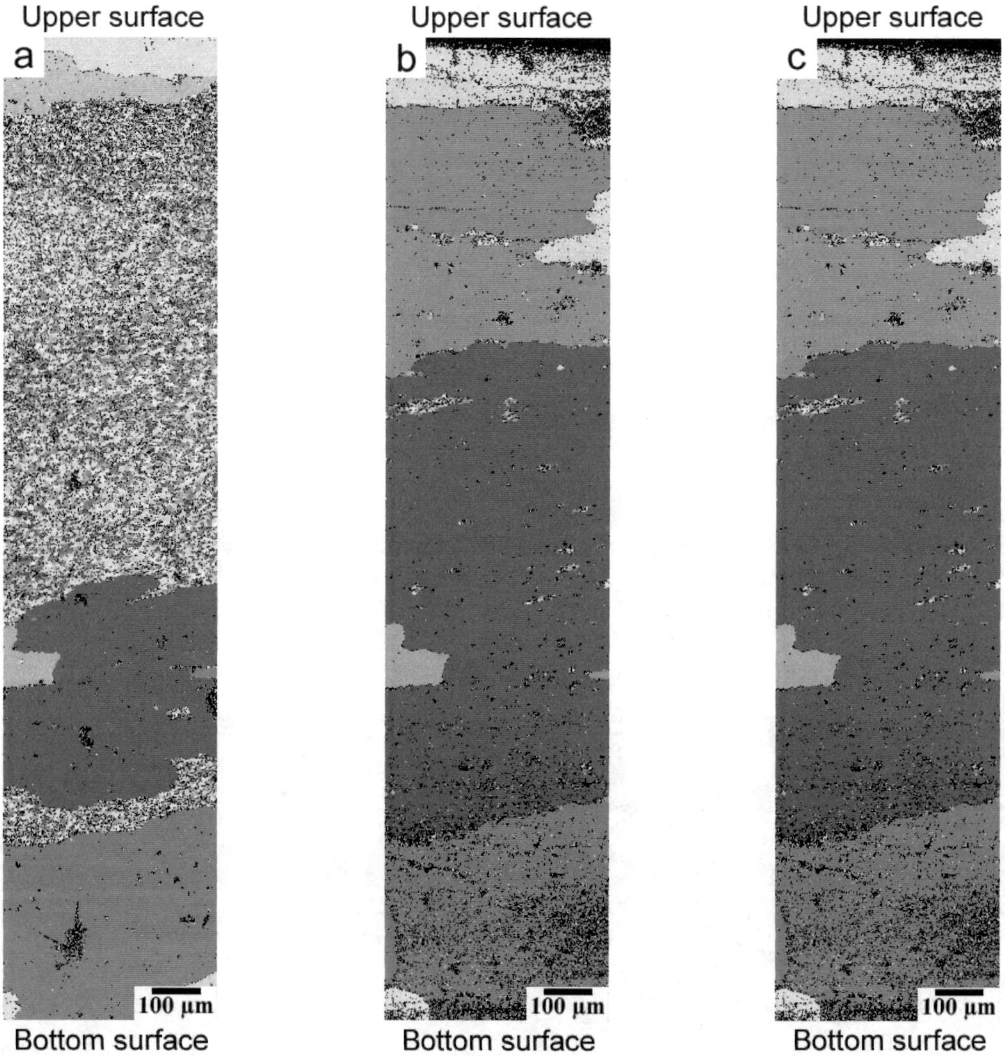


**Figure 2.** Low-magnification EBSD orientation maps in the ND×TD plane of the joint subjected to post-weld annealing at 540 °C for 1 min (a), 4 min (b), or 10 min (c). The surface of the joint at the second pass is marked.

The abnormal grain growth was observed at a very early annealing stage (figure 2a). Remarkably, the abnormally coarse grains were found to preferentially nucleate in the near-surface layer of the stir zone. With further annealing, however, the abnormal grains also evolved in its central section (figure 2b). The catastrophic growth of the latter grains eventually consumed the entire stir zone (figure 2c).

It is evident therefore that the double-sided FSW was not effective for suppression of the abnormal grain growth and thus the initial idea of this work was not correct. Moreover, considering the preferential nucleation of abnormal grains in the shoulder-affective region, the double-sided FSW appears to be even deteriorating for the thermal stability of the welded material.

## 4. Conclusions

In this work, the stability of the double-sided FSWed material against the abnormal grain growth was studied. It was found that the abnormally-coarse grains preferentially nucleated in the shoulder-affected region of the stir zone. Therefore, the thermal stability of the double-sided FSWed material was concluded to be relatively low.

**Acknowlegements**
This study was financially supported by the Russian Science Foundation, grant No. 19-49-02001. The authors are grateful to the staff of the Joint Research Center, "Technology and Materials" Belgorod State National Research University for their assistance with the mechanical and structural characterizations.

## References

[1] Threadgill P L, Leonard A J, Shercliff H R and Withers P J 2009 *Int. Mater. Rev.* **54** 49
[2] Mishra R S, De P S and Kumar N 2014 *Friction Stir Welding and Processing* (Cham: Springer)
[3] Cho J H, Han S H and Lee C G 2016 *Mater. Lett.* **180** 157
[4] Yang C, Ni D R, Xue P, Xiao B L, Wang W, Wang K S and Ma Z Y 2018 *Mater. Charact.* **145** 20
[5] García-Bernal M A, Mishra R S, Verma R and Hernández-Silva D 2015 *Mater. Sci. Eng.* A **636** 326
[6] Mironov S Y 2015 *Met. Sci. Heat Treat.* **57** 40
[7] Baghdadi A H, Rajabi A, Selamat N F M, Sajuri Z and Omar M Z 2019 *Mater. Sci. Eng.* A **754** 728
[8] Malopheyev S, Vysotskiy I, Kulitskiy V, Mironov S and Kaibyshev R 2016 *Mater. Sci. Eng.* A **662** 136
[9] Vysotskiy I, Malopheyev S, Mironov S and Kaibyshev R 2018 *Mater. Sci. Eng.* A **733** 39
[10] Safarkhanian M A, Goodarzi M and Boutorabi S M A 2009 *J. Mater. Sci.* **44** 5452
[11] Esmaeili S, Wang X, Lloyd D J and Poole W J 2003 *Metall. Mater. Trans.* A **34** 751
[12] Wang X, Esmaeili S and Lloyd D J 2006 *Metall. Mater. Trans.* A **37** 2691

IOP Conf. Series: Materials Science and Engineering **672** (2019) 012040 doi:10.1088/1757-899X/672/1/012040

# Friction-stir welding of Al-Mg-Mn-Zr sheets by using a hemispherical tool

**I Vysotskiy, S Malopheyev, S Mironov and R Kaibyshev**

Belgorod State University, Pobeda 85, Belgorod 308015, Russia

E-mail: malofeev@bsu.edu.ru

**Abstract**. A new design of the welding tool for friction-stir welding (FSW) was proposed. The main idea was in using of a hemispherical tool probe. Such a tool is very simple (and cost-effective) in fabrication and provides excellent mixing of the material during FSW, thus avoiding welding defects. In this work, the feasibility of this tool for joining of Al-Mg-Mn-Zr alloy and its effect on the microstructure and mechanical properties of the produced welds were studied.

## 1. Introduction

Friction-stir welding (FSW) is a relatively simple but very effective technology for solid-state joining of structural materials [1–6]. One of the key issues of FSW is the welding tool [1–6]. The tool provides material mixing during the welding process, and, thus governs the microstructure and service properties of the produced welds. To enhance the mixing process, tools of very complex design are sometimes used [3,5]. This facilitates the material flow, but increases the production cost. In the present work, however, an alternative approach in the tool designing was used. Based on the idea that the material flow during FSW is mainly disturbed by abrupt changes in the tool geometry, a spherical design of the tool was suggested to promote very smooth mixing of the material. On the other hand, tools of this design should be relatively simple (and thus cost-effective) in production. Therefore, a welding tool with a hemispherical tool probe was developed. In the previous works, the excellent welding performance of such a tool were shown during FSW of a TWIP-steel [7] and a high-entropy alloy [8]. The purpose of the present study was examination of feasibility of such a tool for joining of the Al-Mg-Mn-Zr alloy and its effect on the microstructure and mechanical properties of the produced welds.

## 2. Material and experimental procedure

The design of the developed tool is shown in figure 1a. Briefly, the tool consisted of a conventional concave-shaped shoulder of 12.5 mm in diameter, and a hemispherical-shaped probe of 2 mm in radius. The tool was fabricated from a W-Co-based alloy. This material is well recommended for joining of relatively high-strength materials, including steels [7-9] and titanium-based alloys.

The program material, i.e. a commercial AA5083 alloy with the chemical composition Al–5.4% Mg–0.5% Mn–0.1% Zr–0.12% Si–0.014% Fe (wt.%), was fabricated by direct chill casting and subsequent homogenization at 360 °C (633 K) for 6 h [10–12]. The homogenized material was subjected to equal-channel angular pressing (ECAP) at ~300 °C and a pressing rate of 3 mm/s to an accumulated true strain of ~12 by using the BCZ processing route. For this purpose, an isothermal die with a rectangular cross-section and an intersection angle of 90° was used. To equilibrate the processing temperature during ECAP, the material was preheated for ~120 min prior to the first pass, whereas the preheating interval between subsequent passes was ~ 5 min. Following the final ECAP pass, the processed material was quenched in water. From the ECAPed material, sheets 10 mm thick were sliced and then rolled parallel to the ECAP direction at ambient temperature to a total thickness reduction of ~80 pct. This was achieved by multiple rolling passes, ~10 pct. of strain each. Following rolling, the produced sheets were annealed at 300 °C for 1 h. This rolled material was defined as the base material.

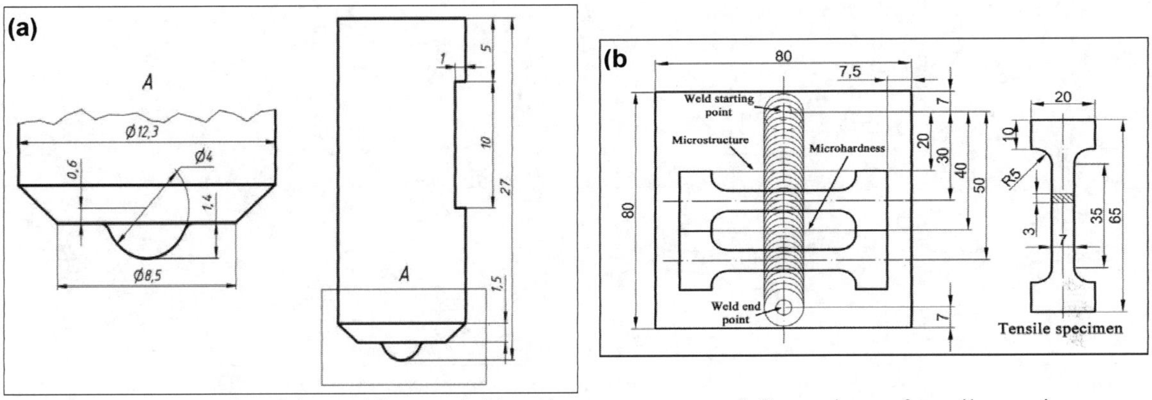


**Figure 1**. A schematic of the welding tool (a) and arrangement and dimensions of tensile specimens (b).

The obtained 2-mm-thick sheets were butt-welded parallel to the rolling direction using an AccuStir 1004 FSW machine. FSW was applied at a tool rotation of 500 rpm, a tool travel speed of 150 mm/min, and a tool tilting angle of 2.5°. The principal directions of the FSW geometry were denoted as the welding direction (WD), transversal direction (TD) and normal direction (ND). The obtained friction-stir welded material was defined as a joint.

Microstructural observations were performed mainly by electron backscatter diffraction (EBSD) technique and optical microscopy. The final surface finish was obtained by electro-polishing in a 25 pct. solution of nitric acid in methanol. EBSD analysis was conducted by using a FEI Quanta 600 field-emission-gun scanning electron microscope (FEG-SEM) equipped with TSL OIMTM software. In the EBSD maps shown, the low-angle boundaries (LABs) (2°<θ<15°) and the high-angle boundaries (HABs) (≥15°) were depicted as red and black lines, respectively.

To examine the mechanical properties, axial tensile tests to failure were applied. The tensile specimens of the joint were centered at the weld centerline and included all characteristic microstructural zones of FSW. The dimensions of the specimens are shown in figure 1b. The tensile tests were conducted at ambient temperature and a constant cross-head velocity corresponding to a nominal strain rate of 10-3 s-1 by using an Instron 5882 universal testing machine. Other experimental details were reported elsewhere [10,12].

## 3. Results and discussion

### *3.1. Microstructure*

A typical appearance of the microstructure of the base material in the ND×RD and ND×TD planes is shown in figure 2. The microstructure could be described in terms of a mixture of relatively-coarse pancake-shaped grains and fine equiaxed grains, thus being somewhat bimodal. Nevertheless, the mean grain intercept length was measured and found to be ~10 μm in the RD and ~6 μm along the ND.

Low-magnification optical observations of the transversal cross-section of the produced joint revealed no volumetric defects, thus confirming excellent performance of the developed welding tool. A typical microstructure evolved in the center of the stir zone is shown in figure 3. From a comparison with figure 2 it is clear that FSW led to significant microstructural changes. Specifically, the stir-zone microstructure was dominated by completely recrystallized fine grains with a mean size of ~1.4 μm and a fraction of HABs as high as 88% (figure 3).

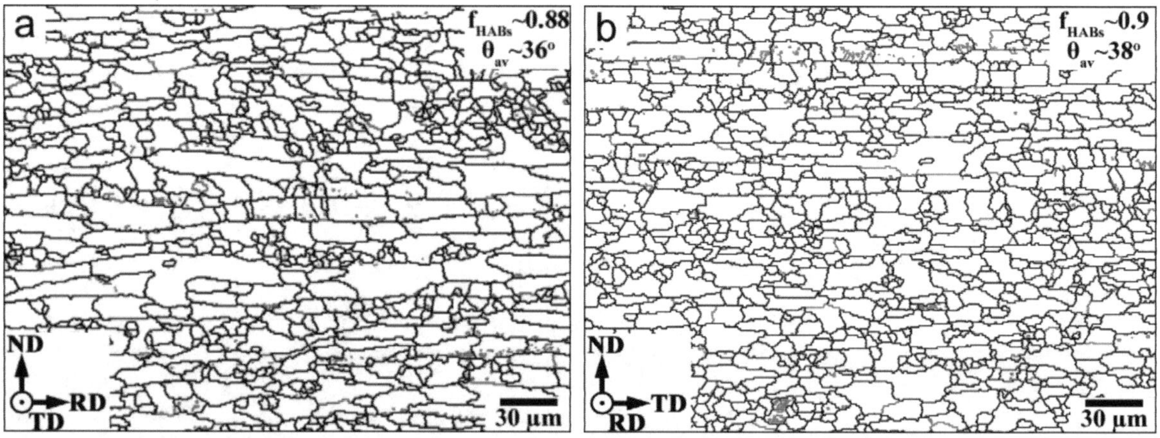


**Figure 2.** EBSD maps showing typical microstructure of *base material* in the ND×RD plane (a) and the ND×TD plane. The $f_{HABs}$ and $\theta_{av}$ in the insets in the top right corner indicate the HABs fraction and the average misorientation, respectively.

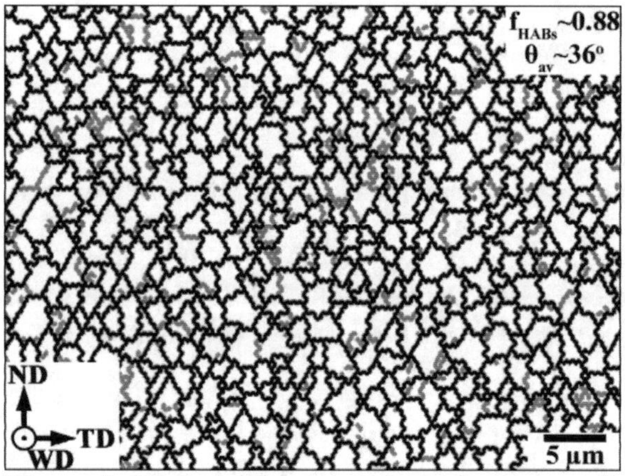


**Figure 3.** Typical microstructure in the center of the *stir zone*. The $f_{HABs}$ and $\theta_{av}$ in the insets in the top right corner are the HABs fraction and the average misorientation, respectively.

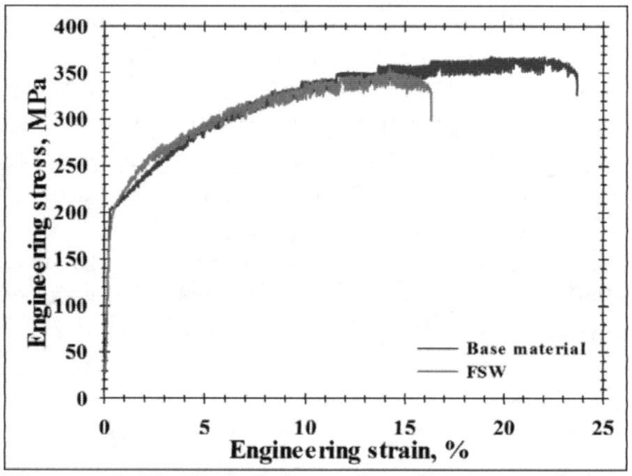


**Figure 4.** Typical deformation diagrams of the base material and the joint recorded during transverse tensile tests.

**Table 1.** Mechanical properties of the *base material* and the *joints*.

| Material condition | Yield strength, MPa | Ultimate tensile strength, MPa | Ductility, % | Joint efficiency for UTS, % | Failure location |
|---|---|---|---|---|---|
| Base material | 205 | 370 | 23 | - | - |
| FSW | 205 | 355 | 16 | 96 | Heat-affected zone |

### *3.2. Tensile properties*

Typical deformation diagrams recorded during the transversal tensile tests of the base material and the joint are shown in figure 4. The relevant results of the tests are summarized in table 1. A characteristic feature of all deformation diagrams was the repeating oscillations indicating the serrated character of the material flow. This phenomenon is characteristic of Al-Mg alloys and is known as the Portevin-Le Chatelier (PLC) effect [10,13].

The produced joint demonstrated excellent mechanical properties. Specifically, the joint efficiency for the yield strength was 100% whereas that for the ultimate tensile strength was to 96% (figure 4, table 1). On the other hand, the joint exhibited relatively low ductility. This effect was associated with strain localization in the softest microstructural region of the weld, i.e. heat-affected zone, which led to premature failure of the weld.

## 4. Conclusions

In this work, the performance of a newly-developed welding tool with a hemispherical probe was examined. To this end, this tool was used for FSW of an Al-Mg-Mn-Zr alloy and its effect on the microstructure and mechanical properties was studied. The produced joint had no volumetric defects, and the efficiency of the joint in terms of strength characteristics was close to 100%. The microstructure generated in the stir zone was dominated by completely recrystallized fine grains with a mean size of ~1.4 μm and a high HAB fraction.

**Acknowlegements**

This study was financially supported by the Russian Science Foundation, grant No. 18-79-10174. The authors are grateful to the staff of the Joint Research Center, "Technology and Materials" Belgorod State National Research University for their assistance with the mechanical and structural characterizations.

## References

[1] Threadgill P L, Leonard A J, Shercliff H R and Withers P J 2009 *Int. Mater. Rev.* **54** 49
[2] Aissani M, Gachi S, Boubenider F and Benkedda Y 2010 *Mater. Manuf. Process.* **25** 1199
[3] Palanivel R, Koshy Mathews P, Murugan N and Dinaharan I 2012 *Mater. Des.* **40** 7
[4] García-Bernal M A, Mishra R S, Verma R and Hernández-Silva D 2016 *Mater. Sci. Eng.* A **670** 9
[5] Elangovan K and Balasubramanian V 2008 *Mater. Des.* **29** 362
[6] Kumar K, Kailas S V, Srivatsan T S 2011 *Mater. Manuf. Process.* **26** 915
[7] Torganchuk V, Vysotskiy I, Malopheyev S, Mironov S and Kaibyshev R 2019 *Mater. Sci. Eng.* A **746** 248
[8] Shaysultanov D, Stepanov N, Malopheyev S, Vysotskiy I, Sanin V, Mironov S, Kaibyshev R, Salishchev G and Zherebtsov S 2018 *Mater. Charact.* **145** 353
[9] El-Batahgy AM, Miura T, Ueji R and Fujii H 2016 *Mater. Sci. Eng.* A **651** 904
[10] Malopheyev S, Kaibyshev R 2015 *Mater. Sci. Eng.* A **620** 246
[11] Nikulin I, Kipelova A, Malopheyev S and Kaibyshev R 2012 *Acta Mater.* **60** 487
[12] Malopheyev S, Vysotskiy I, Mironov S, Zhemchuzhnikova D and Kaibyshev R 2018 *AIP Conf. Proc.* **2051** 020186
[13] Lebyodkin M A, Zhemchuzhnikova D A, Lebedkina T A and Aifantis E C 2019 *Results Phys.* **12** 867

IOP Conf. Series: Materials Science and Engineering **672** (2019) 012041 doi:10.1088/1757-899X/672/1/012041

# Structure and strength of the 1570C aluminum alloy after complex SPD processing

**M V Markushev[1], E V Avtokratova[1], O Sh Sitdikov[1], M L Linderov[2], D L Merson[2] and A Vinogradov[3]**

[1] Institute for Metals Superplasticity Problems, Khalturin st. 39, Ufa 450001, Russia
[2] Institute of Advanced Technologies, Togliatti State University, Togliatti, Russia
[3] Department of Mechanical and Industrial Engineering, Norwegian University of Science and Technology-NTNU, N-7491 Trondheim, Norway

E-mail: mvmark@imsp.ru

**Abstract.** The effect of severe plastic deformation (SPD), realized via multi-directional forging (MDF) at a constant temperature of 325 °C, as well as via a combination of MDF with subsequent room temperature rolling or high pressure torsion, on the microstructure, static and fatigue strength parameters of an ingot of non-age-hardenable commercial aluminum alloy 1570C (Al-5Mg-0.18Mn-0.2Sc-0.08Zr) was studied. It is found that the most effective treatment to enhance the balance of the alloy properties involves a sequence of hot MDF and cold rolling due to the development of a work hardened ultrafine-grain structure with uniformly distributed nanosized aluminides of transition metals. The role of precipitates in the alloy structure- and property-formation is discussed.

## 1. Introduction

Non-age-hardenable Al-Mg alloys with complex additions of transition metals (TM) belong to the class of advanced structural materials with a high potential for a wide range of industrial applications due to the unique balance of their service properties [1-3]. The behavior of the alloys is conventionally conditioned by the chemical (phase) composition, the parameters of the cast microstructure and their changes during thermomechanical processing (TMP). Commercial TMPs are primarily aimed at the control of the dislocation structures, which limits the fabrication of annealed and work-hardened semi-products only. For going beyond these limits, the modern strategies are also aimed at tailoring the grain size distributions and the grain boundary structure - altogether combined under the term "structuring". The latter is frequently called "deformation (nano)structuring" referring to a large group of processing methods and schemes, involving severe plastic deformation (SPD) to strains e>>1 as a pivotal component capable of imparting the nanosized phase components into a wide range of structural materials [4,5]. When applied to aluminum alloys, the SPD approach requires the sufficient content of TMs and dispersion of secondary TM aluminides, which stabilize the dynamically and/or statically recrystallized/polygonized microstructures formed during processing [6].

It has been well documented, that the most capable grain refinement technique is high pressure torsion (HPT), which is usually performed at ambient temperature to produce the crystallite size of the nanometer scale in the small disc-shape samples of various metals and alloys [4,5,7]. Meanwhile, for the efficient processing of the ultrafine grain (UFG) bulk billets with a grain size less than few micrometers, multi-directional forging (MDF) is extremely attractive among other SPD techniques due to its excellent scalability and cost-effectiveness [7].

Owing to the Hall-Petch type relations between the strength and grain size [5,6], the deformation nanostructuring increases the strength of pure metals and solid solutions. However, the significance of SPD is less obvious for the complex alloyed systems due to a number of simultaneously affected structural factors. For instance, the loss in their dispersion hardening under TMP can be much higher than the positive effects gained from work hardening or grain size strengthening. Thus, the point is to develop the TMP effectively exploring most of the known strengthening mechanisms in a single route. The present study is aimed at comparative analysis and assessment of the efficiency of TMP based on MDF and its combinations with conventional cold rolling (CR) and HPT to enhance the static and cyclic performance of the structural Al-Mg-TM alloys at room temperature via their nanostructuring and strengthening during SPD processing.

## 2. Material and procedure

A commercial ingot of the 1570C alloy (Al-5Mg-0.18Mn-0.20Sc-0.08Zr, wt.%) was solution treated at 360 °C for 6 hours and furnace-cooled. The rod billets of ∅80×150 mm were then MDF processed in the isothermal conditions for 12 cycles at 325 °C with a total equivalent strain of $e$ ~12. Cold rolling with a total reduction of ~80% was performed at room temperature on plate shape billets of 10 mm thick. HPT was also performed at room temperature with 10 revolutions and a pressure of 6 GPa applied to the disk-shape samples ∅20×2 mm cut from the MDFed billets.

The microstructure of the alloy was investigated by conventional methods of optical, transmission and scanning electron microscopy (OM, TEM and SEM) and by X-ray analysis. Most parameters of the microstructure (grain size, fractions of the low- and high-angle boundaries (LABs and HABs), mean misorientation angle) were determined by the SEM-EBSD analysis. The size and number fraction of secondary phases were measured using the computerized analysis of TEM images with not less than 2000 precipitates. The microhardness of the alloy was determined at 10 sec loading with 1N. The static and cyclic (dynamic) tension (fatigue ratio R = 0.1, and frequency ν = 60 Hz) tests were performed at ambient temperature of the flat mechanically pre-polished dog-bone and corset samples with a gauge part of 1.5×3×6 $mm^3$ and a minimum cross-section of 1×0.9 $mm^2$, respectively.

## 3. Results and discussions

The TEM analysis showed that the abnormally supersaturated by TM aluminum solid solution was decomposed during the homogenization annealing, and $Al_3(Sc, Zr)$ compact precipitates were formed with a mean diameter of 10-15 nm and a volume density of $1\times10^4$ $\mu m^{-3}$. Owing to the delta-zero contrast, most of them were found to be coherent to the matrix. Their size distribution was unimodal, and the volume distribution was quite uniform with the frequently observed grain boundary precipitate free zones (PFZs). The alloy matrix is constructed by equiaxed grains with a mean size of 25 μm. The grain boundary misorientation spectrum reveals the prevalence of high-angle boundaries (table 1).

It has been shown that the chief process of the grain refinement of the alloy under MDF is the continuous dynamic recrystallization [3]. New fine subgrains and grains, surrounded by LABs and HABs, were formed in the vicinity of the initial grain boundaries with PFZs volumes involved. An increase in the strain resulted in the rise of the crystallites misorientations and the gradual replacement of the coarse-grain microstructure by the fine-grain one (figure 1a). One can notice that even after 12 forging cycles, the microstructure was still partially recrystallized with near 2 μm subgrains inside the fragments of the initial grains surrounded by slightly coarser new grains.

Under HPT, the fine-grain structure was transformed into a typical non-equilibrium nanofragment one comprising of nearly equiaxed crystallites of 150 nm in diameter, which were surrounded predominantly by LABs and characterized by the highly increased dislocation density and the lower lattice parameter (table 1 and figure 1). In spite of the specific TEM contrast from the work hardened matrix, the TM aluminides were easy to observe in the majority of nanofragments. Meanwhile, the decomposition of the aluminum solid solution with the β-phase precipitation was not detected. Thus, it

**Table 1.** Structure parameters of the alloy 1570C.

| Condition | Grain/Subgrain size, μm | Mean misorientation angle, degree | Fraction of HABs, % | Lattice parameter, Å | Dislocation density**, $m^{-2}$ |
|---|---|---|---|---|---|
| Initial | 25/- | 40 | 90 | - | - |
| MDF | 2.2/2.0 | 29 | 67 | 4.073±0.001 | $5.0\times10^{12}$ |
| MDF + HPT | -/0.15* | - | - | 4.069±0.003 | $5.1\times10^{14}$ |
| MDF + CR | 2.4/0.4* | 14 | 27 | - | - |

* derived from TEM analysis, ** X-ray analysis

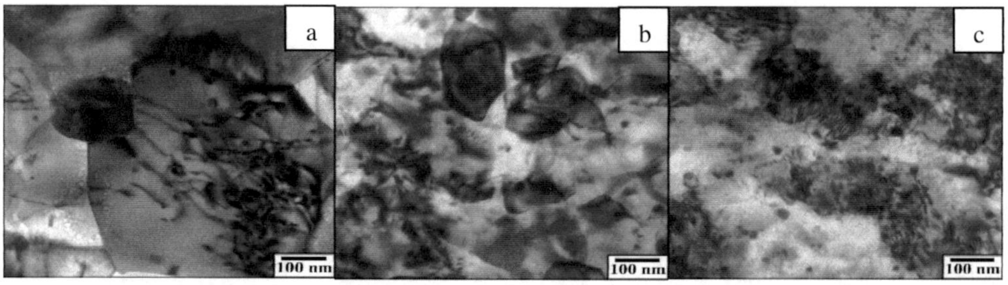


**Figure 1.** 1570C alloy TEM structures processed via MDF (a), MDF+HPT (b) and MDF+CR (c).

could be concluded that the decrease in the lattice parameter under HPT is mainly caused by the changes in amounts of crystal defects, and not by the level of solid solution alloying.

The effect of cold rolling on the microstructure of the matrix of the MDF processed alloy was somewhat less pronounced than that of HPT. The fine-grain structure obtained by MDF was also transformed into the severely hardened one with the high dislocation density and non-equilibrium grain boundaries (table 1 and figure 1). However, the cellular character of the structure with the cell size of about three times larger than the fragment size in the HPT-ed specimens, i.e. quite far beyond the nano-range, was preferably noticed. The TM aluminides were also detected and near homogeneously distributed in the aluminum matrix as in the HPT processed condition.

The mechanical tests reveal good correlation between the microstructures discussed above and the alloy static and dynamic strength parameters presented in table 2. The work-hardening effect originating from the increasing dislocation density is considered to be the main factor controlling the mechanical behavior of the alloy. This conclusion follows from several consistent observations highlighting, in particular, the insignificance of the grain size effect. In particular, the high temperature MDF did not result in any alloy strengthening despite the remarkable grain refinement (table 1). This finding can be only explained as a result of compromised grain boundary strengthening and the alloy softening due to the reduced dispersion strengthening because of the coagulation of TM aluminides under MDF. Besides, the alloy behavior after MDF is also conditioned by the formation of near completely recrystallized equilibrium microstructure where the work hardening effect is absent, and, consequently, the lowest static and fatigue strength is observed along with the largest ductility. Further rolling at ambient temperature was accompanied by a noticeable increase in the strength of the alloy. Particularly, the yield stress increased remarkably while the ductility reduced by a factor of two or so (table 2). Nevertheless, the elongation to failure remains reasonably high (of 18%), which is worth noting for the significantly strengthened material. The increasing monotonic strength results in the concomitant increase of the fatigue life in the high-cycle regime. In line with common trends, such a behavior was reasonably anticipated from SPD processing leading to the UFG microstructure [5,8]. A strong enhancement of the hardness and static strength was also reasonably expected for the MDF + HPT processed alloy due to the extreme grain refinement. However, such a pronounced effect was found below the yield point owing to the nullified ductility. The alloy brittleness accounts for the

**Table 2.** Room temperature mechanical properties of the alloy 1570C.

| Condition | Vickers Hardness, HV | Yield Stress, YS, MPa | Ultimate Tensile Strength, UTS, MPa | Elongation to failure, El, % | Fatigue limit, $\sigma_F$, MPa[a] |
|---|---|---|---|---|---|
| Initial | 110 | 240 | 355 | 28 | 210 |
| MDF | 105 | 235 | 360 | 38 | 230 |
| MDF + CR | 155 | 510 | 570 | 18 | 290 |
| MDF + HPT | 240 | - | 930 | - | 260 |

[a] N= $4\times10^6$ cycles, R=0.1

fatigue limit reduction (table 2), despite the increasing static strength. Since the fatigue is a substantially local phenomenon, and the fatigue limit is governed by the "weakest link" in the microstructure, this is most likely associated with the accumulation of defects in the course of cold SPD, leading to the local increase in the internal stresses, which might be critical for premature fatigue crack initiation.

The observed mechanical behavior permits us concluding that the excellent balance of the static and dynamic properties is obtained after TMP combining hot forging and cold rolling. After this treatment, the alloy exhibits a uniform work-hardened (ultra)fine-grain microstructure with homogeneously distributed nanosized TM aluminides, which is favorable for fatigue performance. It should be also noted that the obtained mechanical properties are outstanding, considering the strength of the alloy comparable with that of high-strength, age-hardenable alloys of the 7xxx series, while their ductility and fatigue resistance are remarkably enhanced.

## 4. Conclusions

The MDF processing of the alloy 1570C at 325 °C up to a total equivalent strain of $e$ ~12 led to a significant grain refinement with a grain size reduction by an order of magnitude down to the UFG range. The grain size reduction was found favorable for ductility, however the Hall-Petch strengthening was simultaneously compensated by softening of the materials resulted from a decrease in the disperse strengthening due to the TM aluminides coagulation. The TMP, involving hot MDF followed by room temperature straining, led to remarkable strengthening of the alloy, which was primarily caused by work-hardening and associated with the dislocation accumulation and the evolution of dislocation structures. The TMP comprising MDF and cold rolling is found preferable for the enhancement of both the static and dynamic properties of the alloy. The observed effect is explained by the formation of a uniform UFG structure with retained strain hardening ability and ductility, which appears to be in sharp contrast with the severely work hardened and brittle nanostructure obtained by hot MDF and cold HPT.

## Acknowledgments

Authors wishing to acknowledge the Russian Science Foundation for the financial support (grant №16-19-10152-P).

## References

[1] Filatov Y, Elagin V and Zakharov V 2000 *Mater. Sci. Eng.* A **280** 97
[2] Avtokratova E, Sitdikov O, Markushev M and Mulyukov R 2012 *Mater. Sci. Eng.* A **538** 386
[3] Sitdikov O *et al.* 2018 *J. Alloys Compd.* **746** 520
[4] Valiev R and Langdon T 2006 *Prog. Mater. Sci.* **51** 881
[5] Vinogradov A and Estrin Y 2013 *Acta Mater.* **61** 782
[6] Avtokratova E *et al.* 2016 *J. Alloys Compd.* **673** 182
[7] Markushev M 2011 *Letters on Mater.* **1** 36
[8] Vinogradov A 2007 *J. Materials Science* **42** 1797

IOP Conf. Series: Materials Science and Engineering **672** (2019) 012042 doi:10.1088/1757-899X/672/1/012042

# Effect of preliminary and post-SPD aging on nanostructuring and strengthening of the HPT processed 2024 aluminum alloy

**M V Markushev, S V Krymskiy, E V Avtokratova and O Sh Sitdikov**

Institute for Metals Superplasticity Problems, 39 Khalturin St., Ufa 450001, Russia

E-mail: mvmark@imsp.ru

**Abstract.** Structural and mechanical behaviors of the 2024 commercial aluminum alloy, differed by heterogeneity of the initial structure, were examined after severe plastic deformation and further annealing. The superior strengthening of the pre-quenched alloy under high-pressure torsion (HPT) at room temperature was caused by the development of a more equiaxed nanostructure of the matrix with a lower fragment size than in pre-aged T6 condition. Post-HPT annealing at 100 °C led to extra strengthening of the pre-quenched alloy only due to the age hardening effect. The nature of the alloy behavior is discussed.

## 1. Introduction

Recent data have shown high potential for improving the service properties of metallic materials due to novel approaches to their structure/phase control through imparting the size less than 100 nm to the main structural components. Such nanostructuring is of particular interest for commercial materials owing to intense studies of the phenomenon of their cold severe plastic deformation (SPD) to effective strains of $e>1$ [1]. To date, a wide range of metals and alloys has been nanostructured and several industrial applications have been announced. However, the nature and structural factors responsible for the effective formation of nanostructures of different types with a guaranteed balance of properties are still not so clear. As for age-hardenable aluminum alloys, these points are mainly controlled by the structure of the matrix and of the secondary phases. The effect of the latter is quite ambiguous and less studied. For example, it was established [2,3] that the preliminary formation and/or increase in the volume fraction of disperse phases resulted in processing of a greater fraction of nanograins of a lower size under SPD. In contrast, it was shown in [4,5] that the formation of nanosized precipitates of high densities in high-strength alloys can completely suppress recrystallization of the matrix.

One of the topical issues for the majority of wrought Al alloys is the estimation of effectiveness of post-SPD annealing in property control and aging response of the SPDed alloys. Along with the structural strengthening owing to the formation of new grains and subgrains, and an increase in the density of dislocations due to SPD, the mechanical behavior of the alloy with further annealing will be also determined by the balance of the disperse hardening and aluminum solid solution strengthening. Though the contribution of the latter to the strength of the SPDed materials is comparatively small, alloying elements should provide noticeable post-deformation hardening of the matrix with the formation of nanosized precipitates owing to decomposition of the supersaturated Al solid solution obtained under preliminary quenching or during SPD. Besides, the strengthening effect of precipitates in the severely deformed matrix should depend on their morphology as may occur in a non-conventional sequence of phase transformations [6]. Therefore, the development of the optimum annealing regimes to realize the aging effect in the SPDed alloy is a rather important task, which requires an understanding of the nature of its structuring and strengthening. The aim of the study is to analyze the effects of preliminary and post-SPD aging on the structure and strength of the widely used commercial aluminum alloy subjected to cold high-pressure torsion (HPT).

## 2. Material and procedure

A hot-pressed rod of the 2024 alloy of a standard chemical composition (Al-4.2Cu-1.3Mg-0.5Mn, wt.%) was used as the starting material. Disk-shaped samples with a diameter of 20 mm and a thickness of 2 mm, cut out from the rod across its axis, were solution treated at 505 °C and quenched in water. Some of them were then aged at 190 °C for 10 hrs (T6 temper). SPD of the quenched (Q) and artificially aged (Q+AA) samples was realized through the HPT with 10 revolutions under a pressure of 6 GPa at room temperature. Post-SPD annealing was performed at 100 °C basing on the data of the previous study [7], showing that annealing at higher temperatures led to intense grain and precipitate coarsening and strong softening of the alloy. The microstructure was analyzed using transmission electron microscopy (TEM). The objects were prepared by electropolishing at -28 °C in 20% solution of $HNO_3$ in $CH_3OH$. Microhardness was determined at a load of 1N. Tensile tests were performed on an Instron 5982 at room temperature on samples with a gage part of $1 \times 1 \times 3$ mm$^3$, and the axis coincided with the line tangent to the circle with half the radius of the deformed discs.

## 3. Results and discussion

TEM analysis showed that the starting material in the Q condition had a coarse-fibered microstructure with homogeneously distributed compact Mn-rich T-phase ($Al_{20}Cu_2Mn_3$) precipitates of 50-200 nm in size inside the fibers (figure 1a). Further artificial aging (Q+AA state) resulted in the decomposition of preliminary supersaturated by Mg and Cu aluminum solid solution, with the formation of the main strengthening needle-like (lamellar) S-phase precipitates (figure 1e).

Due to HPT, the well-developed, non-equilibrium nanostructure of the Al matrix with an average size of near equiaxed crystallites (grains and subgrains) of about 75 nm was formed in the as-quenched alloy (figure 1 b,f). A visible difference in the structure was found in the SPDed artificially aged alloy (figure 1 c,g): it was consisted of predominantly elongated cells (fragments) ~200 nm in size. Judging by the character of both structures, one can conclude that continuous dynamic recrystallization played a dominant role in their formation, and the precipitates had quite a strong effect on its occurrence. The nature of this effect was reported in [4,5]: an increase in the number density of precipitates by one or two orders of magnitude, involving dispersoids and main strengthening phases, could lead to a strong suppression of the nanostructuring of the Al matrix via fragmentation and/or recrystallization, owing to increasing the homogeneity of plastic flow and preventing dislocation rearrangements. As a result, the dislocation pile-ups were formed instead of fragments, and the SPD structure in the pre-aged alloy was less misoriented, having a more uniform distribution of dislocations, cells, and dislocation boundaries and a lower fraction of nanocrystallites. Further post-HPT annealing did not lead to visible grain coarsening (figure 1d) amid the activation of two main processes – recovery of the structure of the matrix and decomposition of the Al solid solution.

The microstructural changes caused by the processing had a strong effect on the mechanical behavior of the alloy. Thus, the hardness of the alloy in the solid-solution hardened Q state was 125 HV and increased to ~170 HV after the T6 treatment because of the precipitation hardening induced by the formation and growth of zones and metastable S phases. After HPT, the hardness in both alloy conditions increased to practically the same value of ~270 HV, testifying a strong dependence of the SPD strengthening on the initial structural heterogeneity of the alloy (figure 2). That is, the strain hardening of the pre-quenched alloy was more intense, and the strength of the preliminarily aged alloy was close to the limit reached under SPD in the Q alloy due to the formation of a more developed nanostructure. Such behavior cannot be explained from a common point of view based on the well-known structural effects, involving the Hall-Petch strengthening and dispersion hardening. This behavior is most likely caused by the interaction of dislocations with the clusters containing Cu and

IOP Conf. Series: Materials Science and Engineering **672** (2019) 012042 doi:10.1088/1757-899X/672/1/012042

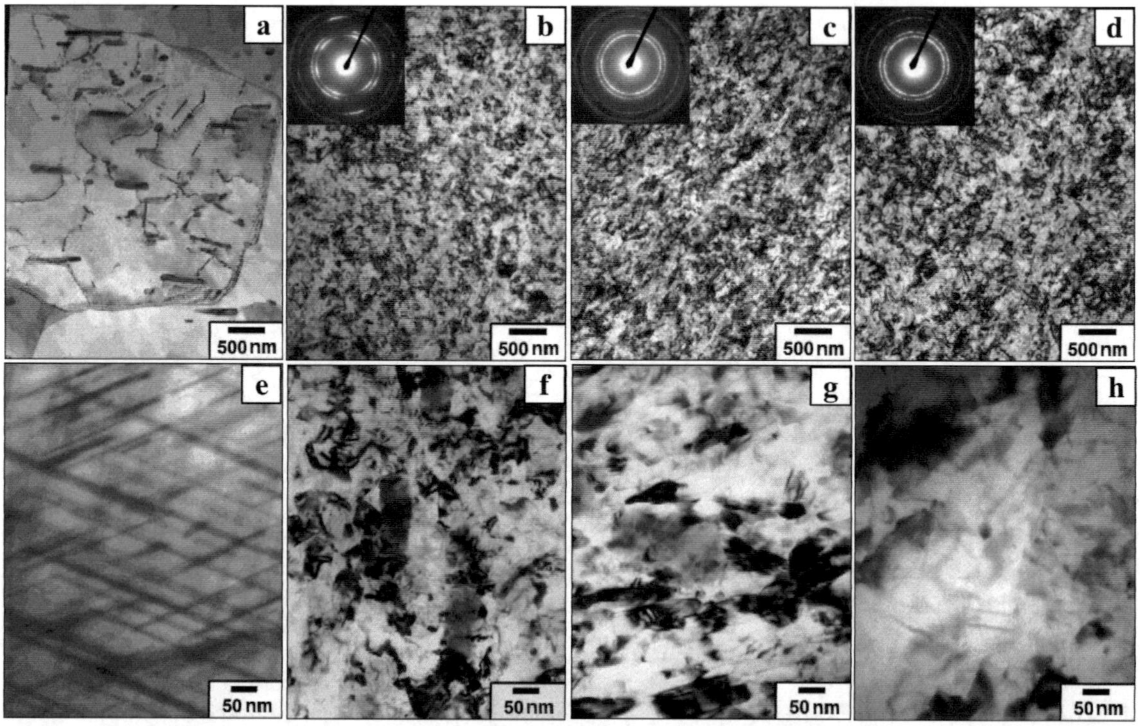


**Figure 1.** TEM structures of the 2024 alloy in the pre-quenched (a,b,f,d,h) and further aged at 190 °C for 10 hrs (c,e,g), before (a,e) and after HPT (b,c,f,g), and post-HPT aging at 100 °C, 100 hrs (d,h).

Mg atoms, namely, the fragments of zones and metastable phases that were formed during SPD. This point of view was advanced and grounded in [8] and our results support it.

The same situation was found in tensile tests: the yield stress and ultimate tensile strength, measured for both initial conditions after HPT, were also close (table 1). Moreover, their ductility values were also equal. It is also necessary to note that the tensile strength of the SPDed alloy were twice as high as those of the age-hardened initial rod, and also those after the "removal" of the SPD hardening effect by further T6 heat treatment. However, in the latter case, the elongation to failure was an order of magnitude higher than in the SPDed alloy and twice higher than in the rod. This effect was caused by the homogenization of the spatial distribution and refinement of the excess phases under HPT [9].

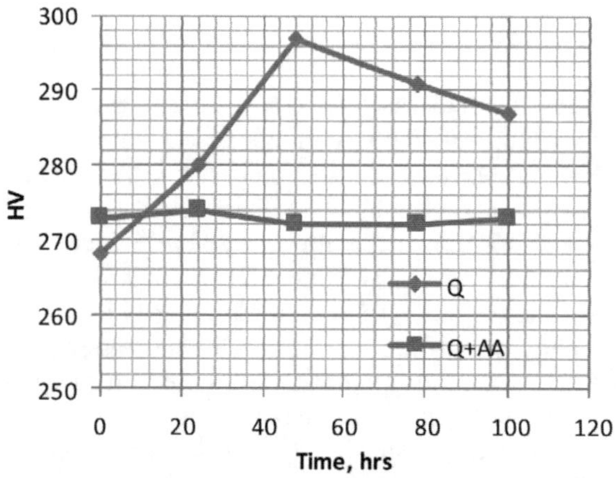


**Figure 2.** Microhardness of the 2024 alloy vs time of post-SPD annealing at 100 °C.

**Table 1.** Room temperature tensile strength parameters of the 2024 alloy.

| Condition | YS, MPa | UTS, MPa | El, % |
|---|---|---|---|
| T6 rod | 400 | 450 | 7.0 |
| Q+HPT | 1020 | 1055 | 1.8 |
| Q+AA+HPT | 1000 | 1075 | 1.8 |
| Q+HPT+T6 | 455 | 550 | 13.8 |

Post-SPD annealing at 100 °C led to the alloy extra hardening in the Q condition only with the maximum reached at 50 hours (figure 2). This behavior was caused by the effect of disperse strengthening due to artificial aging. It should be noted that the products of aging in the form of lamellar zones and precipitates of the main strengthening phases were hardly distinguished by TEM and found only in the recovered areas with a more equilibrium structure (figure 1h). The morphology of these precipitates had no obvious differences from the conventional ones, except for their lower length, limited by the inter-grain boundary spacing. Therewith the loss in the alloy strength owing to the elimination of structural defects, and also a decrease in the level of alloying of aluminum solid solution, were compensated by disperse strengthening with even more than 10% of hardness increase.

The data obtained, especially the absence of aging response in the pre-aged alloy, allow us to conclude that HPT, even to high strains, does not result in sense dissolution of main strengthening phases and sense changing of the alloying of the matrix, frequently declared in the literature.

## 4. Conclusions

1. The room temperature–HPT of a commercial hot-pressed rod of the 2024 Al alloy with differed heterogeneity of the initial structure, performed by 10 revolutions at 6 GPa, led to a more equiaxed and developed nanostructure of the fragmented type in the pre-quenched alloy. In the pre-aged at 190 °C for 10 hrs condition (T6 temper), a mixed (sub)grain/cellular structure, preferably composed of coarser elongated crystallites separated by dislocation boundaries, was formed. HPT resulted also in superior strain hardening with the higher impact also obtained in the pre-quenched alloy.
2. In contrast to the pre-aged alloy condition, post-SPD annealing of the pre-quenched alloy at 100 °C led to its extra hardening. Therewith, the losses in the alloy strength owing to the elimination of structural defects, and also a decrease in alloying of the aluminum solid solution, were compensated by precipitation hardening with even more than 10% of strength increase. The data obtained allow concluding that the absence of aging response in the SPDed pre-aged alloy is conditioned by the absence of sense changes in the contents of the main alloying elements in the aluminum matrix due to severe straining, and dynamic dissolution of secondary phases, frequently discussed in the literature.

## Acknowledgments
The work was performed under the grant No. 16-19-10152-P of Russian Scientific Fund on the basis of Collective center of IMSP RAS "Structural and physics-mechanical analyses of materials".

## References

[1] Valiev R Z, Estrin Y, Horita Z, Langdon T G, Zehetbauer M J and Zhu Y 2016 *JOM* **68** 1216
[2] Rabinovich M Kh, Markushev M V and Murashkin M Yu 1997 *Mater. Sci. Forum* **243** 591
[3] Barlow C Y, Hansen N and Liu Y L 2002 *Acta Mater.* **50** 171
[4] Markushev M V, Avtokratova E V, Krymskiy S V, Sitdikov O Sh 2018 *J. Al. Comp.* **743** 773
[5] Markushev M V, Burdastykh Yu L, Krymskiy S V and Sitdikov O Sh 2017 *Lett. Mat.* **7** 101
[6] Sha G, Wang Y B, Liao X Z, Duan Z C, Ringer S P, Langdon T G 2009 *Acta Mater.* **57** 3123
[7] Markushev M V *et al.* 2018 *IOP Conf. Series: Mater. Sci. Eng.* **447** 012008
[8] Chen Y, Gao N, Sha G, Ringer S P and Starink M J 2015 *Mater. Sci. Eng.* A **627** 10
[9] Krymskiy S V *et al.* 2011 *Perspective Mat.* **12** 387 [in Russian]

IOP Conf. Series: Materials Science and Engineering **672** (2019) 012043 doi:10.1088/1757-899X/672/1/012043

# Abnormal grain growth in fine-grained aluminum produced by friction-stir welding

**S Mironov and R Kaibyshev**

Belgorod National Research University, Pobeda 85, Belgorod 308015, Russia

E-mail: mironov@bsu.edu.ru

**Abstract**. In this work, fundamental aspects of abnormal grain growth, which typically occurs during annealing of friction-stir welded (FSW) aluminium, were studied. To this end, crystallographic orientations of abnormally-coarse grains were systematically examined. Statistical analysis showed that those were not random. Specifically, it was found that the abnormal grain growth resulted in ~30°<111> rotation of FSW-induced B/-B {112}<110> texture. Accordingly, the abnormal character of the grain growth was interpreted in terms of increased mobility of 30°<111> boundaries.

## 1. Introduction

Friction stir welding or processing (FSW/P) is sometimes considered as a relatively new technique for grain refinement of structural alloys [1, 2]. In this approach, a material undergone very large strain at elevated temperature and relatively high strain rate, and this typically gives rise to the formation of a fine-grained microstructure with a mean grain size in the range from 1 to 10 μm [2]. Due to the complex character of the material flow during FSW, the resulting crystallographic textures are often weak, and thus the processed material is characterized by a relatively high fraction of high-angle boundaries. As a result, such materials normally exhibit excellent superplastic properties [1, 2]. Unfortunately, the FSWed materials had relatively low stability against abnormal grain growth [2-6]. This undesirable effect leads to significant degradation of mechanical characteristics. Accordingly, significant efforts have been undertaken recently to elucidate the mechanism of this phenomenon [2-6]. The extensive research has shown that the grain-growth behavior of the friction-stir welded (FSWed) materials could be often described in terms of Humphreys' cellular model, i.e. arises from an imbalance between the grain-growth pressure and the pinning force exerted by second-phase particles or low-angle boundaries [7, 8].

In an attempt to provide a further insight into this phenomenon, crystallographic orientations of the abnormally-growing grains were measured in the present study. To this end, the electron backscatter diffraction (EBSD) technique was used.

## 2. Material and experimental procedure

The program material used in the present study was a commercial aluminum alloy, supplied as 2-mm-thick cold-rolled sheets. The received sheets were FSWed in a bead-on-plate configuration at a spindle rate of 2000 rpm and a feed rate of 10 mm/s. The welding tool was manufactured from tool steel and consisted of a shoulder of 12 mm in diameter and a cylindrical threaded M5 probe of 1.7 mm in length. To maintain consistency with the FSW literature, the reference directions used in this work were denoted throughout as welding direction (WD), normal direction (ND), and transverse direction (TD). After FSW, the produced welds were section perpendicular to the WD for microstructural observations. The selected microstructural samples were annealed at temperatures ranging from 400 to 500 °C for the soaking times ranging from 10 min to 1 hour. To preserve the high-temperature microstructure, the annealed specimens were quenched in iced water. The microstructures and crystallographic textures produced after FSW and the post-weld heat-treatment were examined by using EBSD. To this end, a Hitachi S-4300SE field-emission gun scanning electron microscope equipped with TSL OIM$^{TM}$ EBSD system was employed. To eliminate spurious boundaries induced by orientation noise during EBSD mapping, a lower-limit boundary misorientation cut-off 2 degree was used. A 15-degree criterion was applied to differentiate low-angle boundaries (LABs) from high-angle boundaries (HABs).

## 3. Results and discussion

A sample-scale EBSD map taken from the cross-section of the friction-stir welded material is shown in figure 1. In the map, the grains are coloured according to their crystallographic orientations (the standard colour code triangle is given in the bottom right corner of the map). From the map, it was clear that distribution of the crystallographic texture was very inhomogeneous across the stir zone. This effect is well known from the FSW literature being usually attributed to the rotation of the welding tool and the respective variation of shear direction across the stir zone [9].

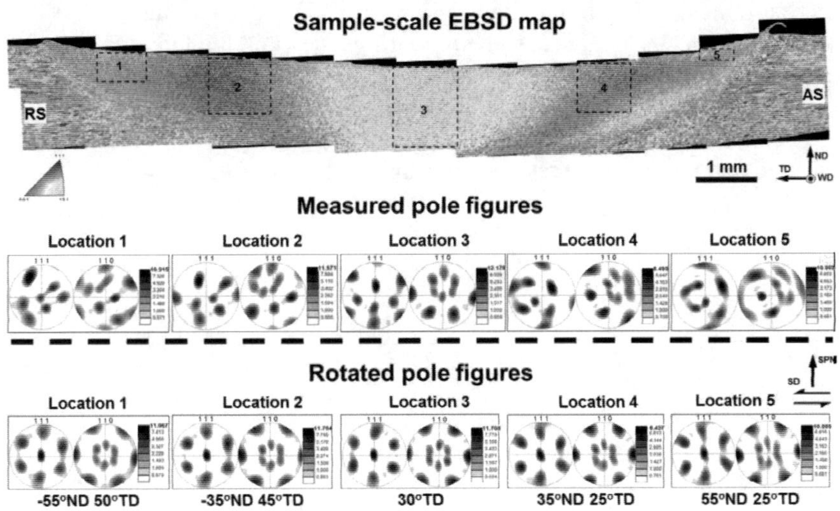


**Figure 1**. Sample-scale EBSD map taken from friction-stir welded material and 111 and 110 pole figures showing texture distribution in various sections of the stir zone. RS and AS are the retreating side and the advancing side, respectively. SPN and SD denote the shear plane normal and the shear direction, respectively. See text for details.

To provide a deeper insight into the evolved texture, orientation data were extracted from selected regions in the EBSD map, arranged as 111 and 110 pole figures, and also shown in figure 1. As expected the measured texture substantially varied. However, after transforming the measured textures into a standard reference frame for the simple-shear deformation (i.e. the local shear direction is horizontal and the local shear normal is vertical), all rotated pole figures showed a typical B/-B {112}<110> texture component (figure 1). Such a texture is often reported to develop during FSW of aluminium alloys [9].

To investigate the grain structure produced during FSW, high-resolution EBSD maps were taken from several locations within the stir zone. It was found that evolved microstructures were dominated by low-aspect-ratio grains of ~1 μm in size (not shown). Remarkably, the measured HAB fraction was found to vary from 63 to 77%, thus being relatively high. It is often believed that such a microstructure should be stable against the abnormal grain growth.

The annealing behaviour of the FSWed material is shown in figure 3. It is clear that the microstructure evolution was dominated by the abnormal grain growth.

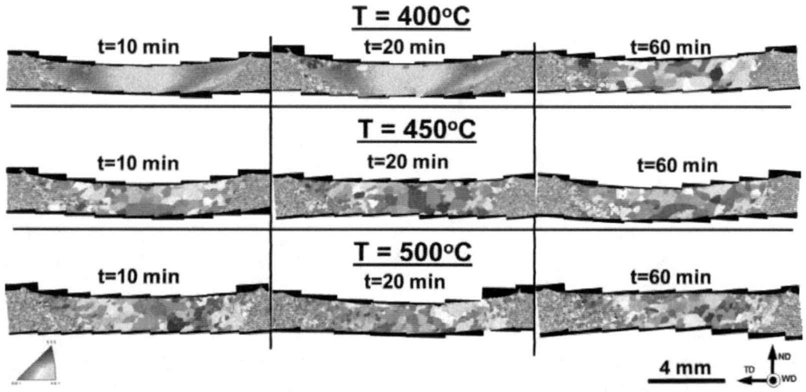


**Figure 2**. Low-resolution EBSD maps showing the annealing behavior of friction-stir welded materials at different temperatures. In the maps, grains are colored according to their crystallographic orientation relative to the WD. In all cases, the retreating side is on the left, and the advancing side is on the right. See text for details.

In order to get insight into this phenomenon, crystallographic orientations of the abnormally-coarse grains were measured and the obtained results were summarized in figure 3.

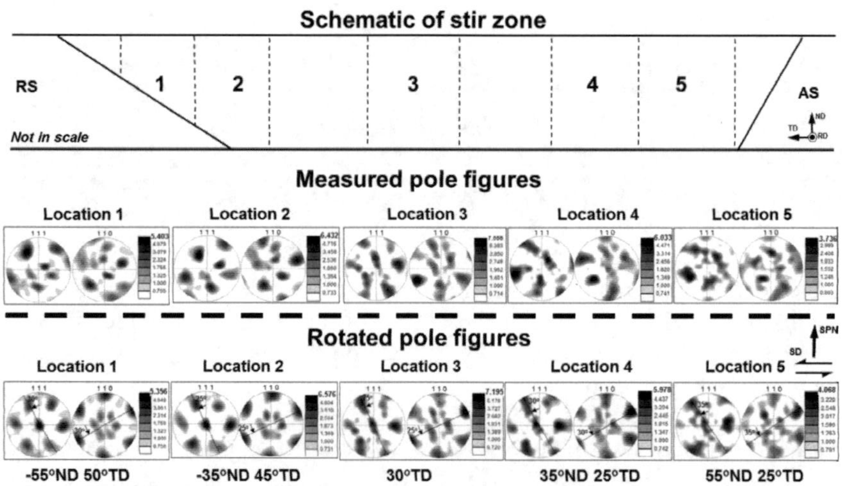


**Figure 3**. Schematic of the weld cross-section of fully annealed material together with 111 and 110 pole figures showing the texture distribution within the stir zone. SPN and SD denote the shear plane normal and the shear direction, respectively. See text for details.

To interpret the textural data taken from different locations within the stir zone, the measured pole figures were rotated appropriately to bring them into a standard reference frame for simple shear, similar to that applied to the as-welded material in figure 2. The applied rotations were noted beneath the rotated pole figures in figure 3.

IOP Conf. Series: Materials Science and Engineering **672** (2019) 012043 doi:10.1088/1757-899X/672/1/012043

From the rotated pole figures in figure 3, it was clear that the crystallographic orientations of the abnormal grains were not random. Moreover, from a comparison of the rotated pole figures in figures 1 and 3, it could be deduced that the abnormal grain growth resulted in a ~30-degree rotation of the as-FSWed texture around the <111> axis.

To the best of the authors' knowledge, this effect has never been reported so far for the FSWed materials. On the other hand, it is well known for annealing of cold-rolled cubic metals, being usually attributed to the theory of orientation growth [10]. According to this concept, the formation of annealing texture is associated with increased mobility of grain boundaries with a particular misorientation. Specifically, the boundaries with ~30-40°<111> misorientation are well accepted to have relatively high migration mobility in faced-centred cubic metals. As a result of the preferential migration of such boundaries, the rolling texture may experience a ~30-40°<111> rotation during grain growth, similar to that observed in the present study.

The obtained result suggests that the abnormal grain growth in FSWed aluminium was associated with increased migration mobility of ~30°<111> boundaries.

## 4. Conclusion

In this work, the EBSD technique was applied to directly measure the crystallographic orientations of abnormally-growing grains in friction-stir welded aluminium. The most important results derived from this work were as follows.

Crystallographic orientations of the abnormally-coarse grains were not random. Statistical analysis showed that these could be described in terms of the as-welded B/-B {112}<110> texture rotated by ~30 degrees around the <111> axis.

The above result suggests that the abnormal grain growth may be associated with increased migration mobility of ~30°<111> boundaries in face-centred cubic metals.

**Acknowledgments**

The authors would like to acknowledge the financial support from the Russian Science Foundation, grant No. 19-49-02001

## References

[1] Mishra R S, Mahoney M W, McFadden S X, Mara N A and Mukherjee A K 1999 *Scripta Mater.* **45** 163
[2] Mishra R S and Ma Z Y 2005 *Mater. Sci. Eng.* R **50** 1
[3] Hassan Kh A A, Norman A F, Price D A and Prangnell P B 2003 *Acta Mater.* **51** 1923
[4] Charit I, Mishra R S 2008 *Scripta Mater.* **58** 367
[5] Mironov S, Masaki K, Sato Y S and Kokawa H 2012 *Scripta Mater.* **67** 983
[6] Vysotskiy I, Malopheyev S, Mironov S and Kaibyshev R 2018 *Mater. Sci. Eng.* A **733** 39
[7] Humphreys F J 1997 *Acta Mater.* **45** 4231
[8] Humphreys F J 1997 *Acta Mater.* **45** 5031
[9] Fonda R W and Knipling K E 2011 *Sci. Technol. Weld. Join.* **16** 288
[10] Humphreys F J, Hartherly M 2004 *Recrystallization and Related Phenomena* (Oxford: Elsevier)

IOP Conf. Series: Materials Science and Engineering **672** (2019) 012044 doi:10.1088/1757-899X/672/1/012044

# Mechanical behavior and impact toughness of the ultrafine-grained VT8M-1 alloy

**Iu M Modina, V V Polyakova, G S Dyakonov, A V Polyakov, I P Semenova and A G Raab**

Institute of Physics of Advanced Materials, Ufa State Aviation Technical University, 12 K. Marx st., Ufa, 450008, Russian Federation

E-mail: modina_yulia@mail.ru

**Abstract**. This paper reports on a study of the relationship between the microstructure, mechanical behavior and impact toughness of the two-phase Ti alloy VT8M-1 (Ti-5.7Al-3.8Mo-1.2Zr-1.3Sn) in the ultrafine-grained (UFG) state, which was produced by two different techniques – equal-channel angular pressing (ECAP) and rotary swaging (RS). It is shown that both processing techniques enable forming a UFG structure and enhanced mechanical properties in the alloy. The UFG structures formed in the billets are characterized by different shapes and sizes of the α- and β-phases. The relationship between the structure, strength, ductility and impact toughness of the alloy processed by RS and ECAP is discussed.

## 1. Introduction

Two-phase titanium alloys are widely used for producing complex-loaded parts of a gas-turbine engine (GTE). Therefore, high requirements are imposed on them not only in terms of strength, fatigue resistance, but also ductility-related properties, such as fracture toughness and impact toughness, which characterize a material's resistance to brittle fracture.

To date, one of the most promising approaches to drastically modify and enhance the mechanical properties of metals and alloys is to produce ultrafine-grained (UFG) states using various severe plastic deformation (SPD) techniques [1]. This type of treatment, applied to two-phase Ti alloys, enabled considerably increasing both strength and endurance limit, implementing high-speed and/or low-temperature superplasticity [2,3]. However, recent studies show that the formation of UFG structures in metals and alloys often leads to a decrease in both fracture toughness and impact strength [4]. Many investigators attribute this trend to the decline in the capacity of UFG metals to strain hardening [4]. As a result, the practical application of high-strength UFG Ti alloys as engineering materials can be severely limited.

In this work, two different techniques were used to form the UFG structure in the VT8M-1 alloy under study: equal-channel angular pressing (ECAP) and rotary swaging (RS). RS consists in the high-frequency deformation of metals with anvils according to the quasi-scheme of all-round compression and it is new to the processing of titanium alloys [5,6]. RS can be seen as an alternative to ECAP methods, and a combination of the two approaches is possible, which will allow creating new technologies for producing UFG materials on a commercial scale. To compare the potential of two alternative methods, the present paper is devoted to the study the relationship between the microstructure, mechanical behavior and impact toughness of the VT8M-1 alloy subjected the two techniques.

## 2. Material and experimental procedure

The experiments were conducted using the VT8M-1 alloy (VSMPO-AVISMA Corporation, Verkhnaya Salda, Russia) in the form of hot-rolled rods with a diameter of 70 mm. The alloy had the following chemical composition (in wt.%): 5.7% Al, 3.8% Mo, 1.2% Zr, 1.3% Sn and Ti – base. The initial structure of the rod made of the VT8M-1 alloy was a mixed globular-plate structure. The average size of the primary α-phase was 5 μm, and its fraction in the alloy was 50%, the thickness of α-lamellae was 0.2 μm [7]. To ensure the conditions of plastic deformation of hard-to-deform Ti alloys, preliminary heat treatment (HT) was performed by heating the alloy to a temperature of 940 °C, subsequent quenching in water, annealing for one hour at 700 °C, and cooling in air [7]. The UFG state was formed via two different techniques. Four ECAP passes via $B_C$ route at a temperature of 750°C (e=2.7 with a strain rate of 4 mm $s^{-1}$) were applied to produce rods with a diameter of 30 mm [8]. A rod with a diameter of 70 mm was successively processed by RS to a diameter of 30 mm with intermediate annealing at 750 °C (e ~ 1.7, the strain rate was ~300 mm $s^{-1}$).

The metallographic study was conducted using a JEOL JEM-2100 microscope (TEM) and a JEOL JSM 6390 microscope (SEM). The stress-strain curves were recorded using an initial strain rate of 1.0 x $10^{-3}$ $s^{-1}$ with an Instron universal testing machine. The samples with cross-sectional dimensions of 10 × 10 mm and a V-notch were tested using an Instron CEAST 9350 impact tester at room temperature in accordance with ASTM E23-16. At least three samples were used per experimental point.

## 3. Results and their discussion

### *3.1. Microstructure and mechanical properties of VT8M-1 after SPD processing*

The microstructure after HT (figure 1a) is a globular plate-like structure, which consists of 0.12 μm thick α-phase lamellae divided by β-phase interlayers, and 25% of the globular primary α-phase with an average grains size of 2.7 μm.

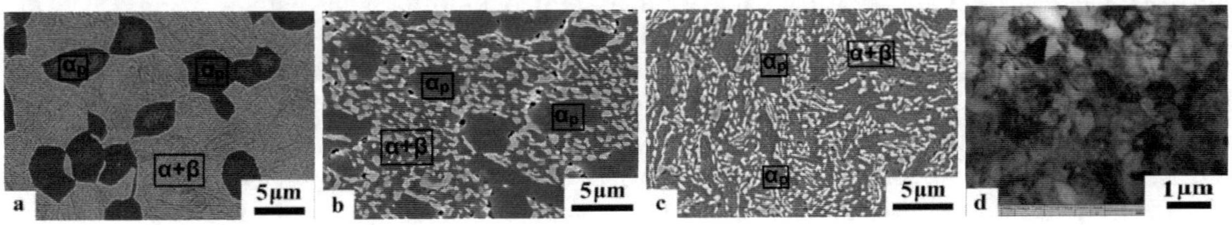


**Figure 1.** Microstructure of the VT8M-1 alloy: (a) after HT; (b) after HT+ECAP; (c) after HT+RS; (d) after HT+RS (TEM).

According to the SEM data (figures 1b and 1c), in the structures after ECAP and RS, the lamellar constituent is almost fully transformed into the equiaxed one, the grain size of the primary α-phase is ~3 μm. Part of the grains of the primary α-phase after RS is elongated along the deformation direction, their length is 7 μm (figure 1c), whereas after ECAP the grain shape of the primary α-phase is close to the equiaxed one (figure 1b).

According to the TEM data, after ECAP, the average size of grains and subgrains in the (α+β) areas was 0.48 μm, and after RS it was 0.25 μm (figure 1d). The features of the UFG structure formation resulting from ECAP are considered in more detail in [8].

Table 1 shows that the UFG structure formation by ECAP and RS leads to an increase in the ultimate tensile strength (UTS) and yield stress (YS) (by 10% for ECAP and 20% for RS). The difference in the UTS values of the investigated states is primarily conditioned by a finer structure in the alloy after rotary swaging (0.25 and 0.48 μm for RS and ECAP, respectively).

In particular, a higher value of uniform elongation in the alloy after RS (4%) as compared to the ECAP-processed condition (1.3%) (table 1) can be attributed to a lower dislocation density inside grains/subgrains, and also to a more equilibrium state of the boundaries [7]. The total elongation was practically the same and was found to be within the error range (11 and 10%, respectively).

**Table 1.** Mechanical properties of the VT8M-1 alloy before and after SPD processing.

| Condition | UTS, MPa | YS, MPa | Elong., % | Uniform elongation, % |
|---|---|---|---|---|
| **As-received** | 1040±20 | 970±20 | 15±2 | 5.0±0.9 |
| **HT+ECAP** | 1150±10 | 1080±20 | 10±1 | 1.3±0.5 |
| **HT+RS** | 1230±10 | 1150±15 | 11±1 | 3.8±0.5 |

### *3.2. Impact toughness of VT8M-1 after SPD processing*

Table 2 presents the results of impact tests of the samples with different types of microstructure. A tendency is observed for an almost 2-fold decrease in absorbed energy for UFG VT8M-1 as compared to the CG alloy, which is primarily conditioned by the increase in yield stress both after ECAP and after RS (table 1). Such a tendency is typical for many metals [4]. The absorbed enegry values after impact tests of the ECAP-processed and RS-processed samples are practically identical and on average amount to ~0.25 MJ/m$^2$ (table 2). The RS-processed samples exhibit a smaller spread in the Kcv values as compared with the ECAP-processed samples.

The appearance of the fracture surfaces of the VT8M-1 alloy (figure 2) after testing is typical for titanium alloys, it has a flat area, which is characterized by a fibrous zone and shallow fracture dimples, and also has a shear lip, but also some differences are observed related to the features of the microstructure of the alloy. The rupture area in the RS-processed sample, in contrast to the CG and ECAP-processed samples, is located at an angle of 45° with respect to the fracture plane (figure 2g), which may be related to inhomogeneity and the formation of a specific texture by rotary swaging [6].

In the case of fracture in the CG alloy, the microrelief consists of comparatively large (~6 μm in size, table 2) and deep, slightly elongated dimples, which are typical for conventional CG Ti alloys (figure 2c). On the fracture surfaces after ECAP and RS (figures 2f and 2i), the dimples are smaller with sizes ~ 4 and 2 μm, respectively (table 2), and the fraction of dimples with a size of less than 1 μm is much larger, which indicates the predominantly intercrystalline character of the fracture, in contrast to the CG sample, where the transcrystalline character of the fracture prevails. Thus, the conducted study shows that producing an ultrafine grain size in the VT8M-1 alloy by ECAP and RS leads to a decline in the impact toughness Kcv from 0.5 MJ/m$^2$ in the CG state to 0.25 MJ/m$^2$ due to an increase in yield strength. At the same time, the RS-processed sample with a higher yield stress in comparison to the ECAP-processed condition (1150 and 1080 MPa, respectively) exhibits the same level of impact toughness. Apparently, this could be attributed to the higher values of the uniform elongation of the samples after RS (4%) in comparison to the ECAP-processed samples (1.3%), which characterizes the greater ability of the material to strain hardening [4]. Achieving in the UFG VT8M-1 alloy a combination of high strength (UTS 1230 MPa) and impact toughness (0.25 MJ/m$^2$) creates prospects for its successful application as a structural material in aviation and mechanical engineering.

**Table 2.** Impact toughness of the VT8M-1 alloy in different structural states.

| Condition | Average grain size, D, μm | Kcv, MJ/m$^2$ | Dimple size, μm |
|---|---|---|---|
| **As-received** | 5±1 | 0.510±0.010 | 6.40±0.38 |
| **HT+ECAP** | 0.48±0.08 | 0.254±0.040 | 3.86±0.18 |
| **HT+RS** | 0.25±0.05 | 0.246±0.005 | 2.35±0.18 |

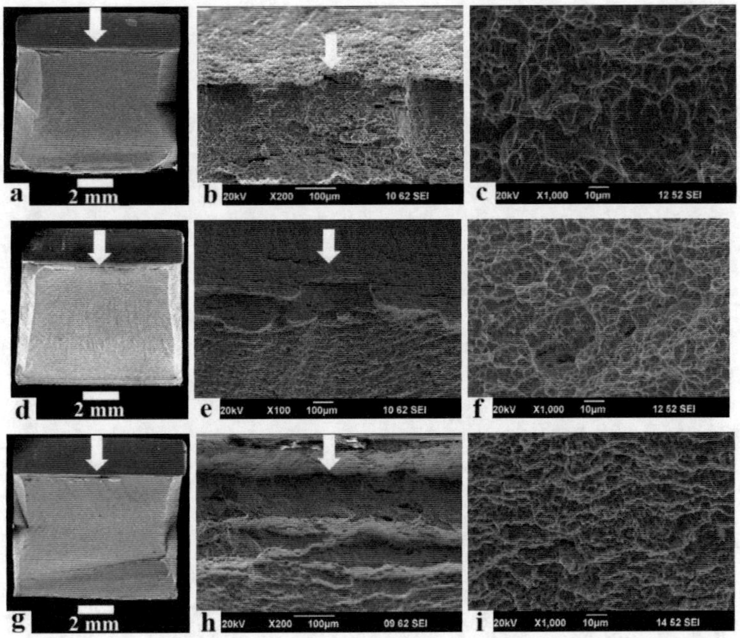


**Figure 2.** Fracture surface of the samples after testing: CG alloy (a, b, c), HT+ECAP (d, e,f); (g, h, i) HT+RS: (a, d, g) general view of fracture; (b, e, h) failure zone; (c, f, i) flat area. The arrows shows the onset and direction of fracture propagation.

## 4. Conclusions

The following conclusions can be drawn from the conducted studies of the mechanical behavior and impact toughness of the UFG VT8M-1 alloy produced by two different techniques – RS and ECAP:

1. The formation of the UFG structure with a finer grain size during RS as compared to ECAP (0.25 and 0.48 µm, respectively) results in a higher ultimate tensile strength (1230 and 1150 MPa, respectively) and unifrom elongation (4 and 1.3%, respectively).

2. It is shown that the impact toughness of the alloy in the UFG state produced by the two techniques (RS and ECAP) decreases almost 2-fold in comparison to the coarse-grained state due to an increase in yield stress. The UFG RS-processed samples, having a higher yield stress and uniform elongation than the ECAP-processed samples, exhibit identical Kcv values (0.25 MJ/m$^2$).

## Acknowledgements

This work was supported by the Ministry of Education and Science of the Russian Federation in the scope of the State Assignment. The authors are grateful to I. Ramazanov for performing the impact toughness tests at the Center for Common Use of the Research and Educational Center "Nanostructured Materials and High Technologies" of Ufa State Aviation Technical University.

## References

[1] Valiev R Z *et al.* 2015 *Mater. Res. Lett.* **4** 1
[2] Semenova I P, Raab G I and Valiev R Z 2014 *Nanotechnologies in Russia* **9** 311
[3] Zherebtsov S V *et al.* 2012 *Mater.Sci. Eng.* A **536** 190
[4] Pippan R and Hohenwarter A 2016 *Mater. Res. Lett.* **4** 127
[5] 2005 *ASM Handbook: Metalworking: Bulk Forming* vol 14A, ed S L Semiatin (ASM International) pp 179–182
[6] Ren D, Su H, Zhang H, Wang J, Jin W and Yang R 2019 *Acta Metal. Sinica* **55** 480
[7] Modina Iu M *et al.* 2019 *IOP Conf. Ser.: Mater. Sci. Eng.* **461** 012056
[8] Semenova I P *et al.* 2018 *Adv. Eng. Mater.* **20** 1700813

IOP Conf. Series: Materials Science and Engineering **672** (2019) 012045 doi:10.1088/1757-899X/672/1/012045

# Strengthening mechanisms of ultrafine grained copper alloys

**A Morozova, R Mishnev, A Belyakov and R Kaibyshev**

Laboratory of Mechanical Properties of Nanostructured Materials and Superalloys, Belgorod National Research University, Belgorod, Pobeda str. 85, Russia

E-mail: morozova_ai@bsu.edu.ru

**Abstract**. Cu-Cr-Zr alloys are characterized by an excellent combination of strength and electrical conductivity and are applied as structural materials in various electrical/electronic devices. The microstructure design of Cu-Cr-Zr alloys, the effect of thermomechanical treatment based on severe plastic deformation, and physical mechanisms providing high strength and electrical conductivity are investigated. The main structure-property relations in Cu-Cr-Zr alloys are considered. The dissolution of the dispersed particles during severe plastic deformation and an increase in the dislocation density, as well as the effect of heat treatment after deformation on the mechanical and electrical properties are discussed.

## 1. Introduction

Nowadays, Cu-Cr-Zr alloys attract the attention of researchers and engineers due to the unique combination of strength and electrical conductivity. Many works concentrate on investigation of the effect of severe plastic deformation (SPD) at room temperature on the microstructure and properties of the alloys, because deformation at room temperature produces smaller grain size and higher dislocation density in comparison with warm deformation. However, deformation at elevated temperature can provide optimal properties of Cu-Cr-Zr alloys. SPD at elevated temperature can promote dynamic aging, i.e., particle precipitation during deformation that can improve strength by dispersion strengthening and increase electroconductivity due to the decomposition of a supersaturated solid solution. Moreover, the kinetics of dynamic recrystallization is accelerated with an increase in the deformation temperature, so, grain refinement can be realized after relatively small strain. In addition, deformation at elevated temperature can be carried out under low pressure, decreasing the wear of the forming device. Subsequent aging after SPD improves electrical conductivity in Cu-Cr-Zr alloys.

The aim of the present work is to study the microstructure evolution, the mechanical properties, the strengthening mechanisms and the electrical conductivity of Cu-Cr-Zr alloys subjected to SPD at elevated temperature and aging.

## 2. Experiment

The investigated alloys were Cu-0.1%Cr-0.1%Zr (hereafter 0.1Cr-0.1Zr) and Cu-0.3%Cr-0.5%Zr (hereafter 0.3Cr-0.5Zr). 0.1Cr-0.1Zr was forged at 800 °C to a total strain of 2, then annealed at 920 °C during 1h with subsequent water quenching and aged at 550 °C during 4 h. 0.3Cr-0.5Zr was annealed at 920 °C for 0.5h with subsequent water quenching and aged at 450 °C during 1 h. The alloys were subjected to equal channel angular pressing (ECAP) at 400 °C via route $B_C$ to 1, 2, 4, 8, 12 passes. Then, 0.1Cr-0.1Zr was annealed at 500 °C for 1 h and 0.3Cr-0.5Zr was annealed at 450 °C for

1 h. The microstructures were investigated using a Nova NanoSEM 450 scanning electron microscope (SEM) equipped with an EBSD analyzer and a Jeol Jem 2100 transmission electron microscope (TEM). The average grain size and the fraction of ultrafine grains (the grains with a size of below 2 micron) were estimated using the OIM Software. The dislocation density was determined by counting the number of intersections of individual dislocations located within the grains/subgrains interiors with the foil surfaces in at least 5 arbitrary selected TEM images. The specimens for SEM and TEM investigations were prepared by a Tenupol 5 with an electrolyte consisted of 25% $HNO_3$ and 75% $CH_3OH$ at a temperature of -20 °C under a voltage of 10 V. The mechanical properties were evaluated using tensile tests at room temperature at a starting strain rate of $2\times10^{-3}$ $s^{-1}$. The electrical conductivity was measured by the eddy current method using a Constant K6 device.

## 3. Results and discussion

### *3.1. Microstructure evolution of Cu-Cr-Zr alloys*

After aging, the Cu-Cr-Zr alloys contained nanoscale particles. Fine ellipsoid Cr-rich precipitates with a bcc-lattice were 10-15 nm in the long direction and 5-7 nm in the transverse direction in 0.1Cr-0.1Zr. Spherical fcc-Cr particles with a diameter of about 4-5 nm precipitated in 0.3Cr-0.5Zr. The Cu-Cr-Zr alloys consisted of coarse grains with a size of 100-120 μm and 180-200 μm in 0.1Cr-0.1Zr and 0.3Cr-0.5Zr, respectively.

Plastic deformation at elevated temperature was accompanied by grain refinement (figure 1). After the 1st ECAP pass, new strain-induced low-angles boundaries (LAB) formed in the Cu-Cr-Zr alloys. Parallel long high-angle boundaries (HAB) developed in coarse initial grains and formed deformation microbands. An increase in strain led to the appearance of new grains with a size of less than 1 μm inside the deformation microbands. The formation of new grain resulted from an increase in the misorientations of LAB during deformation. After 8 ECAP passes, a uniform ultrafine grain structure with a grain size below 1 μm was formed.

The kinetics of grains refinement can be estimated using the modified Johnson-Mehl-Avrami-Kolmogorov (JMAK) equation [1]:

$$F_{UFG} = 1 - \exp(-ke^{n}) \qquad (1),$$

where $F_{UFG}$ is the fraction of ultrafine grains, e is the logarithmic strain, k is the constant characterizing incubation period for grain refinement, and n is the kinetic constant.

The kinetic constant, n, was almost the same for 0.1Cr-0.1Zr and 0.3Cr-0.5Zr of 1.99 and 1.72, respectively. The incubation period of grain refinement of 0.1Cr-0.1Zr was longer than for 0.3Cr-0.5Zr alloy, so the grain size was smaller in 0.3Cr-0.5Zr alloy after the same ECAP strain. The change in the average grain size was good approximated with following equitation [1]:

$$D = D_{UFG}\,(1 - \exp(-ke^{n}))^{-0.5} \qquad (2),$$

where D is the current grain size, $D_{UFG}$ is the average size of ultrafine grains.

The dislocation density in 0.3Cr-0.5Zr was very high for copper alloys and achieved about $10^{15}$ $m^{-2}$ after the 8th ECAP pass. In 0.1Cr-0.1Zr, smaller dislocation density of about $(2\text{-}4)\times10^{14}$ $m^{-2}$ was observed. The change in the average dislocation density, ρ, can be expressed by the following equations [2]:

$$\rho = (1 - F_{UFG})\,\rho_{wh} + F_{UFG}\,\rho_{rec} \qquad (3a),$$

$$\rho_{wh} = h/r + (h/r - \rho_0)\exp(-re) \qquad (3b),$$

$$\rho_{rec} = h/r' + (h/r' - \rho_0)\exp(-r(e - e_C)) \qquad (3c),$$

where $\rho_{wh}$ is the average dislocation density in work hardened grains, $\rho_{rec}$ is the average dislocation density in recrystallized grains, h is the athermal work hardening rate, r is the dynamic recovery rate in work hardened grains, r' is the dynamic recovery rate in recrystallized grains, $\rho_0$ is the initial dislocation density $e_C$ is the onset recrystallization strain.

IOP Conf. Series: Materials Science and Engineering **672** (2019) 012045 doi:10.1088/1757-899X/672/1/012045

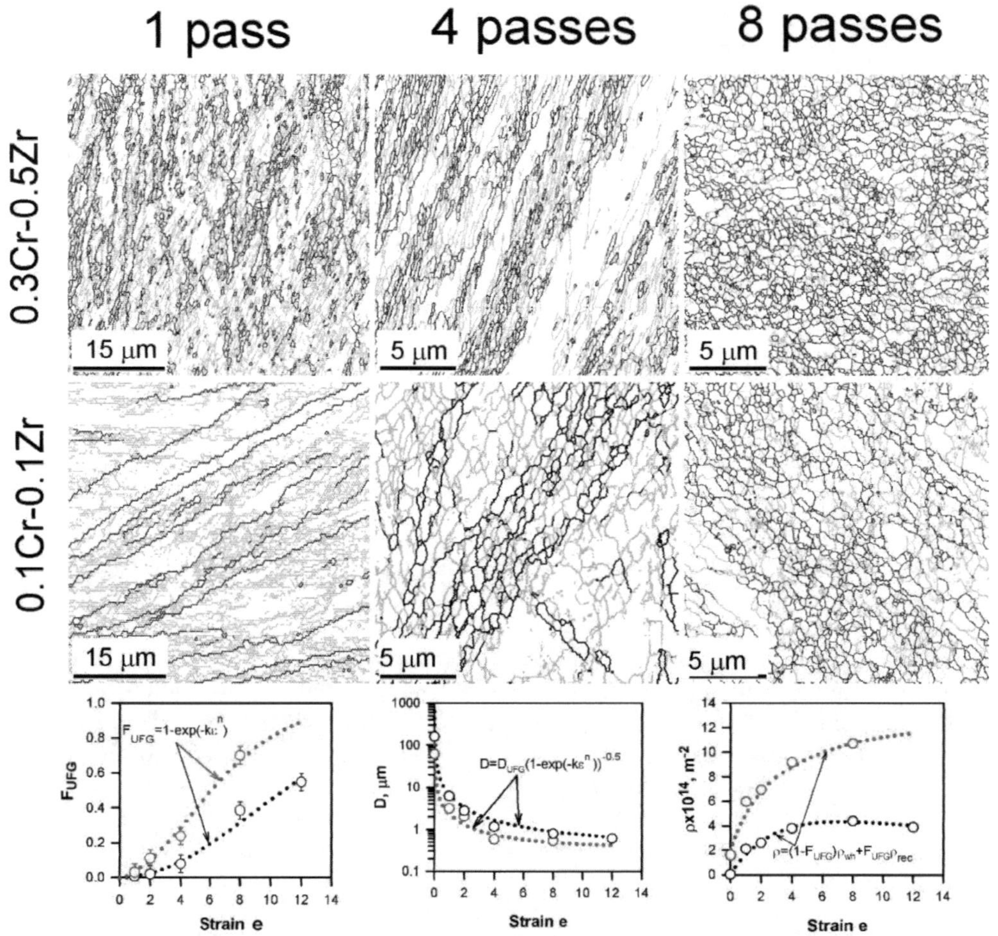


**Figure 1.** Microstructure evolution during ECAP in the Cu-Cr-Zr alloys. Black and brown lines represent HAB and LAB, respectively. Grain refinement and dislocation density in 0.1Cr-0.1Zr shown by black points and those in 0.3Cr-0.5Zr by red points.

The rate of dynamic recovery was higher, and the work hardening rate was lower in 0.1Cr-0.1Zr than in 0.3Cr-0.5Zr. The difference in the kinetics of grain refinement, dynamic recovery and work hardening can result from the difference in the structure, shape and size of Cr-particles in the initial states. Also, in the 0.3Cr-0.5Zr alloy, additional precipitation of particles during deformation was observed, leading to an increase in the volume fraction of particles from 0.0015 to 0.0027.

### *3.2. Properties of Cu-Cr-Zr alloys*

ECAP led to an increase in the strength and a decrease in the elongation (figure 2). In 0.1Cr-0.1Zr after 8 passes, the ultimate tensile strength achieved 430 MPa. Significant strengthening was observed after the 1$^{st}$ ECAP pass. Then, the strength increased slowly with further straining. In 0.3Cr-0.5Zr the strength increased gradually with straining. 8 ECAP passes resulted in the ultimate tensile strength above 700 MPa. The elongation in the Cu-Cr-Zr alloys degraded with strengthening, although after 4 ECAP passes, the change in plasticity slowed down. The elongation was about 15% and 20% for 0.1Cr-0.1Zr and 0.3Cr-0.5Zr, respectively. In 0.1Cr-0.1Zr, the deformation led to a slight decrease in the electrical conductivity. In contrast, pressing of 0.3Cr-0.5Zr was accompanied by an increase in the electrical conductivity from 47% IACS to 65% IACS that could result from the precipitation of Cr-particles.

IOP Conf. Series: Materials Science and Engineering **672** (2019) 012045 doi:10.1088/1757-899X/672/1/012045

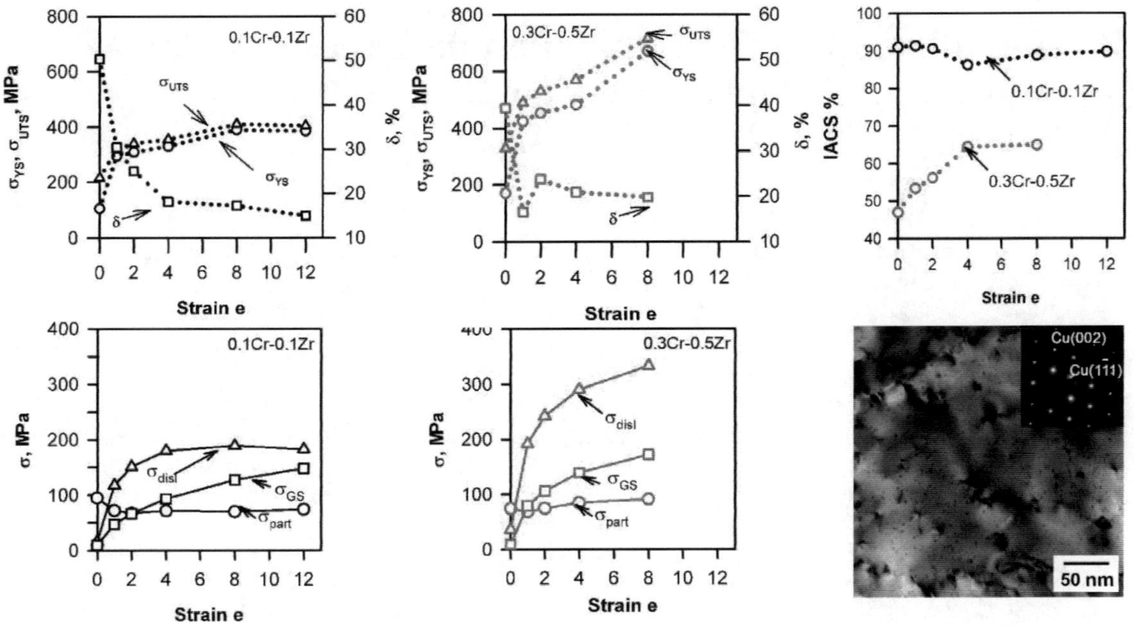


**Figure 2.** Effect of strain on the yield strength ($\sigma_{YS}$), ultimate tensile strength ($\sigma_{UTS}$), elongation (δ), electrical conductivity (IACS %), and particle ($\sigma_{part}$), dislocation ($\sigma_{disl}$) and grain size ($\sigma_{GS}$) strengthening of the Cu-Cr-Zr alloys, and the fine microstructure of 0.3Cr-0.5Zr after 4 ECAP passes.

The contribution of the different strengthening mechanisms to the total yield strength can be estimated as [1]:

$$\sigma_{YS} = \sigma_0 + \sigma_{part} + \alpha M G b \rho^{0.5} + k_y D^{-0.5} \quad (4),$$

where $\sigma_{YS}$ is the yield strength, $\sigma_{part}$ is the Orowan strengthening, α=0.24, $k_y$=0.12 are constants, M is the Taylor factor, G is the shear modulus.

The dislocation strengthening ($\alpha MGb\rho^{0.5}$) dominated in the studied deformation range. The particle strengthening ($\sigma_{part}$) was nearly the same with the grain size strengthening ($k_yD^{-0.5}$) after small strains. In the 0.3Cr-0.5Zr alloy, the dislocation density and the dislocation strengthening were significantly higher than in the 0.1Cr-0.1Zr alloy that can be associated with the high volume fraction of particles in the initial state in the 0.3Cr-0.5Zr alloy.

Aging of the Cu-Cr-Zr alloys led to increase in electrical conductivity to 93%IACS and 72% IACS for 0.1Cr-0.1Zr and 0.3Cr-0.5Zr, respectively, with a minor decrease in strength to 400 MPa for 0.1Cr-0.1Zr and to 650 MPa for 0.3Cr-0.5Zr that can be associated with the compensation of a decrease in dislocation strengthening by particle strengthening.

**References**

[1] Morozova A and Kaibyshev R 2017 *Philosophical Magazine* **97** 2053
[2] Quelennec X and Jonas J J 2012 *ISIJ international* **52** 1145

IOP Conf. Series: Materials Science and Engineering **672** (2019) 012046 doi:10.1088/1757-899X/672/1/012046

# Effect of ultrasonic pretreatment on the superplasticity of titanium alloy Ti-6Al-4V processed by ECAP

**A A Mukhametgalina, M A Murzinova, A A Nazarov**

Institute for Metals Superplasticity Problems, Russian Academy of Sciences, 39 Khalturin st., 450001 Ufa, Russia

E-mail: a.mukhametalina@mail.ru

**Abstract**. Improving the superplasticity characteristics of alloys widely used in industry, such as titanium alloy Ti-6Al-4V, is an important practical problem. Processing by the methods of severe plastic deformation can reduce the temperature of superplasticity. However, the material exhibits high flow stresses. In this work we studied the effect of preliminary ultrasonic treatment with an amplitude of oscillating tension-compression stresses of 100 MPa on the characteristics of superplasticity of the Ti-6Al-4V alloy processed by equal channel angular pressing (ECAP). It is shown that high-frequency oscillating stresses resulted in the reduction of the flow stress, the increase in elongation and the strain rate sensitivity coefficient during tensile tests in the conditions of superplasticity.

## 1. Introduction

Metals and alloys with ultrafine-grained (UFG) and nano-scale structure exhibit high strength at ambient conditions and high ductility at elevated temperatures [1]. For example, titanium alloy Ti-6Al-4V processed by ECAP demonstrated an elongation of about 240% during isothermal deformation at 600 °C with a strain rate of $10^{-4}c^{-1}$ [2]. At the same time, low temperature superplastic deformation usually occurs at high flow stresses, which limits the possibilities of its practical applications. Ultrasonic treatment (UST) is one of the metal processing techniques, which allows modifying the defect structure of materials. It has been shown that UST of UFG metals results in a relaxation of the nonequilibrium structure of grain boundaries, the release of internal strains, and increases the ductility and impact toughness of these materials [3,4]. Imposing axial ultrasonic vibrations to the samples during superplastic deformation tests was shown to reduce the flow stress and increase the elongation to failure [5]. It is of a significant interest to study the effect of preliminary treatment of UFG alloys by ultrasound on their superplastic deformation. In the present paper, we study the changes in the structure and superplastic properties of the Ti-6Al-4V alloy processed by ECAP due to the UST with a stress amplitude of 100 MPa.

## 2. Experimental materials and procedure

Bars of titanium alloy Ti-6Al-4V were subjected to 6 passes of ECAP via route Bc at a temperature of 600 °C in a tool with an intersection angle of channels 120° ($\varepsilon_{sum} \approx 4$) (marked as ECAP condition). Samples with a length of 40 mm were cut from the ECAP-ed material. Several samples were then subjected to UST with a frequency of 20 kHz. They were tightly clamped in the middle part of a cylindric steel waveguide with a length of 128 mm that is equal to the half-wavelength for this frequency, so that a standing wave is formed in the waveguide. The amplitude of oscillating tension-compression stresses in the mid cross section of the samples was equal to 100 MPa. These samples will be further marked as "ECAP+UST samples", while the ones tested directly after ECAP as "ECAP samples".

Dog-bone samples with a gauge length of 10 mm, a width of 3.4 mm and a thickness of 1.5 mm were used for tensile tests. The tensile tests were carried out in isothermal conditions at a temperature of 600 °C with initial strain rates ща $10^{-4} – 10^{-3}$ $s^{-1}$ in an Instron 5982 electromechanical-testing machine. Relative elongation to failure δ, true stress σ and true strain ε on the stage of the uniform elongation, and the strain rate sensitivity coefficient *m* at $\varepsilon = 0.05$ were determined.

The structure of the samples was studied in their longitudinal section by scanning electron microscopy (SEM) using a TESCAN MIRA 3 LMH FEG microscope in the regimes of back-scattered electron imaging (BSE) and electron back-scattered diffraction (EBSD) and by transmission electron microscopy (TEM) using a JEM 2000EX microscope.

## 3. Experimental results and discussion

The microstructure of the investigated samples is shown in figure 1. No differences in the microstructure of the samples subjected to ECAP and ECAP+ UST were detected. Elongated primary α-grains with sizes in the longitudinal and cross directions 15±2 and 6±1 μm, respectively, and a mixture of fine crystals with sizes less than 1 μm of the secondary α- and β-phases were observed. Volume fractions of the primary α-phase and β-phase were about 25% and 9%, respectively. Reliable EBSD structural maps were not obtained, probably due to the high level of internal stresses in the ECAP and ECAP+UST samples (figure 1b). Room temperature mechanical properties of the ECAP and ECAP+UST samples were nearly the same: the ultimate strength and elongation to failure were 1250 MPa and 12%, respectively.

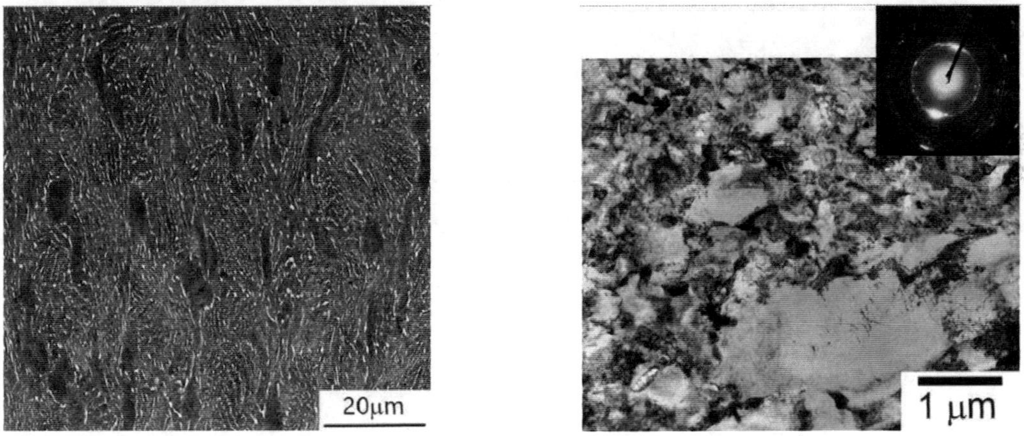


a b

**Figure 1.** BSE (a) and TEM (b) images of microstructure of Ti-6Al-4V alloy subjected to ECAP (a) and ECAP+UST (b).

Preheating of the samples before tensile tests at 600 °C leads to the development of recovery and recrystallization processes [6]. Fine misoriented fragments of the α-phase were observed in the BSE images (figure 2 a, c) and in the orientation maps (figure 2 b, d). Apparently, the UST promoted to the development of relaxation processes and facilitated the formation of new grain boundaries during preheating of the ECAP-ed alloy Ti-6Al-4V. The fractions of high-angle boundaries were 0.38 and 0.53 in the preheated samples in ECAP and ECAP + UST conditions, respectively.

This difference in the microstructure provided differences in the properties of the investigated samples during superplastic deformation. The results of tensile tests are presented in figures 3-5. The shapes of σ - ε curves were not typical for the classical superplastic flow (figure 3). The presence of strain softening stages in σ - ε curves assumes that certain relaxation processes occurred in the alloy during testing. UST resulted in a decrease in the intensity of softening processes during superplastic deformation. A steady state flow stage was observed during tensile tests with initial strain rates of $1.7 \cdot 10^{-4}$ $s^{-1}$ and lower; at higher strain rates, the stage of strain localization followed after the strain softening stage.

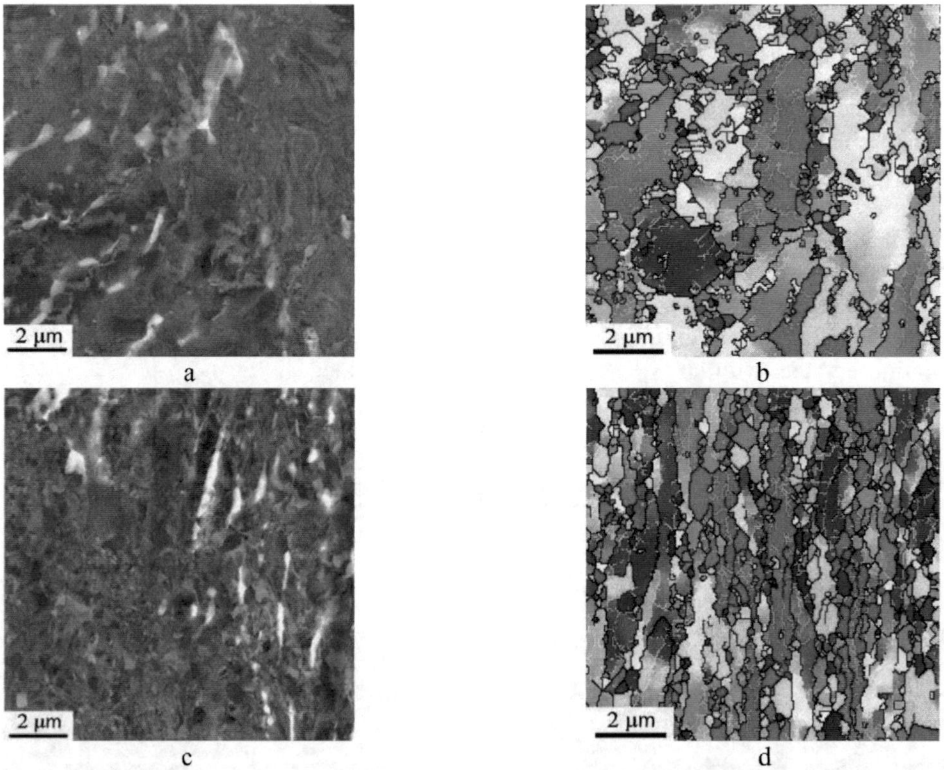


**Figure 2.** Microstructure of the preheated samples of the Ti-6Al-4V alloy subjected to ECAP (a, b) and ECAP+UST (c, d); (a,c) – BSE images, (b,d) – orientation maps.

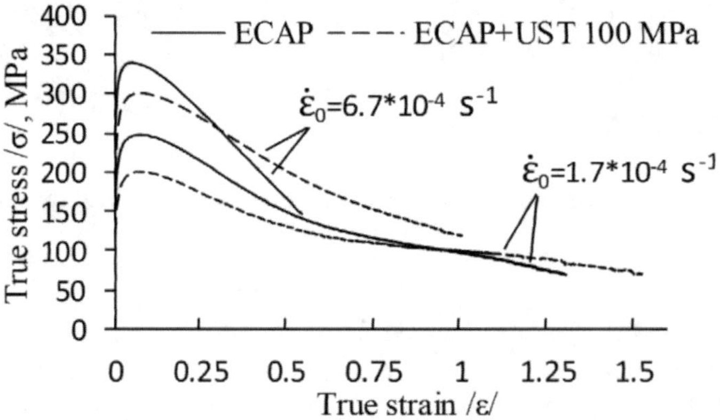


**Figure 3.** Typical true stress – true strain curves of tensile deformation of UFG Ti-6Al-4V after ECAP and ECAP+UST.

ECAP-ed Ti-6Al-4V alloy subjected to UST demonstrated lower values of the flow stress in initial stages of deformation (figures 3, 4) and higher values of strain rate sensitivity (figure 5) and elongation to failure (figure 4).

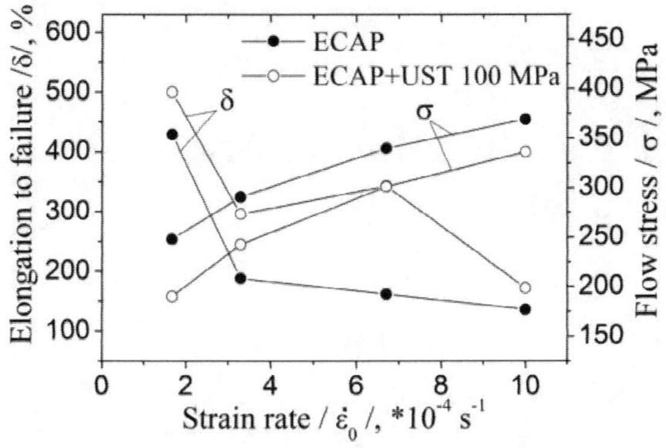


**Figure 4.** Effect of UST on the flow stress and elongation to failure of UFG Ti-6Al-4V alloy

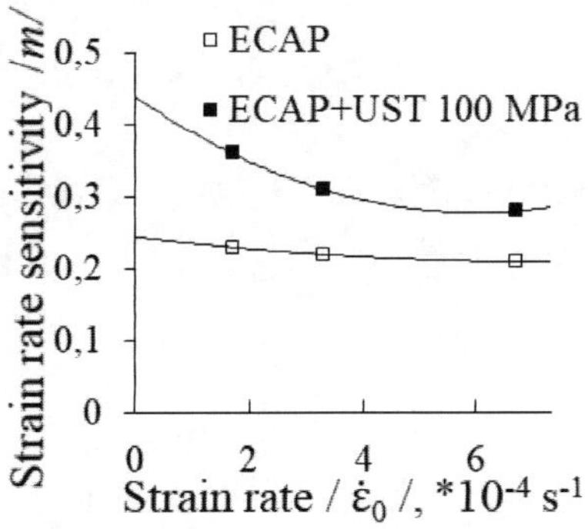


**Figure 5.** Effect of UST on the strain rate sensitivity coefficient, *m*

More detailed structural studies are needed to understand the nature of the effect of ultrasonic treatment on the superplastic behavior of the Ti-6Al-4V alloy. In particular, the complete, incomplete and pseudopartial wetting of grain boundaries by the second solid phase discussed in [7] can explain the observed effect of the UST on high temperature deformation of the alloy.

## 4. Conclusions

Structural studies demonstrated that UST accelerated the recovery and recrystallization processes in the ECAP-ed alloy Ti-6Al-4V during the preheating of the samples at 600 °C. This led to the formation of a finer-grained structure compared to an alloy that was not subjected to UST. It was shown that high-frequency oscillations with a stress amplitude of 100 MPa resulted in the reduction of the flow stress, increase of elongation and the strain rate sensitivity coefficient during tensile tests in the conditions of low-temperature superplasticity.

## Acknowledgments

The present work was supported by the Russian Science Foundation (grant No. 16-19-10126). Experimental studies were carried out on the facilities of shared services center of IMSP RAS "Structural and Physico-Mechanical Studies of Materials".

## References

[1] Valiev R Z, Zhilyaev A P and Langdon T G 2013 *Bulk Nanostructured Materials: Fundamentals and Applications* (Hoboken: Wiley)
[2] Valiev R Z, Islamgaliev R K and Semenova I P 2007 *Mater. Sci. Eng.* A **463** 2
[3] Samigullina A A, Nazarov A A , Mulyukov R R, Tsarenko Yu V and Rubanik V V 2014 *Rev. Adv. Mater. Sci.* **39** 48
[4] Samigullina A A, Mulyukov R R, Nazarov A A, Mukhametgalina A A, Tsarenko Y V and Rubanik V V 2014 *Lett. Mater.* **4** 52
[5] Myshlyaev M M, Shpeizman V V, Klubovich V V, Kulak M M and Lyu Y 2015 *Phys. Solid State* **57** 2039
[6] Humphreys F J and Hatherly M 2004 *Recrystallization and Related Annealing Phenomena*, 2nd ed (Amsterdam: Elsevier)
[7] Gornakova A S, Straumal B B, Nekrasov A N, Kilmametov A and Afonikova N S 2018 *J. Mater. Eng. Perf.* **27** 4989

IOP Conf. Series: Materials Science and Engineering **672** (2019) 012047 doi:10.1088/1757-899X/672/1/012047

# The microstructure and properties of the alloy Ti-5Al-0.5V subjected to ECAP and ultrasonic treatment

**A A Mukhametgalina, M A Murzinova and A A Nazarov**

[1] Institute for Metals Superplasticity Problems, Russian Academy of Sciences, 39 Khalturin st., 450001 Ufa, Russia

E-mail: a.mukhametalina@mail.ru

**Abstract**. This work presents the results of a study of the mechanical properties of a titanium alloy Ti-5Al processed by equal channel angular pressing (ECAP) and ultrasonic treatment (UST). The UST was carried out at a frequency of 20 kHz with an amplitude of oscillating tension-compression stresses of 100 MPa. It was shown that the alloy subjected to UST exhibited low temperature superplastic properties. Its elongation exceeded a value of 300%, the coefficient *m* was 0.3 at a strain rate of $1.7\times10^{-4}$ 1/s.

## 1. Introduction

ECAP is an effective processing method, which results in a significant improvement of the technological and service properties of metals and alloys [1]. At room temperatures, the alloys subjected to ECAP possess high strength but low ductility due to their highly nonequilibrium structure. At high temperatures, these alloys exhibit low-temperature superplasticity.

Ultrasonic treatment (UST) affects the defect structure of metals and alloys that results in a change of their mechanical properties. Depending on the conditions of UST, the latter can lead either to an increase or to a decrease of internal stresses [2]. Also, the effect of UST depends on a material, its lattice type, alloy composition, initial microstructure etc. Most of the studies of the effect of ultrasound on the structure and properties of materials to date have been carried out on commercially pure fcc metals, while the studies on metals with ahcp lattice are scarce. In the present work, we report on comparative studies of the properties of the titanium alloy Ti-5Al-0.5V subjected to ECAP and ECAP followed by UST with a frequency of 20 kHz and an amplitude of tension-compression stress of 100 MPa.

## 2. Experimental materials and procedure

The titanium alloy Ti-5Al-0.5V was chosen for this study. Rods of the alloy were subjected to 6 passes of ECAP in the die with an the angle between channels of 120° by Bc rout at a temperature of 600 °C followed by extrusion at a temperature of 400 °C to a cross-sectional diameter of 16 mm. Then the samples were cut to cylinders with a length of 40 mm.

Some cylinders were treated by UST with an amplitude of oscillating tension-compression stresses of 100 MPa according to the scheme described in [3]. A standing wave with a sinusoidal stress distribution was excited in the waveguide. The antinode of compression-tension stresses was located in the centre of the sample; their amplitude was equal to 100 MPa.

Samples with a gauge length of 10 mm, a width of 3.4 mm and a thickness of 1.5 mm were used for tensile tests. The tensile tests were carried out at room and elevated temperatures in an Instron 5982 electromechanical-testing machine. The tensile tests in the conditions of superplasticity were performed with initial strain rates of 1.7, 3.3 and 6.7 times $10^{-4}$ $s^{-1}$ in isothermal conditions at 600 °C. The relative elongation to failure δ, the true stress σ and the true strain ε on the stage of homogeneous elongation, and the strain rate sensitivity coefficient *m* at $\varepsilon = 0.01$ were determined.

The Vickers microhardness of the samples was measured by an AFFRI DM8A testing instrument with a load on the indenter of 300 g and an exposure time of 10 s.

X-ray diffraction (XRD) analysis of the samples was carried out on a DRON-4 apparatus in the cross section of the cylinder. The Cu $K_\alpha$ radiation was used with Bragg-Brentano focusing. A Soller collimator for primary beam and a graphite monochromator for the diffracted beam were used. The experimental data were processed by means of the «Maud» (Materials Analysis Using Diffraction) software

The microstructure of the samples was studied using a TESCAN MIRA 3 LMH FEG scanning electron microscope (SEM) in the regime of back-scattered electron imaging (BSE) and using a JEM 2000EX transmission electron microscope (TEM).

## 3. Experimental results and discussion

The microstructure of the α titanium alloy after ECAP is presented in figure 1. The ECAP did not result in a transformation of the initial plate-like structure into an equiaxed nanosized structure (figure 1 a). The alloy retained a plate-like microstructure with α-plates having the thickness in between 1 and 10 μm. In the longitudinal section of the sample, plates aligned along its axis dominate. In the cross section, colonies of curved α-plates are observed. The boundaries of the plates are decorated with dispersed particles, the fraction of which is about 3% and the sizes are less than 0.5 μm. According to [4], these are particles of α-phase with an enhanced content of vanadium and reduced content of aluminum.

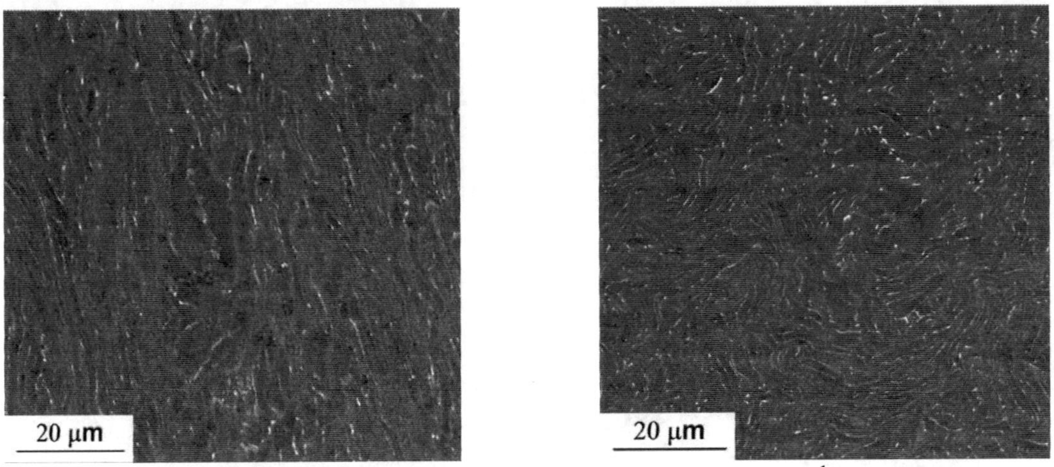


a b

**Figure 1.** BSE images of microstructure of the Ti-5Al-0.5V alloy after ECAP in longitudinal (a) and cross sections (b).

The TEM micrographs of samples after ECAP show that in the bulk of α-plates there are weakly misoriented fragments with sizes of about 1 μm with a high density of dislocations (figure 2 a). The estimates of microstrains and dislocation density obtained by XRD have given the values of approximately $2.6\times10^{-5}$ and $4\times10^{14}$ $m^{-2}$, respectively. The UST did not result in considerable changes of the fine structure of the alloys (figure 2 b), the microstrain and dislocation density values are approximately equal to $2.9\times10^{-5}$ and $5\times10^{14}$ $m^{-2}$.

IOP Conf. Series: Materials Science and Engineering **672** (2019) 012047 doi:10.1088/1757-899X/672/1/012047

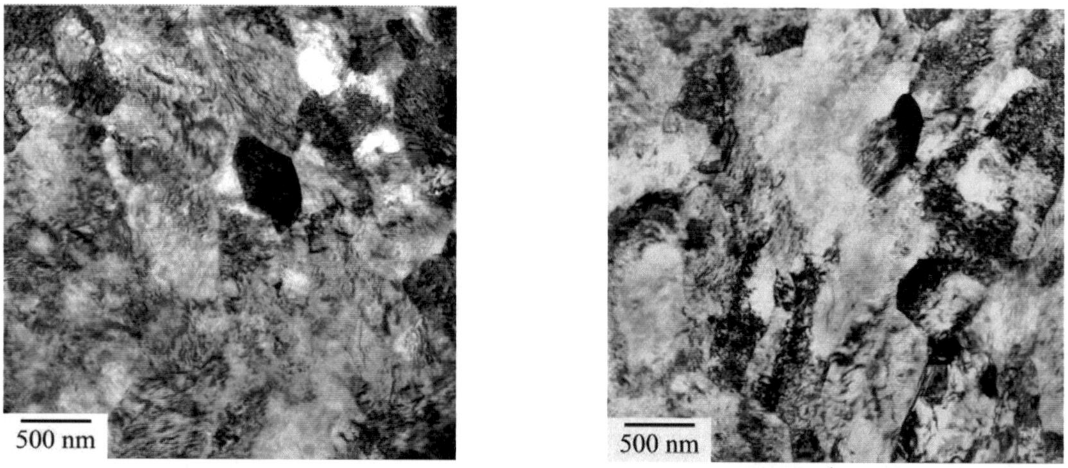


a b

**Figure 2.** TEM images of microstructure of Ti-5Al-0.5V alloy after ECAP (a,) and ECAP+UST (b).

The strength properties of the alloy at room temperature after ECAP and additional UST are similar. The microhardness values in the two states are equal to 3090±130 and 3170±130 MPa, respectively, and the ultimate tensile strength values are equal to 1401±50 and 1433±50 MPa, respectively. These values are typical for the titanium alloys subjected to SPD [1].

The UST did not affect significantly the deformation behavior of the Ti-5Al-0.5V alloy tested in the conditions of low-temperature superplasticity (figure 3 and table 1). The flow stress for all samples attains a maximum value near $\varepsilon \approx 0.01$ and then gradually decreases in a fairly large strain interval. The features of superplasticity are observed only at the strain rates about $10^{-4}$ $s^{-1}$. During deformation with initial strain rate of $1.7\times10^{-4}$ $s^{-1}$, the steady stage flow strain is observed and the strain rate sensitivity coefficient is close to the value of 0.3. Values of the uniform elongation and elongation to failure increase with decreasing strain rate (table 1). The maximum elongation of 342% is demonstrated by the Ti-5Al-0.5V alloy subjected to ECAP followed by UST at a strain rate of $1.7 \times 10^{-4}$ $s^{-1}$.

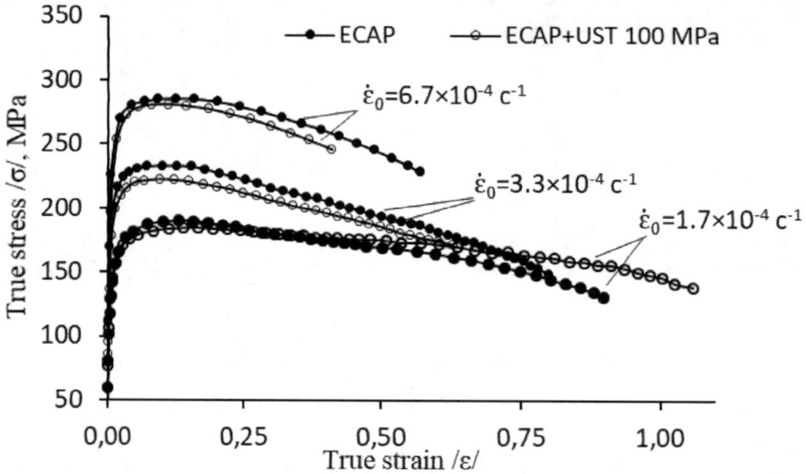


**Figure 3.** True stress – true strain curves of tensile deformation of UFG Ti-5Al after ECAP and ECAP followed by UST.

**Table 1.**

| | Elongation to failure, % | | Uniform elongation, % | | Strain rate sensitivity coefficient /*m*/ | |
|---|---|---|---|---|---|---|
| $\dot{\varepsilon}$, $s^{-1}$ | ECAP | ECAP+UST | ECAP | ECAP+UST | ECAP | ECAP+UST |
| $6.7 \cdot 10^{-4}$ | 143 | 117 | 77 | 51 | - | - |
| $3.3 \cdot 10^{-4}$ | 191 | 184 | 122 | 87 | 0.29 | 0.33 |
| $1.7 \cdot 10^{-4}$ | 257 | 342 | 191 | 203 | 0.30 | 0.29 |

It should be pointed out that α-titanium alloys usually do not exhibit features of superplastic deformation both in the fine grained [5] and submicrocrystalline states [6]. This is explained by the limited number of slip systems in the hcp lattice of titanium and quick growth of α-grains at test temperatures.

## 4. Conclusions

The results of the present study show that the single-phase α-alloy Ti-5Al-0.5V subjected to ECAP and UST can demonstrate superplastic behavior and allow significant elongations of up to about 340% at a temperature of 600 °C and the initial strain rate of $1.7 \times 10^{-4}$ $s^{-1}$.

## Acknowledgments

The present work was supported by the Russian Science Foundation (grant No. 16-19-10126). Electron microscopic studies were carried out on the facilities of shared services center of IMSP RAS "Structural and Physical-Mechanical Studies of Materials".

## References

[1] Valiev R Z, Zhilyaev A P and Langdon T G 2013 *Bulk Nanostructured Materials: Fundamentals and Applications* (Hoboken: Wiley)
[2] Samigullina A A, Nazarov A A , Mulyukov R R, Tsarenko Yu V and Rubanik V V 2014 *Rev. Adv. Mater. Sci.* **39** 48
[3] Samigullina A A, Murzinova M A, Mukhametgalina A A, Zhilyaev A P and Nazarov A A 2018 *Def. Diff. Forum* **385** 53
[4] Lutjering G and Williams J C 2007 *Titanium – Engineering Materials and Processes* (Berlin: Springer)
[5] Kaibyshev O A 1992 *Superplasticity of Alloys, Intermetallides and Ceramics* (Berlin: Springer)
[6] Malysheva S P, Salishchev G A, Galeev R M, Danilenko V N, Myshlyaev M M and Popov A A 2003 *Phys. Metals Metallogr.* **95** 390

IOP Conf. Series: Materials Science and Engineering **672** (2019) 012048 doi:10.1088/1757-899X/672/1/012048

# Microstructure evolution of a Re-containing nickel-based superalloy by hot deformation and cold severe plastic deformation

**Sh Mukhtarov, R Shakhov and A Ganeev**

Institute for Metals Superplasticity Problems RAS, 39, Stepan Khalturin, Ufa, 450001, Russia

E-mail: shamil@anrb.ru

**Abstract**. The microstructure evolutions of the cast rhenium containing nickel-based superalloy subjected to thermomechanical treatment and severe plastic deformation by high pressure torsion were investigated. Thermomechanical treatment led to the formation a fine-grained structure with γ grain size of about 7 μm and both large precipitates of the γ′ phase with a size of 1-5 μm and dispersed precipitates with a size of about 0.2 μm. Microstructure of the superalloy subjected to high pressure torsion up to 5 revolutions was nanocrystalline with partly dissolved of the dispersed γ′ precipitates. The effect of different kinds of deformation and subsequent annealing on the microstructure, grain boundaries and microhardness changes are also discussed.

## 1. Introduction

Improving the chemical composition of nickel-based superalloys for an increase in operating temperature of gas turbine engines leads to a decrease in their deformability. Nickel-based superalloys are multiphase materials, which usually consist of the γ matrix phase, carbides and dispersed hardening phases as γ′ phase, γ″, δ and others [1]. Deformation processing of superalloys can lead to recrystallization of the γ matrix phase, dissolution and precipitation of dispersed phases during high temperatures [1] and the same thing occurs by severe plastic deformation at room temperature followed by annealing [2]. The development of new superalloys goes towards increasing the content of refractory elements [1], which leads to a decrease in the diffusion rate and affects the development of recrystallization. The recently developed superalloy contains of about 2 wt. % of rhenium and its chemical composition is similar to some of the second generation single-crystal superalloys. The deformation behavior, recrystallization of such complex superalloys has not been sufficiently studied. Previous studies [3] showed that a decrease in the size of the recrystallized γ grains of the investigated superalloy to 3-20 μm led to an increase in mechanical properties both at room and at elevated temperatures. The formation of a smaller γ grain size in this superalloy has not been previously obtained. In this regard, it is relevant to study the evolution of the microstructure during deformation of both hot deformation and severe plastic deformation by high pressure torsion at room temperature.

IOP Conf. Series: Materials Science and Engineering **672** (2019) 012048 doi:10.1088/1757-899X/672/1/012048

## 2. Material and methods

The material selected for this study was precipitation hardened with the strengthened phase $\gamma'$ (fcc, $Ni_3Al$) cast nickel-based superalloy Ni-12.5(Al,Ti,Nb,Ta)-37(Cr,Co,W,Mo,Re)-0.17(C,La,Y,Ce,B) (in wt.%). The ingots produced by induction melting had a diameter of 40 mm and a height of 15 mm. The temperature of the full dissolution $\gamma'$ phase was measured via water quenching experiments and correspond to $T_s$=1220±5 °C. The superalloy ingots were subjected to: i) heterogenization annealing in the temperature range of ($T_s$-120)÷($T_s$+20) followed by slow cooling with a rate of 25 °C/min; ii) rapid cooling in air from temperature $T_s$+20. The sample subjected to annealing was compressed to an engineering strain of ε=50% with an initial strain rate of $5\times10^{-4}$ $s^{-1}$ at a temperature of 1175 °C and then cooled in air [3]. The samples with dimension 8 mm in diameter and 0.7 mm in height were cut out of the superalloy. These small discs were processed by HPT up to 5 revolutions at room temperature and annealed at 700 °C for 2 hours in an airtight casing.

The microstructure investigation was carried out using a Tescan Mira-3 scanning electron microscope (SEM) in backscattering electron (BSE) mode and a JEOL JEM-2000EX transmission electron microscope (TEM). Electron backscatter diffraction (EBSD) analysis was performed with a scan-step size of 3 μm from central parts of the deformed sample. The grain boundaries with misorientation more than 15° were assumed as high-angle ones. The grain boundaries having a misorientation angle of less than 2° were excluded from consideration. Vickers microhardness tests were carried out using an AFFRI DM8A digital microhardness tester at a load of 100 g and a holding time of 10 seconds.

## 3. Results and discussion

### *3.1. Microstructure of the superalloy in as-cast condition and subjected to annealing*

The superalloy in the as-cast condition was characterized by a coarse-grained structure with a dendritic segregation and non-equilibrium eutectic (figure 1 a). The dispersed $\gamma'$ phase homogeneously precipitated within the $\gamma$ phase. Carbides are uniformly distributed in the microstructure with a size of about 10 μm. The as-cast superalloy was subjected to heterogenization annealing in order to improve hot workability. The annealing led to the dissolution of the nonequilibrium eutectics and some coagulation of the $\gamma'$ phase (figure 1 b). The annealing also resulted in a partial loss of coherence between the $\gamma'$ phase and the $\gamma$ matrix. Heterogenization annealing led to coagulate of $\gamma'$ particles up to 5-10 μm (figure 1 b).

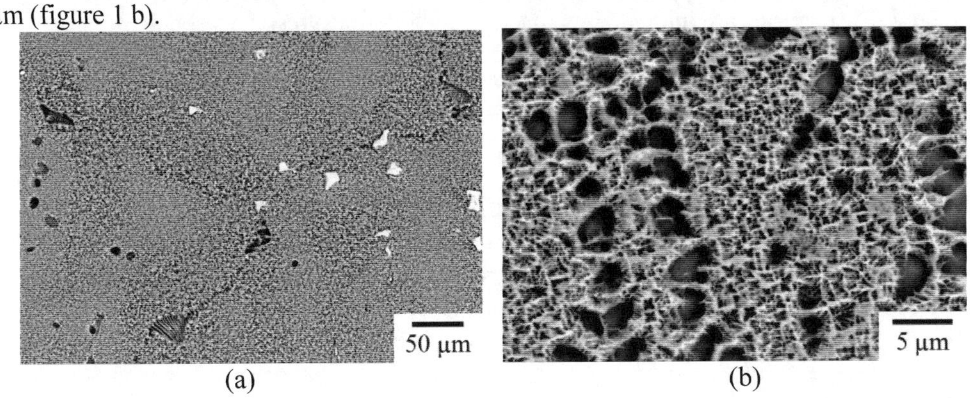


**Figure 1.** SEM images of cast superalloy Ni-12.5(Al,Ti,Nb,Ta)-37(Cr,Co,W,Mo,Re)-0.17(C,La,Y,Ce,B): (a) – as-received condition; (b) – subjected to heterogenization annealing in the temperature range of ($T_s$-120)÷($T_s$+20) followed by slow cooling with a rate of 25 °C/min.

### *3.2. Microstructure evolution of a Re-contain nickel-based superalloy by hot deformation*

Figure 2 shows the microstructure and EBSD map obtained from the central part of the compressed sample of the superalloy. The recrystallization processes occurred extensively at 1175 °C in the central part of the sample. The average size of the recrystallized γ grains in the small central part was down to 7 μm (figures 2 a). In the microstructure there were both coarse γ′ grains with a size of 1-5 μm, which were not dissolved during high temperature heating before hot deformation, and dispersed γ′ precipitates with a size of 0.15-0.25 μm (figure 2 a) that formed during air cooling.

EBSD analysis of the central part of the compressed sample revealed that the first stage of the deformation led to the formation of a developed substructure, which consisted of subgrains (top and bottom part of figure 2 b). An increase in the strain value led to the occurrence of recrystallization along the initial γ grain boundaries, and then near completely recrystallized microstructure was formed. The size of recrystallized γ grains was in the range of 5-50 μm. The large grains contained low-angle boundaries. The main mechanism for the microstructure refinement during compression at 1175 °C was defined as continuous dynamic recrystallization. Therefore, the hot deformation of cast rhenium containing nickel-based superalloy showed the possibility of recrystallization behaviour that was generally similar to that of other nickel-based superalloys.

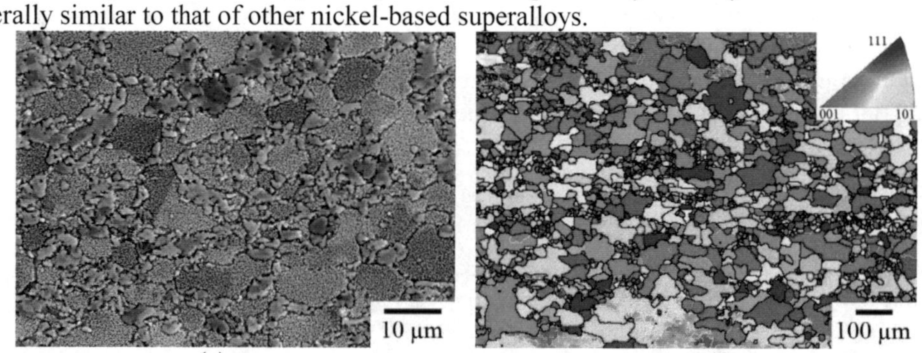


(a) (b)

**Figure 2.** Microstructure and EBSD map of superalloy Ni-12.5(Al,Ti,Nb,Ta)-37(Cr,Co,W,Mo,Re)-0.17(C,La,Y,Ce,B) subjected to heterogenization annealing and following hot deformation at the temperature ($T_s$-25) to strain of ε=50% with an initial strain rate of $5\times10^{-4}$ $s^{-1}$: (a) – BSE image; (b) – normal-direction EBSD (inverse-pole-figure) map, the compression axis is vertical, high- and low-angle grain boundaries are indicated by black and white lines, respectively.

### *3.3. Microstructure evolution of a Re-containing nickel-based superalloy by severe plastic deformation and annealing*

The investigation of cast superalloy subjected to rapid cooling from high temperature showed a coarse grained microstructure with γ grain size of about 300 μm and dispersed γ′ precipitates. The superalloy Ni-12.5(Al,Ti,Nb,Ta)-37(Cr,Co,W,Mo,Re)-0.17(C,La,Y,Ce,B) in three conditions: i) cast condition; ii) cast alloy after heterogenization annealing; and iii) cast alloy after rapid cooling were subjected to HPT up to 5 revolutions. The studies of the microstructure revealed that HPT of samples with different initial conditions led to refined of the microstructure to nanosized. The average grain size was about 30 nm (figure 3 a). Studies of the fine structure revealed that the γ′ phase did not dissolve in all conditions. Figure 3 b shows a TEM image of the sample subjected to heterogenization annealing followed by HPT and annealing at 700 °C for 2 hours. The average γ grain size increased from about 30 to 50 nm independently of the initial condition. Annealing twins were detected within some of the grains.

### *3.4. Effect of HPT and annealing on the microhardness of a Re-containing nickel-based superalloy*

The microhardness of the superalloy Ni-12.5(Al,Ti,Nb,Ta)-37(Cr,Co,W,Mo,Re)-0.17(C,La,Y,Ce,B) in different conditions is represented in figure 4. The microhardness increases up to 6.4-6.7 GPa after

HPT. This is due to the refinement of the microstructure down to the nanocrystalline level. The microhardness increase up to 9.3-9.8 GPa for all conditions after subsequent annealing, is probably associated with the onset of polygonization that occurred before recrystallization [4]. The microhardness of the superalloy subjected to heterogenization annealing following by HPT and annealing at 700 °C is slightly higher than the microhardness of other conditions probably because of the presence of large γ′ precipitates that not dissolved during HPT.

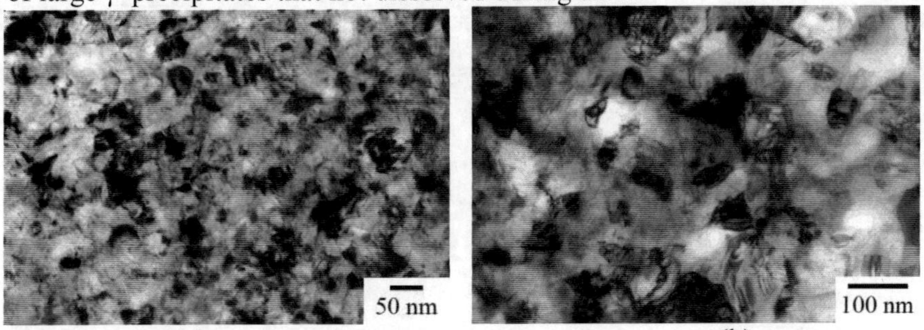


(a) (b)

**Figure 3.** TEM images of the superalloy Ni-12.5(Al,Ti,Nb,Ta)-37(Cr,Co,W,Mo,Re)-0.17 (C,La,Y,Ce,B) subjected to heterogenization annealing following by: (a) – HPT; (b) – HPT and annealing at 700 °C during 2 hours.

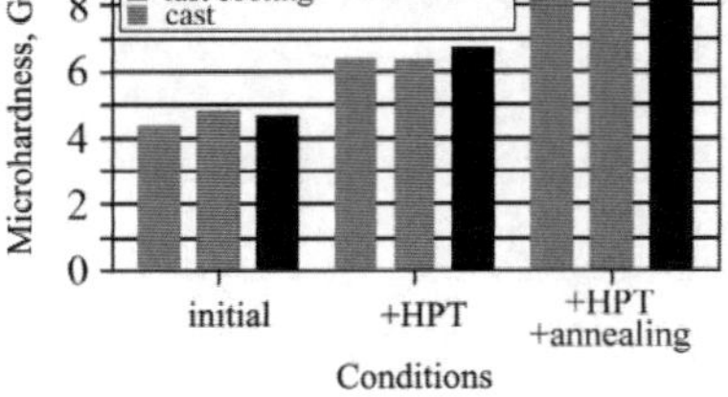


**Figure 4.** Microhardness of the superalloy Ni-12.5(Al,Ti,Nb,Ta)-37(Cr,Co,W,Mo,Re)-0.17(C,La,Y,Ce,B) subjected to different treatments.

## 4. Summary

Hot deformation of the cast superalloy Ni-12.5(Al,Ti,Nb,Ta)-37(Cr,Co,W,Mo,Re)-0.17(C,La,Y,Ce,B) at 1175 °C with an initial strain rate of $5\times10^{-4}$ $s^{-1}$ only on ε=50% led to the formation of the recrystallized microstructure with an average grain size of about 7 μm. HPT of the samples of the cast superalloy with different initial conditions led to refined of microstructure to nanocrystalline. The average grain size was about 30 nm. Annealing at 700 °C for 2 hours of the superalloy with different initial conditions subjected to HPT slightly increased the grain size up to 50 nm. The highest microhardness (9.8 GPa) was achieved in the superalloy subjected to heterogenization annealing following by HPT and subsequent annealing at 700 °C for 2 hours.

**Acknowledgements**

The present work was funded by the Russian Foundation for Basic Research according to research Project #18-08-00997-a. The work was performed using the facilities of the shared services center "Structural and Physical-Mechanical Studies of Materials" at the Institute for Metals Superplasticity Problems of Russian Academy of Sciences.

**References**

[1] Reed R C 2006 *The Superalloys: Fundamentals and Applications* (New York: Cambridge University Press)
[2] Sauvage X and Mukhtarov Sh 2014 *IOP Conf. Ser.: Mater. Sci. Eng.* **63** 012173
[3] Mukhtarov Sh Kh *et al.* 2018 *IOP Conf. Ser.: Mater. Sci. Eng.* **447** 012055
[4] Mukhtarov Sh Kh 2010 *Mater. Sci. Forum* **633-634** 569

IOP Conf. Series: Materials Science and Engineering **672** (2019) 012049 doi:10.1088/1757-899X/672/1/012049

# Microstructure and microhardness studies of Inconel 718, manufactured by selective laser melting and subjected to severe plastic deformation and annealing

**K Sh Mukhtarova[1], R V Shakhov[2], V V Smirnov[1] and Sh Kh Mukhtarov[2]**

[1] Ufa State Aviation Technical University, 12 Karl Marx st., Ufa, 450008, Russia
[2] Institute for Metals Superplasticity Problems of RAS, 39 Stepan Khalturin st., Ufa, 450001, Russia

E-mail: kamilla551@gmail.com

**Abstract**. Microstructure investigations and microhardness measurements have been comparatively performed for the nickel-based superalloy Inconel 718 produced by selective laser melting (SLM) and conventional technology. The superalloy produced by SLM had a finer $\gamma$ grain size and a developed dislocation substructure in contrast to a conventional wrought material. High pressure torsion of the superalloy led to a refinement in the microstructure and an increase in the microhardness in both superalloys. Annealing at 600 °C for 2 hours of the deformed superalloys caused an additional increase in microhardness. The microhardness of the SLM material was slightly higher than that of the conventional superalloy.

## 1. Introduction
Inconel 718 is a nickel-chromium-based superalloy, capable of withstanding high mechanical and thermal loads, widely used in rocket and aircraft engines [1]. The performance characteristics of this alloy depend on the phase composition, as well as on the technology of its production and heat treatment. For example, selective laser melting (SLM) is one of the additive manufacturing methods, utilizing a 3D layer-by-layer fabrication to produce a solid, simple or geometrically complex objects from a series of layers [2]. This process includes many cycles of heating and rapid cooling, which lead to the formation of a complex microheterogeneous structure with the precipitation of dispersed strengthening phases and oxides. Therefore, the microstructure of the SLM material is different from the material produced by conventional methods. The microstructure of the conventional superalloy subjected to severe plastic deformation by high pressure torsion (HPT) was refined to nanosized [3]. HPT can lead to the enhancement of diffusion processes and the dissolution of the precipitates in the alloy. In the present paper, the microstructure and microhardness of the alloy Inconel 718 manufactured by SLM and the conventional alloy, produced by conventional technology, subjected to HPT and annealing, were compared.

## 2. Experimental
The Inconel 718 alloy was produced in Ufa State Aviation Technical University. The chemical composition of the Inconel 718 manufactured by SLM is presented in Table 1. The samples with dimensions 8 mm in diameter and 0.7 mm in thickness were cut out of the SLM alloy. These small

IOP Conf. Series: Materials Science and Engineering **672** (2019) 012049 doi:10.1088/1757-899X/672/1/012049

discs were processed by HPT up to 5 revolutions at room temperature and annealed at 600 °C for 2 hours in an airtight casing.

**Table 1**. The chemical composition of the Inconel 718 alloy manufactured by SLM [wt.%].

| Cr | Al | Ti | Fe | Nb | Mo | Co | B | C | Ni |
|---|---|---|---|---|---|---|---|---|---|
| 19.84 | 0.64 | 1.01 | 18.40 | 5.93 | 3.44 | 0.12 | 0.025 | 0.04 | base |

The microstructural examination was carried out by scanning electron microscopy (SEM) in the backscattering electron (BSE) mode. The electron backscattered diffraction (EBSD) analysis was performed with a scan-step of 0.5 μm and was carried out by Tescan Mira-3. The EBSD analysis was conducting using the Channel 5 processing software. Transmission electron microscopy was carried out by JEOL JEM-2000EX, JEM-2100Plus. The samples and foils were polished by Tenupol-5 at 50V using an electrolyte with the composition: 10% perchloric acid and 90% butanol. Vickers microhardness tests were carried out using an Axiovert 100A with a microhardness tester MNT-10 attachment with a load of 100 g and a holding time - 10 seconds.

## 3. Results and discussion

### *3.1. The microstructure of SLM Inconel 718*

Figure 1a represents the BSE cross section image. γ grains consist of columnar and equiaxed subgrains (figure 1b). The average size of the matrix of γ grains is about 20 μm. The plates of the δ phase and small carbides are distinguished at the γ grain boundaries. Columnar subgrains consist of submicrocrystalline equiaxed subgrains with a size of about 0.5 μm (figure 1b, c). EBSD analysis of the initial microstructure shows that the fraction of high-angle grain boundaries was 26% [4].

(a) (b) (c)

**Figure 1.** The microstructure of SLM Inconel 718: (a, b) - BSE images of coloumnar subgrains and of equiaxed subgrains; (c) - TEM image of submicrocrystalline equiaxed subgrains.

### *3.2. The microstructure of SLM Inconel 718 processed by HPT*

Figure 2 shows BSE and TEM images of SLM Inconel 718 processed by HPT. The γ grain size in the SLM alloy processed by HPT was about 30 nm that is comparable with that obtained in the industrial alloy subjected to HPT [3]. Dispersed particles, probably fine carbides and oxides, are visible in figure 2a. The diffraction pattern of the alloy after HPT (figure 2b) shows the presence of orthorhombic $Ni_3Nb$ (δ phase, (012)), as well as $Ni_3Al$ ($\gamma'$ phase, (222)). It is well known that the volume fraction of $\gamma'$ phase in Inconel 718 is less than 5% and precipitates like $\gamma''$ (δ) can be more than 40%. $\gamma'$ and $\gamma''$ phases are almost completely dissolved in the industrial alloy subjected to HPT [3]. It is likely that the large amount of dispersed precipitates as δ phase, carbides and oxides of the SLM alloy do not dissolved in the γ matrix during HPT.

IOP Conf. Series: Materials Science and Engineering **672** (2019) 012049 doi:10.1088/1757-899X/672/1/012049

### *3.3. The microstructure of SLM Inconel 718 subjected to annealing*

Figure 3 represents the BSE images and the EBSD map of SLM Inconel 718 after annealing at 600 °C for 2 hours. A number of the δ phase precipitates, carbides and oxides increased along the grain and subgrain boundaries (figure 3a, b). The EBSD analysis was obtained (figure 3c) and the fraction of high angle grain boundaries increased during annealing and reached approximately 53%. The microstructure consists of large and fine γ grains. Annealing twins were not detected. The average γ grain size did not change.

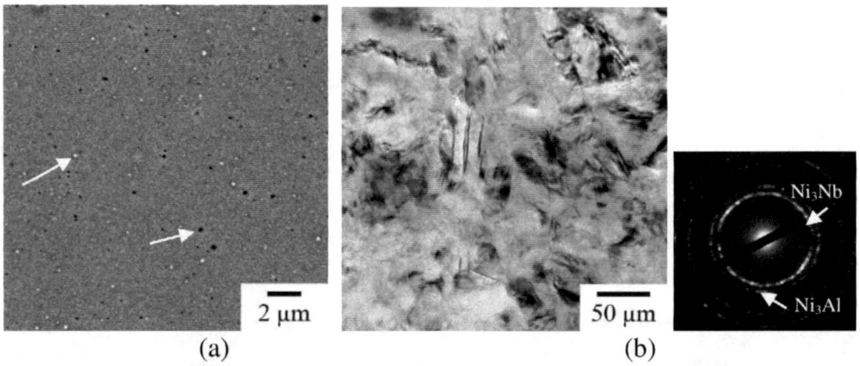


**Figure 2.** Microstructure of SLM Inconel 718 subjected to HPT: (a) - BSE image, arrows show black and white particles; (b) - TEM image with electron diffraction pattern.

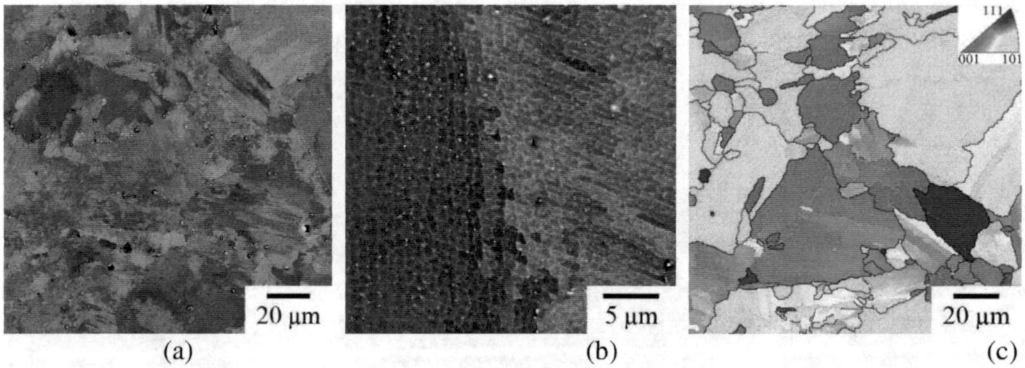


**Figure 3.** The microstructure of SLM Inconel 718 after annealing at 600 °C for 2 hours: (a, b) - BSE images, (c) - EBSD normal-direction (inverse-pole-figure) orientation map.

### *3.4. The microstructure of SLM Inconel 718 subjected to HPT and annealing*

Figure 4 shows the BSE and TEM images of the sample subjected to HPT and annealing at 600 °C for 2 hours. The average γ grain size increased from about 30 to 50 nm. The amount of fine carbides and oxides decreased. Some γ grain boundaries had a sharp contrast. The δ phase precipitates were revealed in the microstructure. The diffraction pattern at figure 4b of the alloy after HPT and annealing showed the presence of orthorhombic $Ni_3Nb$ (δ phase, (412)) and Ni (γ phase, (220)). The γ′ phase disappeared after annealing. Severe plastic deformation and annealing of the SLM alloy led to the structure close to the industrial superalloy subjected to HPT and annealing.

### *3.5. Effect of HPT and annealing on the microhardness of SLM Inconel 718 in comparison with industrial Inconel 718*

The microhardness of SLM Inconel 718 and industrial Inconel 718 in different conditions is represented in figure 5. The increase in microhardness after HPT is related with a refinement of the microstructure down to the nanocrystalline level. The increase in microhardness also after subsequent annealing is probably associated with the additional precipitation by the δ phase and the onset of polygonization that occurred before recrystallization [5]. The microhardness of SLM Inconel 718 is higher than that of the industrial alloy, probably due to the presence of dispersed oxides and carbides in the SLM alloy.

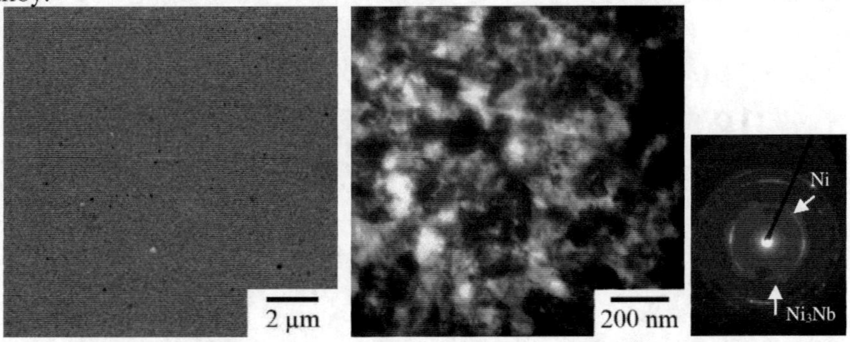


**Figure 4.** The microstructure of SLM Inconel 718 subjected to HPT and annealing at 600 °C for 2 hours: (a) - BSE image, (b) - TEM image with electron diffraction pattern.

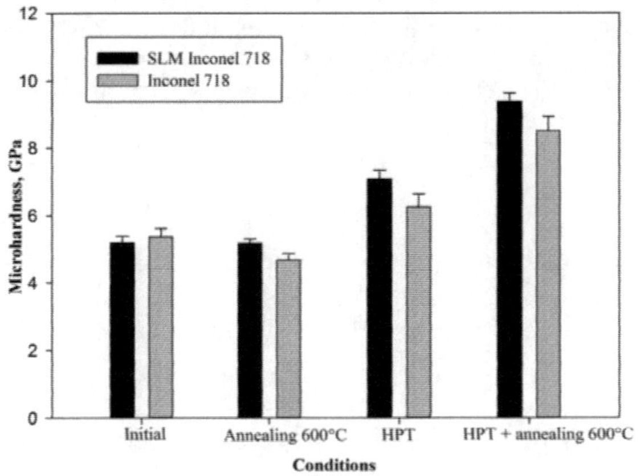


**Figure 5.** Microhardness of SLM Inconel 718 and industrial Inconel 718 subjected to different treatments.

## 4. Conclusions

Annealing of SLM Inconel 718 at 600 °C for 2 hours did not affect the γ grain size and the microhardness. The nanostructure was formed in SLM Inconel 718 after HPT, which was found to the thermally near stable at 600 °C. The highest microhardness (9 GPa) was achieved in SLM Inconel 718 processed by HPT and annealed at 600 °C for 2 hours.

## Acknowledgments

The work was accomplished according to the state assignment of IMSP RAS (AAAA-A17-117041310215-4). Electron microscopic, experimental studies were carried out on the facilities of shared services center of the Institute for Metals Superplasticity Problems of Russian Academy of Sciences "Structural and Physical-Mechanical Studies of Materials".

## References

[1] Schafrik R and Sprague R 2008 *Key Eng. Mater.* **380** 113
[2] Murr L E *et al.* 2012 *J. Mater. Res. Technol.* **1** 42
[3] Sauvage X and Mukhtarov Sh 2014 *IOP Conf. Ser.: Mater. Sci. Eng.* **63** 012173
[4] Shakhov RV and Mukhtarova K Sh 2018 *Letters on Materials* **8** 105
[5] Mukhtarov Sh Kh 2010 *Mater. Sci. Forum* **633-634** 569

IOP Conf. Series: Materials Science and Engineering **672** (2019) 012050 doi:10.1088/1757-899X/672/1/012050

# Strength and electrical conductivity of UFG Cu-Fe alloys subjected to HPT

**K M Nesterov and R K Islamgaliev**

Institute of Physics of Advanced Materials, Ufa State Aviation Technical University, 12 K. Marx st. Ufa, 450008 Russia

E-mails: kmnesterov@mail.ru, rinatis@mail.ru

**Abstract.** This paper investigates the strength and specific conductivity of UFG Cu-Fe alloys processed by high-pressure torsion (HPT). The effect of an ultrafine-grained structure and further heat treatment on the increase of microhardness and specific conductivity is examined. The results of the structural study of the UFG samples by electron back-scatter diffraction (EBSD) are presented.

## 1. Introduction

It is known that pure Cu is widely used in electrical-engineering products as a material with a high conductivity. At the same time, for the application under high mechanical loads, e.g. in electrical connectors or contacts, conductive thermally stable materials with enhanced strength characteristics are required. In such cases where an industry does not require a high electrical conductivity, the manufacturers prefer using cheaper bronzes of brass. In recent years, due to the low cost of iron, copper-iron alloys and composites have been gaining popularity, as they have higher strength and lower electrical conductivity as compared to pure copper [1].

On the other hand, in recent years, special interest is aroused by ultrafine-grained (UFG) materials having an average grain size of below 1 μm that are able to exhibit unique mechanical properties (high strength and enhanced endurance limit) [2, 3]. Besides, it is known that at room temperature grain boundaries make a smaller contribution into the enhancement of electrical resistance as compared to alloying element atoms, which creates conditions for the preservation of the initial values of electrical conductivity in UFG copper materials [4, 5]. All this provides prerequisites for achieving in UFG samples of Cu alloys a combination of high strength and improved electrical conductivity.

In this paper, we examine the effect of the UFG structure of a Cu-Fe alloy on the strength enhancement by means of grain refinement and improvement of electrical conductivity by reducing the concentration of impurities and alloying element atoms as a result of their migration to grain boundaries.

## 2. Materials and research methods

As the initial material for the study, we selected the copper alloys Cu-0.15Fe, Cu-0.5Fe, Cu-2Fe and Cu-3.5Fe. The initial as-cast samples were subjected to homogenizing annealing at a temperature of 1070 °C for 4 hours with cooling in water. The heat treatment of the samples was carried out in a Nabertherm muffle furnace.

IOP Conf. Series: Materials Science and Engineering **672** (2019) 012050 doi:10.1088/1757-899X/672/1/012050

Nanostructuring was implemented by the high-pressure torsion (HPT) method at room temperature. For this purpose, disks with a diameter of 20 mm and thickness of 2 mm were cut off from the initial rods and subjected to high-pressure torsion (6 GPa) for 10 revolutions.

In order to study the thermal stability of the UFG samples, we studied the variation of microhardness and electrical conductivity after annealing in a temperature range of 200-600 °C with 30 minutes' exposure. Microhardness (HV) was tested by the Vickers method using a Micromet 5101 device with a load of 100 g and an exposure time of 10 seconds.

Specific conductivity was measured by the eddy-current method using a VE-27NTs/4-6 device. The measurement range of specific conductivity was 3.5-60 MS/m, the maximum permissible error was ±3%. The data are presented in % IACS (International Annealed Copper Standard).

Structural studies by electron back-scatter diffraction (EBSD) were performed using a scanning electron microscope FEG-SEM TESCAN MIRA 3 LMH, with an accelerating voltage of 20 kV and a scanning step of 50 to 150 nm. To analyze the Kikuchi lines, we used the CHANNEL 5 software by Oxford Instruments. The error in measuring the crystal lattice orientation was not more than 0.6°.

## 3. Results and discussion

The testing of strength reveals that the maximum microhardness value of 1770 MPa was achieved in the UFG Cu-2Fe samples at an annealing temperature of 300 °C (figure 1 a). The maximum values of specific conductivity 74% and 80% IACS were achieved in both the UFG Cu-0.5Fe and Cu-0.15Fe samples at an annealing temperature of 450°C (figure 1 b). On the basis of the obtained data, the alloys Cu-0.5Fe and Cu-2Fe were selected for further structural analysis.

Figure 2 a presents an EBSD image for the Cu-0.5Fe alloy subjected to HPT and annealing a temperature of 350 °C for 30 minutes. A heating temperature of 350 °C was selected because at this temperature one can observe the best combination of strength and electrical conductivity (figure 1). It can be seen that the samples have a UFG structure with an average grain size of 650 nm (table 1). The fraction of high-angle grain boundaries is 39.9%, the fraction of special boundaries Σ3 is 2.45%.

Figure 2 b presents the results of the EBSD analysis of the Cu-2Fe alloy subjected to HPT and annealing at a temperature of 450 °C for 30 minutes. It can be seen that such a processing has produced a UFG structure with an average grain size of 230 nm (table 1). The fraction of high-angle grain boundaries is 37.3%, the fraction of special boundaries Σ3 is 4% (table 1).

It should be noted that there are papers that investigated the structure and properties of the alloys and composites based on Cu-Fe [6-9]. In particular, the papers [7, 8] examined the formation of a supersaturated solid solution in Cu-Fe composites with Fe content 12 at.% [7] and 15 wt.% [8] subjected to HPT, but did not touch upon the studies of electrical conductivity. There is also a paper that examined the strength and electrical conductivity of the Cu-14Fe composite, but in that paper the samples were produced by extrusion and further heat treatment, resulting in UTS=900 MPa and electrical conductivity of 57% IACS [9]. Meanwhile, in the present paper we have demonstrated that the formation of a UFG state enables achieving microhardness in a range of 1750-1500 MPa and electrical conductivity of 45-80% IACS, depending on the Fe content and the heat treatment regimes.

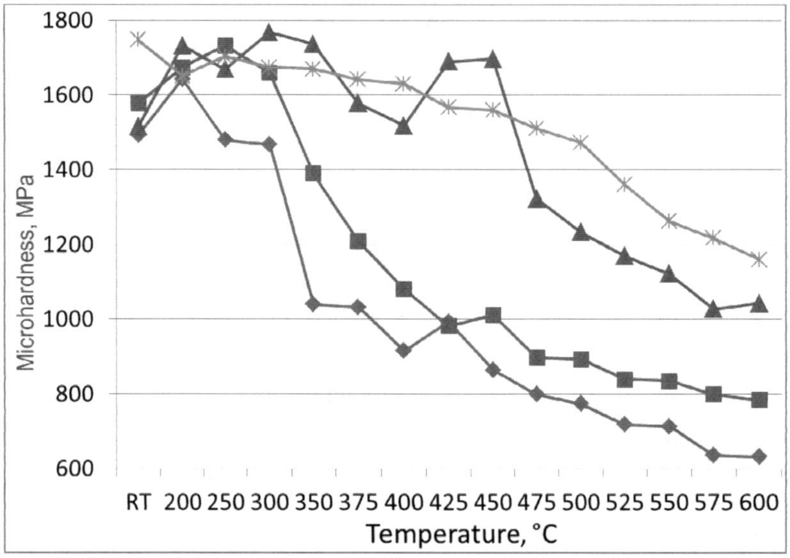


a)

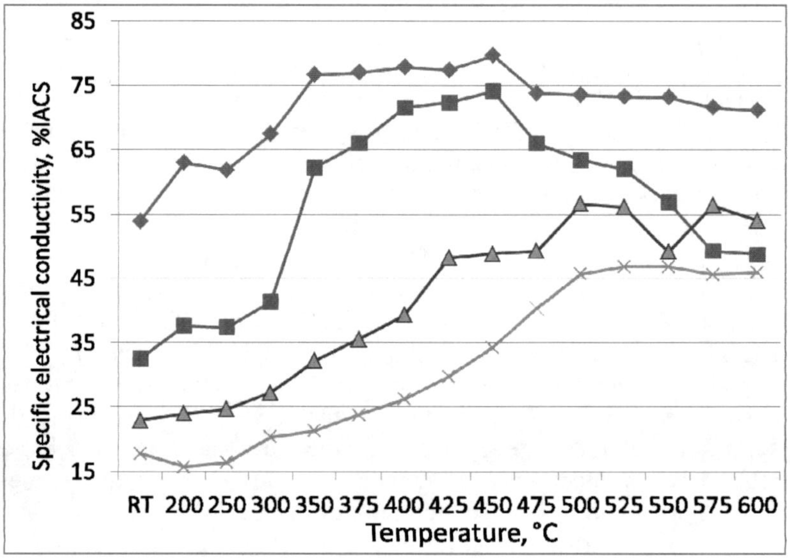


b)

**Figure 1.** a – microhardness, b – specific electrical conductivity of the Cu-Fe alloys under different heat treatments. HPT Cu-0,15Fe HPT Cu-0,5Fe HPT Cu-2Fe HPT Cu-3,5Fe

IOP Conf. Series: Materials Science and Engineering **672** (2019) 012050 doi:10.1088/1757-899X/672/1/012050

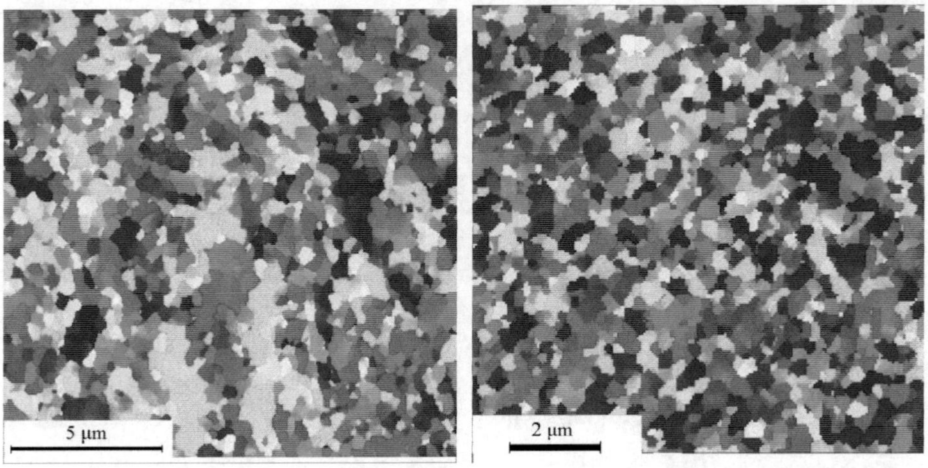


a) b)

**Figure 2.** a – EBSD image for the UFG Cu-0.5Fe alloy, b – EBSD image for the HPT-processed UFG Cu-2Fe alloy.

**Table 1.** Summarized data from the EBSD analysis.

| Alloy | HPT processing | Heat treatment | Average grain size, nm | Fraction of LABs, % | Fraction of HABs, % | Special boundaries Σ3, % |
|---|---|---|---|---|---|---|
| Cu-0.5Fe | HPT 20 °C 10 rev. | HT 350 °C | 653 | 60.1 | 39.9 | 2.45 |
| Cu-2Fe | HPT 20 °C 10 rev. | HT 450 °C | 235 | 62.7 | 37.3 | 4 |

## 4. Conclusions

Based on the results of the studies of strength and electrical conductivity, the following conclusions can be made:

1. HPT processing and further heat treatment in a temperature range of 300-450 °C have produced a UFG structure in the copper alloys Cu-0.5Fe and Cu-2Fe, which contain a fraction of high-angle boundaries 37.3-39.9% and a fraction of special boundaries Σ3 2.45-4%.
2. UFG samples of copper alloys of the Cu-Fe system with Fe content in a range of 0.15-3.5 wt.% can exhibit microhardness in a range of 1750-1500 MPa and electrical conductivity of 45-80% IACS, depending on the Fe content and the heat treatment regimes.

## Acknowledgements

This work was supported by the Russian Foundation for Basic Research under grant No. 20-02-00598.

## References

[1] Xie Z X, Gao H Y, Wang J and Sun B D 2011 *Mater. Sci. Eng.* A **529** 388
[2] Valiev R Z, Islamgaliev R K and Alexandrov I V 2000 *Prog. Mater. Sci.* **45** 103
[3] Wen H, Zhao Y, Nesterov K M, Islamgaliev R K, Valiev R Z and Lavernia E J 2013 *Phil. Mag. Letter.* **8** 481
[4] Islamgaliev R K, Nesterov K M, Bourgon J, Champion Y and Valiev R Z 2014 *J. Appl. Phys.* **115** 194301
[5] Islamgaliev R K, Pekala K, Pekala M and Valiev R Z 1997 *Phys. Stat. Sol.* A **162** 559
[6] Bachmaier A and Pippan R 2013 *International Materials Reviews* **58** 41
[7] Quelennec X, Menand A, Le Breton J M, Pippan R and Sauvage X 2010 *Phil. Mag.* **90** 1179
[8] Bachmaier A, Kerber M, Setman D and Pippan R 2012 Acta Mater. **60** 860
[9] Liu K, Jiang Z, Zhao J, Zou J, Lu J and Lu D 2015 *Metall. Mater. Trans.* A **46** 2255

IOP Conf. Series: Materials Science and Engineering **672** (2019) 012051 doi:10.1088/1757-899X/672/1/012051

# Effect of the UFG structure on the fatigue properties of ferritic/martensitic steel

**M A Nikitina, R K Islamgaliev, A V Ganeev, V D Sitdikov and V S Zhernakov**

Institute of Physics of Advanced Materials, Ufa State Aviation Technical University, 12 K. Marx str., Ufa 450008 Russia

E-mail: nik.marina.al@gmail.com

**Abstract**. This paper presents the results of investigation of the effect of an ultrafine-grained structure on the fatigue properties of ferritic/martensitic steel. Special attention is paid to the study of structural changes after fatigue tests. Scanning electron microscopy (SEM) and transmission electron microscopy (TEM) were employed to study the features of fatigue fracture and microstructural evolution, depending on the distance from the fracture surface. Using X-ray diffraction, we determined the lattice parameter, coherent scattering region, dislocation density, the volume fraction and size of carbide particles. The paper discusses the effect of the ultrafine-grained structure, as well as the volume fraction and size of carbides on the mechanism of fracture during fatigue tests.

## 1. Introduction

It is known that in 9-12%Cr steels during low-cycle fatigue tests at elevated temperatures, softening occurs due to the instability of the dislocation structure, which leads to a change in the dislocation density and carbide coarsening [1, 2]. Besides, the fatigue strength is greatly influenced by the structural features formed after quenching and tempering. For example, it was found that carbide in the form of chains at the boundaries of former austenite grains are the spots for the crack nucleation in ferritic/martensitic steels [3]. Among the methods that have been actively developed in the past two decades and enable a considerable enhancement of the strength properties of metallic materials are severe plastic deformation (SPD) techniques. It is known that the UFG structure formation by means of equal-channel angular pressing (ECAP) leads to an increase in microhardness and the ultimate tensile strength in various metals and alloys [4], and could result in an increase of the fatigue endurance limit. For instance, in [5] it was demonstrated that the UFG structure formation occurring in the process of ECAP, alongside with a more uniform distribution of strengthening particles, also contributed to an increase in fatigue endurance limit in a number of Al-, Mg- and Ti-based light alloys.

Ferritic/martensitic steel 12Cr-2W-2Ni-0.5Mo, used in power mechanical engineering and operating at temperatures of up to 600 °C, contains a large number of carbide particles. It was found that in the process of SPD, these particles become refined and their volume fraction increases [6, 7]. In [8], it was shown that the particles of $Me_{23}C_6$ carbides precipitating at grain boundaries inhibit the dislocation motion, enhancing the low-cycle fatigue behavior. However, currently there is little data from the study of the fatigue properties and the effect of carbide particles in UFG ferritic/martensitic steels. The aim of the present paper is to study the features of fatigue fracture in the UFG ferritic/martensitic steel 12Cr-2W-2Ni-0.5Mo.

## 2. Experimental procedure

As the initial material, we used samples of ferritic/martensitic steel 12Cr-2W-2Ni-0.5Mo (table 1). Prior to ECAP processing, we performed a standard treatment – quenching from a temperature of 1050 °C in oil and tempering at 800 °C. Processing by ECAP was performed using 6 passes on the route $B_c$ with the angle of intersection of the channels equal to 120 degrees at a temperature of 550 °C on samples with a diameter of 20 mm and a length of 100 mm [6,7].

**Table 1.** Chemical composition, wt.%.

| C | Si | Mn | Ni | S | P | Cr | Mo | W | V |
|---|---|---|---|---|---|---|---|---|---|
| 0.14 | 0.31 | 0.32 | 1.63 | 0.009 | 0.02 | 11.53 | 0.45 | 1.66 | 0.18 |

To find the fatigue endurance limit, we used cylindrical samples with a gage diameter of 3 mm. The fatigue tests of the standard samples were performed under a symmetric loading cycle at room temperature with 12 samples per each condition. The cycle asymmetry coefficient $R = \sigma_{min}/\sigma_{max}$ was considered as equal to -1.

For X-ray and TEM analysis, disks with a diameter of 3 mm were cut near the fracture surface. The first disk was cut at a distance of 0.3 mm from the fracture surface, followed with a step of 0.5 mm. To calculate the lattice parameter and the sizes of the coherent scattering regions, we used diffraction patterns obtained with a Rigaku Ultima IV diffractometer employing the Bragg-Brentano focusing method (goniometer geometry). The microstructure was characterized using a JEM-2100 transmission electron microscope. Thin foils for TEM investigations were punched from the slices of the samples. Then they were mechanically ground to a thickness of 0.15 mm, and finally double-jet electropolished to perforation using an electrolyte based on n-butyl alcohol. To study the relief surface of the fractured samples, and also to further study the structure, a JSM-6390 scanning electron microscope with an accelerating voltage of 30 kV was employed [7].

## 3. Results

The microstructural study by TEM revealed that the structure after quenching was martensitic, with an average band width of 520 nm and a length of 3-4 μm. Tempering resulted in the precipitation of the second-phase particles about 40-60 nm in size, predominantly at the boundaries of former martensite plates [6,7]. ECAP treatment led to the formation of an equiaxed UFG structure with an average grain size of up to 0.6 μm with high-angle misorientations and an average particle size of up to 90 nm. In this case, a uniform arrangement of particles is observed both at the boundary and in the body of the grain. The analysis revealed the presence of $Cr_{23}C_6$, $Fe_3C$, WC, $Ni_3C$ particles [7]. The X-ray phase analysis demonstrated that SPD leads not only to a decrease in the size of coherent scattering regions (CSR) and an increase in the dislocation density, but also a decrease in the lattice parameter and an increase in the volume fraction of carbide particles [7]. After ECAP, in comparison with the tempered sample, one can observe an increase in the ultimate tensile strength from 950 MPa to 1100 MPa and the yield stress from 790 MPa to 1040 MP without significant changes in ductility [7].

Figure 1 shows the fatigue curves in the initial state (tempering) and after ECAP processing. It can be seen that after ECAP processing, the fatigue endurance limit at $10^7$ cycles increases by 20% from 472 MPa to 570 MPa. The enhancement of the fatigue properties in the samples with a UFG structure is related to grain refinement, as well as with an increase in the volume fraction of second-phase particles [7]. In the low-cycle region (up to $10^5$ cycles), there are practically no differences in the fatigue curves between the tempered and the UFG samples. However, in the high-cycle region, the larger the number of cycles to failure, the more significant is the advantage of the UFG material.

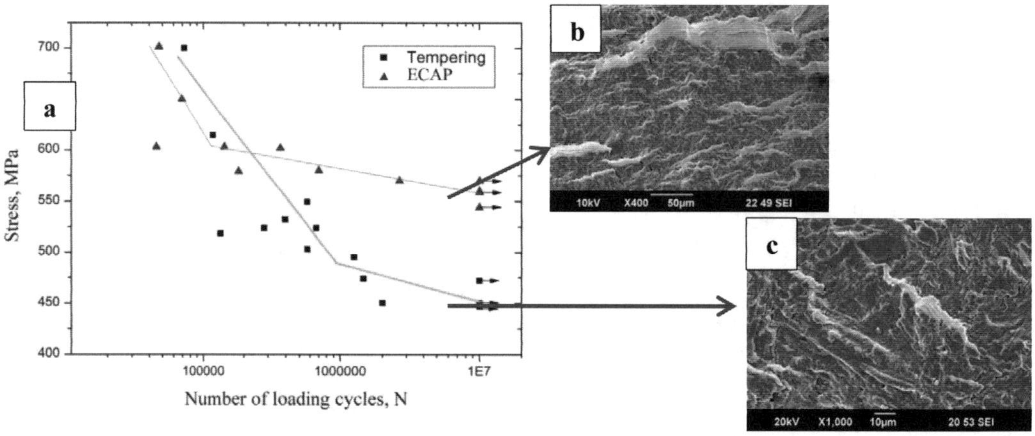


**Figure 1.** (a) Dependence of the number of cycles to failure, N, on the stresses in a cycle; microrelief of the fracture in the zone of stable crack growth: (b) in the UFG state, (c) in the CG state.

The basic difference in the fracture behavior, based on the data of fractographic analysis, is that in the ECAP-processed samples, the zone of stable crack growth is practically twice as large as the similar zone in the samples after tempering. In the UFG state, we observe the formation of wider (up to 50 μm) fatigue striations where the inhibition of secondary cracks occurs (figure 1 b, c).

In the samples after the standard treatment (tempering), we observe a line arrangement of carbide particles along the boundaries of former austenite grains (figure 2), which is a factor that reduces the fatigue endurance limit in ferritic/martensitic steels [3]. In the ECAP-processed samples, in the immediate vicinity of the fracture zone, we observe finer carbide particles. Besides, the dislocation density immediately in the fracture zone grows by a factor of 1.5, but with distance from the fracture zone, the values gradually decrease to the initial state before testing.

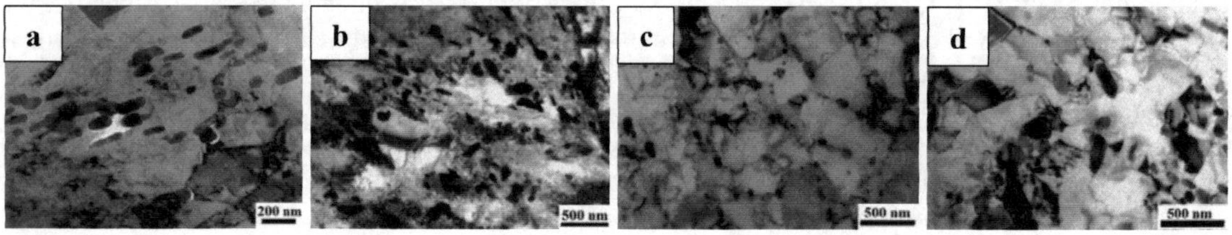


**Figure 2.** Microstructure (a, b) of the tempered samples and (c, d) of the ECAP-processed samples after fatigue tests: (a, c) – in the fracture zone, (b, d) at a distance of 0.5 mm from the fracture zone.

X-ray diffraction analysis shows that in the fracture zone in the UFG state a dynamic aging process takes place, as a result of which the lattice parameter decreases (table 2).

**Table 2.** XRD and TEM results after fatigue testing of the tempered and the ECAP-processed samples.

| State | | Lattice parameter, A | Lattice strain, % | Crystallite size, nm | Disl. dens., $10^{15}$ m$^{-2}$ | Precipitate size, nm (TEM) |
|---|---|---|---|---|---|---|
| Coarse-grained sample | | | | | | |
| Tempering | | 2.87801(5) | 0.22(9) | 122(31) | 0.96(9) | 250±30 |
| Distance from the fracture zone, mm | 0.3 | 2.878206(6) | 0.166(9) | 81(6) | 2.11(8) | 150±14 |
| | 0.8 | 2.878186(7) | 0.154(6) | 96(8) | 2.04(7) | 160±12 |
| | 1.3 | 2.878159(8) | 0.131(8) | 114(11) | 1.43(9) | 200±14 |
| | 1.8 | 2.878148(9) | 0.123(5) | 127(15) | 1.12(6) | 220±20 |
| | 2.3 | 2.878133(9) | 0.112(7) | 131(18) | 1.06(11) | - |
| Ultrafine- grained sample | | | | | | |
| ECAP | | 2.87409(4) | 0.13 | 36(4) | 3.87(16) | 90±8 |
| Distance from the fracture zone, mm | 0.3 | 2.874197(5) | 0.227(5) | 24(4) | 5.72(6) | 60±7 |
| | 0.8 | 2.874184(8) | 0.219(8) | 27(3) | 5.21(5) | 80±9 |
| | 1.3 | 2.874146(9) | 0.211(7) | 31(2) | 4.86(5) | 90±10 |
| | 1.8 | 2.874122(6) | 0.205(8) | 34(2) | 4.27(4) | 95±12 |
| | 2.3 | 2.874110(7) | 0.193(9) | 37(3) | 3.94(5) | - |

## 4. Conclusions

The formation of the UFG structure enabled increasing the fatigue endurance limit by 20% as compared to the standard treatment (from 472 MPa to 570 MPa). Smaller second-phase particles and their uniform distribution along the grain boundaries and in grain interiors inhibited crack propagation, thereby increasing the fatigue endurance limit. The main differences in the fracture behavior of the 12Cr-2W-2Ni-0.5Mo steel in the ultrafine-grained and coarse-grained samples were, first, the different areas of the crack nucleation zone (in the CG state, the zone width was 0.7 mm, and in the UFG state – 1.4 mm), and second, the size and distribution of carbide particles.

**Acknowledgements**

The authors are grateful to the Russian Ministry of Science and Higher Education for support within the project part of the program for universities through project No. 16.2061.2017/4.6. M.A. Nikitina is grateful for the support for the fabrication of the UFG samples of the ferritic/martensitic steel and their studies by TEM from RFBR under project No. 18-38-00649. The X-ray investigations of the ferritic/martensitic steel were supported by the Russian Science Foundation under project No.19-19-00496 (V.D. Sitdikov).

**References**

[1] Giordana M, Giroux P, Alvarez-Armas I, Sauzay M, Armas A and Kruml T 2012 *Mater. Sci. Eng.* A **550** 103
[2] Song X, Yang G, Zhou S, Fan H, Yang S, Zhu J and Liu Y 2008 *Mater. Sci. Eng.* A **483–484** 211
[3] Batista M, Hereñú S and Alvarez-Arma I 2014 *Procedia Engineering* **74** 228
[4] Valiev R, Islamgaliev R and Alexandrov I 2000 *Prog. Mater. Sci.* **45** 103
[5] Estrin Y and Vinogradov A 2010 *International Journal of Fatigue* **32** 898
[6] Sitdikov V, Islamgaliev R, Nikitina M, Sitdikova G, Alexandrov I and Wei W 2019 *Phil. Mag.* **99** 73
[7] Islamgaliev R, Nikitina M, Ganeev A and Sitdikov V 2019 *Mater. Sci. Eng.* A **744** 163
[8] Jing H, Luo Z, Xu L, Zhao L and Han Y 2018 *Mater. Sci. Eng.* A **731** 394

IOP Conf. Series: Materials Science and Engineering **672** (2019) 012052 doi:10.1088/1757-899X/672/1/012052

# Effect of severe plastic deformation on the physical and mechanical properties of the ultrafine-grained β-alloy Ti-15Mo

**V V Polyakova, S A Gatina, A A Polyakov and I P Semenova**

Institute of Physics of Advanced Materials, Ufa State Aviation Technical University, 12 K. Marx street, Ufa, 450008, Russian Federation

E-mail: vnurik@gmail.com

**Abstract.** The influence of severe plastic deformation on physical and mechanical properties of the pseudo-β-alloy Ti-15Mo alloy has been investigated.

## 1. Introduction

The combination of a low elastic modulus, good corrosion resistance and unique biocompatibility determines the wide use of titanium and its alloys in medicine as functional materials for designing implants of a wide range of applications [1]. Lately, much attention has been paid in the world to the development of pseudo-β-titanium alloys, which have the ability to change their elastic modulus depending on the phase composition [2]. Among the commercial Ti alloys certified in medicine, the Ti-15Mo alloy (ASTM F2066) is intended for applications in orthopedics, traumatology, and cardiovascular surgery. The Ti-15Mo alloy has the lowest elastic modulus (about 80 GPa) in the single-phase β-state. However, it is characterized by unsatisfactory strength (UTS$\leq$ 800 MPa, $\sigma_{-1}\leq$400 MPa) (ASTM F 2066). Conventional methods of improving the strength of pseudo-β-titanium alloys by thermal or thermomechanical treatment, including rolling or drawing in combination with aging, normally lead to a reduction of ductility, an inevitable increase in the elastic modulus due to the precipitation of the second phases (α and $\omega_{iso}$), as well as the formation of a sharp texture and consequently, strong anisotropy of properties [3]. One of the effective methods for increasing fatigue strength while preserving ductility in metals and alloys in the formation of an ultrafine-grained (UFG) structure in them by severe plastic deformation (SPD) that enables very large strains to be achieved at relatively low temperatures under high applied pressures [4]. Therefore, increasing the fatigue strength of the pseudo-β alloy Ti-15Mo while preserving its relatively low elastic modulus is a relevant problem in the area of design of biomedical materials and permanent joint replacements.

In this work, the possibility of increasing the strength and fatigue properties of the Ti-15Mo alloy without a significant increase in the elastic modulus due to the formation of the UFG structure in it using equal-channel angular pressing (ECAP) is demonstrated.

## 2. Material and experimental procedure

The object of the study was rods with a diameter of 15 mm from the pseudo-β-titanium alloy Ti-15Mo in the two-phase (α+β)-state [5]. The UFG state was produced by thermomechanical treatment that included: β-solution quenching in water from a temperature of 810 °C (heating time 30 minutes) and subsequent ECAP processing (4 passes via route Bc with a channels intersection angle of 135°) at a temperature of 250 °C, with a total strain of $\varepsilon = 1.91$ [6]. A metallographic study was conducted using a JEOL JEM-2100 microscope (TEM) and a JEOL JSM 6390 microscope (SEM). The stress-strain curves were recorded using an initial strain rate of $1.0 \times 10^{-3}$ s$^{-1}$ with an Instron universal testing machine. The tension-compression fatigue tests were conducted with an Instron 8801 machine at room temperature under a symmetric loading cycle R = -1 with a frequency of 30 Hz at $10^7$ cycles to failure, in compliance with the international standard for cyclic tests ASTM E466, on smooth samples of a round section having a gauge diameter of 3 mm. The elastic modulus of the alloy was determined by nanoindentation using a CETRN universal microtribometer with a load of 100 mN, with a loading-unloading time of 15 s. The nanoindentation procedure was in compliance with ISO 14577.

## 3. Results and their discussion

### *3.1. Microstructure*

For medical products that require resistance to intensive loads, the Ti-15Mo alloy is used in its stronger ($\alpha+\beta$) state with an increased elastic modulus (figure 1, a). To form a single-phase β-structure, the initial material was exposed to a temperature of 810 °C for 30 minutes followed by water quenching. This temperature was selected on the basis of the results of X-ray phase analysis and differential scanning calorimetry [7].

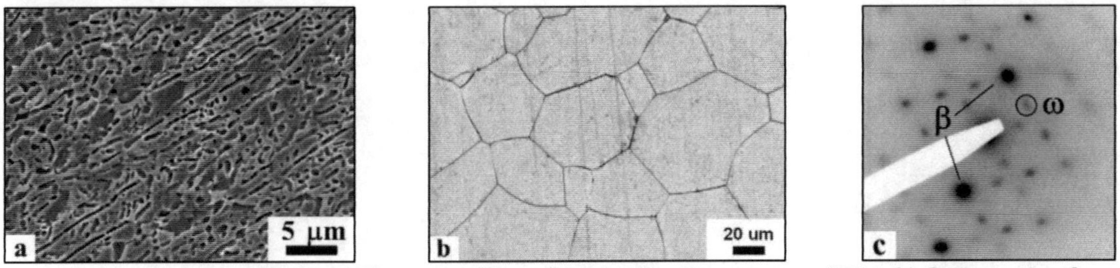


**Figure 1.** Coarse-grained Ti-15Mo alloy: a) initial ($\alpha+\beta$)-structure, SEM; b) β-structure after quenching, SEM; c) microdiffraction pattern taken from a β-grain, TEM.

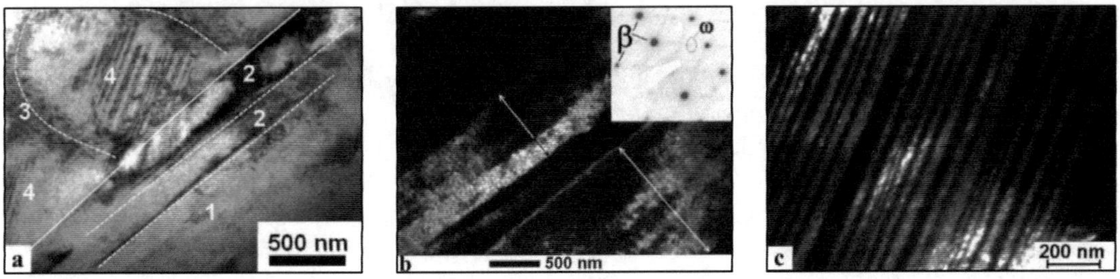


**Figure 2.** ECAP-processed Ti-15Mo (250 °C, 4 passes, 135°): a) typical structural features. b) mesobands consisting of microbands, microdiffraction pattern; c) microtwins. TEM.

After quenching, the alloy is characterized by a polycrystalline bimodal structure consisting of recrystallized β-grains with average sizes of 35±2 µm and 15±3 µm (figure 1, b). The volume fraction of grains larger than 35 µm in size is 55%. The microdiffraction pattern of the Ti-15Mo alloy (figure 1, c) taken from the axis <110> of the bcc lattice of the β-phase, in addition to the β-phase reflections, is also characterized by the presence of secondary reflections of the athermal $\omega_{ath}$-phase that correspond to the positions 1/3 and 2/3 of {112}β. The diffusionless formation of the $\omega_{ath}$-phase is typical for pseudo-β-titanium alloys quenched from temperatures above the polymorphic transition temperature [8].

By analyzing the microstructure of the ECAP-processed samples, several typical structural elements can be distinguished (figure 2, a). In the structure, closed (loop-like) configurations of dislocation boundaries (figure 2, a-3), as well as extensive mesobands (figure 2, a-1), representing a system of parallel dipoles with disrupted dislocation boundaries, are observed. Inside the mesobands, microbands are observed (figure 2, a-2). A weak contrast between them indicates small misorientation angles (figure 2, b). Also, packs of microtwins are observed in the structure, the width rof which ranges from 10 to 50 nm (figure 2, a-4, c).

In general, after ECAP, the microstructure is characterized by ultrafine structural elements in the form of grain/subgrain fragments with a cross-sectional size of less than 1 μm, oriented along the axis rod of the rod. In addition, weakly-expressed reflections of the ω-phase can be seen in the diffraction patterns (figure 2, b). Due to the low intensity of the reflections, an image of the ω-phase could not be obtained.

### 3.2. Mechanical properties

The tensile curve in the initial (α+β)-state is characterized by the presence of a sharp yield point attributed to a sharp increase in the number of mobile dislocations of the β-grain at the onset of plastic flow, which is typical for metals with a bcc lattice [9]. Quenching led to an increase in the ductility of the alloy from 17±2 to 32±3 % and to a decrease in the ultimate tensile strength (UTS) from 1020±10 to 790±5 MPa due to the growth of the β-grain and the complete dissolution of the strengthening α-phase, i.e. the transition of the alloy from the two-phase state to the single-phase state (figure 3, a).

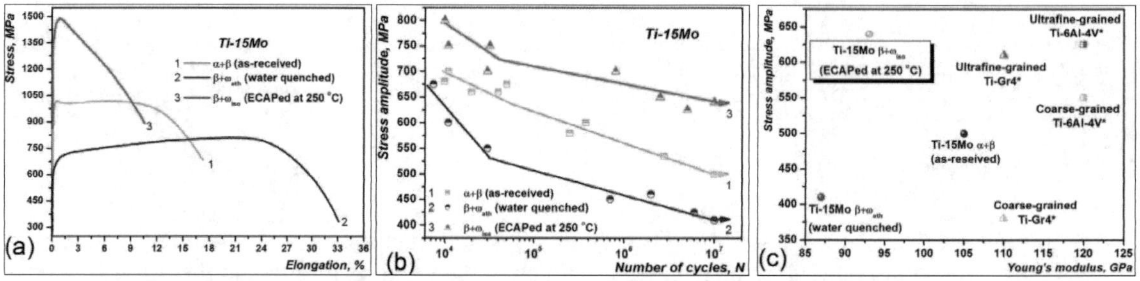


**Figure 3.** Physical and mechanical properties of the Ti-15Mo alloy: a) tensile curves; b) tension-compression cyclic tests; c) endurance limit vs elastic modulus of biomedical Ti alloys.* [11],** [12].

The formation of an ultrafine structure in the Ti-15Mo alloy by ECAP resulted in a significant enhancement of the mechanical properties (figure 3, a). The UTS increased from 790 ± 5 to 1450 ± 20 MPa, and the elongation to failure decreased from 32 ± 3 to 10 ± 1 %. At the same time, worth noting is a visible reduction in uniform elongation on the tensile curve of the processed alloy, which indicates an early strain localization, characteristic of many metallic materials produced by SPD and associated with impeded nucleation of dislocations in the structure [4]. Figure 3, b shows the fatigue curves of the Ti-15Mo alloy in different structural states, obtained during tension-compression tests. Two main areas can be distinguished on the fatigue curves: low-cycle and high-cycle fatigue. In the initial (α+β)-state, the endurance limit of the Ti-15Mo alloy is 500 ± 10 MPa at $10^7$ cycles to failure. The formation of a single-phase β-state in the alloy led to a decrease in the endurance limit to 410 ± 10 MPa, which is conditioned by the exclusion of the strengthening α-phase from the structure. The ultrafine structure formation by ECAP led to an increase in the endurance limit of the alloy, which increased by more than 50 % as compared to the quenched condition, and amounted to 640 ± 10 MPa at $10^7$ cycles to failure (figure 3, b).

### 3.3. Stress-elasticity modulus ratio

Figure 3 shows the relationship between the endurance limit and elastic modulus for the Ti-15Mo alloy and other alloys widely used as biomaterials, in particular, Grade 4 Ti, Ti-6Al-4V ELI after a conventional treatment and SPD [10]. In the initial α+β-state, the alloy has an elastic modulus of 105±7 MPa, influenced by the presence of a fraction of the α-phase in the structure, its elastic modulus for Ti-15Mo being 124 GPa [11]. After quenching, the value of E was 87±5 GPa. As it was shown in [9], the elastic modulus of the Ti-15Mo alloy is strongly influenced by the stability of the β-phase and the stability of the ω-phase particles that grow with increasing processing temperature. The athermal $\omega_{ath}$-phase, present in the alloy after quenching, is the pre-precipitate of the isothermal $\omega_{iso}$-phase that formed via the mechanism of atoms displacement in the lattice of the β-phase [9], and therefore its effect on the elastic modulus value of Ti-15Mo is not so large. A slight increase in the alloy's elastic modulus to 93±5 GPa was observed after ECAP processing at a temperature of 250 °C. Apparently, the main reason could be a very small amount of the $\omega_{iso}$-phase after the processing. Besides, some contribution into the preservation of the low values of Young's modulus is made by the formation in the structure of a high density of non-equilibrium grain boundaries with introduced dislocations, as it was demonstrated for the nanostructured TNTZ alloy [12].

## 4. Conclusions

1. This paper reveals the microstructural features of the Ti-15Mo alloy processed by SPD. It is shown that the processing of the alloy by equal-channel angular pressing in the β-state at a temperature of 250 °C leads to the formation of a complex ultrafine structure with a structural element size of less than 1 μm, providing a significant enhancement of the mechanical (UTS from 800 to 1450 MPa) and fatigue properties ($\sigma_{-1}$ from 420 to 640 MPa) of the alloy.
2. The elastic modulus of the Ti-15Mo alloy after ECAP at T=250 °C slightly increases from 87 to 93 GPa as compared to the single-phase β-state after quenching, which could be due to the increase in the density of the non-equilibrium boundaries of the β-phase grains/subgrains and the formation of a relatively small amount of the $\omega_{iso}$-phase.
3. Comparison of the study results for the ECAP-processed Ti-15Mo alloy and the similar parameters of Grade 4 Ti and Ti-6Al-4V ELI visibly demonstrates that Ti-15Mo with an ultrafine structure can occupy a worthy place among other biometallic materials, and its application is promising for the manufacture of products for osteosynthesis and orthopedics.

**Acknowledgements**
This work was supported by RFBR grant № 18-32-00585.

## References

[1] 2004 *Engineering Materials for Biomedical Applications* ed S H Teoh (World Scientific)
[2] Kuroda D *et al.* 2005 *Mat. Sci. Eng.* C **25** 312
[3] Bowen A W 1977 *J. Mater. Sci.* 1355
[4] Valiev R Z *et al.* 2016 *JOM* **68** 1216
[5] ATI 15Mo™ Titanium Alloy IMI, Technical Data Sheet –PA 15222-5479 U.S.A., 2014. – 4 p.
[6] Segal V M 1995 *Mater. Sci. Eng.* A **197** 157
[7] Gatina S A, Semenova I P, Leuthold J and Valiev R Z 2015 *Adv. Eng. Mater.* **17** 1699
[8] Devaraj A *et al.* 2009 *Scripta Mater.* **61** 701
[9] Honeycombe R W K 1984 *The Plastic Deformation of Metals* (Maidenhead: Edward Arnold Publ.)
[10] Semenova I P *et al.* 2012 *JOM* **47** 7777
[11] Bowen A K 1980 On the strengthening of metastable β-titanium alloy by ω- and α-precipitation *Titanium'80 Science and Technology* (Warrendale: Metallurgical Society of AIME) pp 1317–1326
[12] Xie K Y *et al.* 2013 *Mater. Sci. Eng.* C **33** 3530

IOP Conf. Series: Materials Science and Engineering **672** (2019) 012053 doi:10.1088/1757-899X/672/1/012053

# Features of a severe plastic deformation method for threaded steel reinforcing bars used in construction

**A G Raab and G I Raab**

Ufa State Aviation Technical University, 12 K. Marx str., Ufa 450008 Russia

E-mail: agraab@mail.ru

**Abstract.** Enhancing the mechanical properties of products by developing new technologically advanced SPD methods, as the most efficient for strengthening of materials due to deep refinement of the initial structure, is an urgent task. For this purpose, a method was developed for the continuous forming of threaded rods with the required properties, using successive operations of rolling and transverse shear. The produced rods with a threaded profile can be used in construction as reinforcing bars for the reinforcement of concrete structures. The studies were conducted using mathematical modeling and physical experiments. We analyzed the stress-strain state of the produced billets. In one processing cycle, the strength of the initial billet from low-carbon steel 10 was raised from 350 to 530-550 MPa and simultaneously a threaded profile was formed. The obtained parameters comply with the Russian state standard GOST for steel reinforcing bars of the A500 class used in construction.

## 1. Introduction

For producing construction reinforcing bars, the hot rolling of steel rods with special rolls of complex design is predominantly used. It is evident that the operating, engineering and energy costs of such a technology are very high [1]. In the present study, we propose an alternative technology for producing construction reinforcing bars (rebars) that employs the principles of severe plastic deformation (SPD) [2-4] and is based on the use of successive forming operations of rolling and transverse shear (by free torsion) (figure 1). As the initial billets, we used the most common round rolled sections (rods, coils). In the process of rolling, a square cross section is formed from a round section, the initial billet becomes strengthened and during the second operation it is subjected to shear. It is known that shear deformation is less energy-consuming than other types of deformation (compression, tension). Shear provides not only the forming of a threaded profile with an inclination angle of 45-60°, but also additional deformation and strengthening of a rod. In this connection, the task is to verify experimentally the proposed technology using laboratory devices and a physical experiment. To solve the set task, at the first stage it is necessary to evaluate the stress-strain state of the deformed billets using computer simulation, which will reduce the costs at the experiment stage and provide scientific and technical information on the process parameters.

## 2. Material and experimental procedures

To perform the experiment by means of numerical simulation, we used the software package DEFORM-3D. Using a CAD software Kompas-3D, we designed the 3D models of the deforming tool and the billet under study.

As the material for the investigation, we selected from the Deform-3D material library the low-carbon steel AISI1010, an analogue of steel 10.

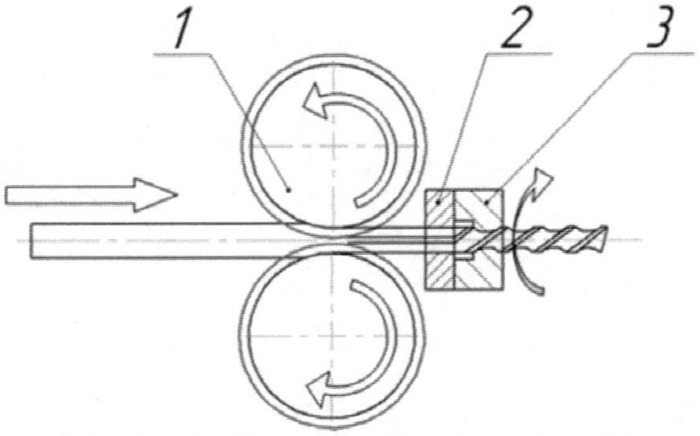


**Figure 1.** Principle of the method of rolling with forming: 1- rolling rollers, 2-guide, 3-twisting mechanism [5].

The following boundary conditions were used in the simulation:
1) The initial material is ductile and isotropic.
2) The experimental temperature is 20 °C, deformational heating is not considered.
3) The tool is rigid.
4) The initial material's diameter is 10 mm, and the length is 500 mm.
5) The simulation includes 500 steps.
6) The billet is divided into 70000 elements.
7) The coefficient of friction between the billet and the tool, calculated according to Zibel's formula, is 0.12.
8) The number of passes - 1 pass
9) The size of the bar after deformation - 8.6 mm in diameter
10) Sample twist angle - 45 ° +/- 2 °
11) Linear speed during deformation - 1.8 m / min

The simulation results were used to find the variation in the intensity of accumulated plastic strain and the average stresses of the billets produced by the proposed method.
To produce the experimental samples of reinforcing bars, we developed and manufactured a laboratory die-set (figure 2).

**Figure 2.** Experimental die-set for implementing the operation of transverse shear and the appearance of the produced samples.

### 3. Experimental results and discussion
The computer simulation results show that during the deformation by rolling, the largest strain is accumulated in the surface layers of the billet, where strain reaches a value of e=1.4. Further deformation

by transverse shear raises the level of accumulated strain which reaches values of about e=2 in the surface layers of the billet (figure 3). As a rule, such a level of cold non-monotonic strain ensures a rather intensive refinement down to the ultrafine sizes [6].

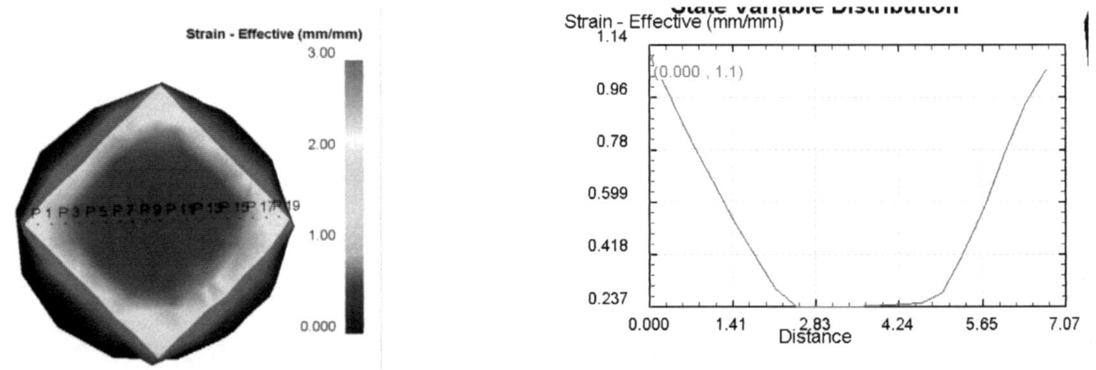


a

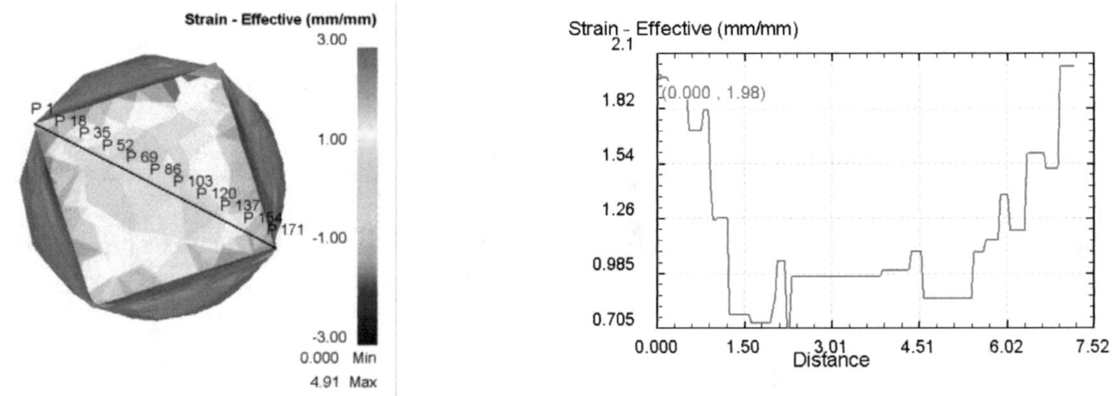


b

**Figure 3**. Strain state of the billet after the operations of rolling and transverse shear: a – the pattern and diagram of accumulated strain distribution after rolling, b – the pattern and diagram of accumulated strain distribution after transverse shear.

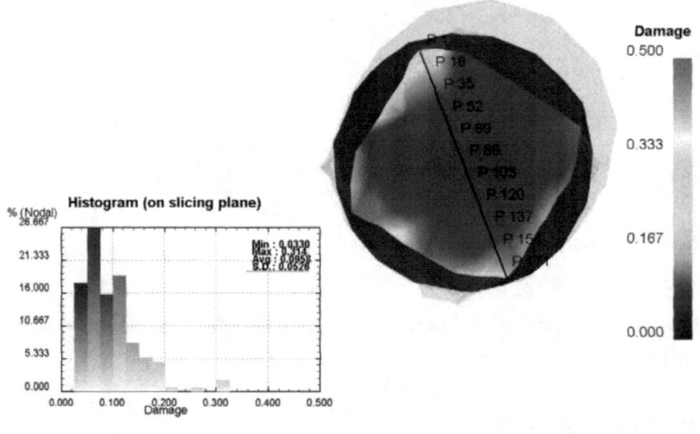


**Figure 4.** Pattern of damage distribution across the section of the deformed rod.

Studies of the damage parameter, conducted in the software system Deform-3D on the basis of the Cockroft-Latham criterion, reveal its low values not exceeding 0.3, which indicates a rather large operational safety margin of the produced rods (figure 4).

Studies of the mechanical properties of the produced billets demonstrate that after rolling the billet is strengthened to 420-430 MPa, and after transverse shear – to 530-550 MPa.

## 4. Conclusions

1. With the use of computer simulation, we studied an end-to-end process for producing threaded reinforcing bars of the A500 class, analyzing the stress-strain state parameters and the damage parameters of the deformed billets during their processing by the strengthening and forming operations of rolling and transverse shear. A gradient strain distribution has been found in the bulk of the finished product. The strain accumulated in the peripheral layers reaches e~2, while the central layers it reaches e~0.7.
2. The obtained damage parameter based on the Cockroft-Latham criterion does not exceed 0.3, which indicates a rather large operational safety margin of the produced reinforcing bars.
3. In one processing cycle, the strength of the initial billet from low-carbon steel 10 was increased from 350 to 550 MPa, and simultaneously the required threaded profile with an inclination angle of 50° was formed. Both parameters comply with the Russian state standard GOST for construction reinforcing bars of the A500 class.

## Acknowledgments

The authors gratefully acknowledge the financial support from the Russian Foundation for Basic Research under project No. 17-08-00720/19.

## References

[1] Rybyov I A 2006 *Materials Science in Construction* (Moscow: Akademiya Publishing Center) [in Russian]

[2] Valiev R Z, Zhilyaev A P and Langdon T G 2014 *Bulk Nanostructured Materials: Fundamentals and Applications* (Hoboken: Wiley STM)

[3] Utyashev F Z and Raab G I 2013 *Deformation Methods for Fabrication and Processing of Ultrafine-Grained and Nanostructured Materials* (Ufa: Gilem) [in Russian]

[4] Zhilyaev A P, Nurislamova G V, Bae-Kyun K, Baró M D, Szpunar J A and Langdon T G 2003 *Acta Mater.* **51** 753

[5] Raab G I and Raab A G Method of strengthening and forming screw reinforcing bar, Russian Federation patent No. 2640705, publication date 11.01.2018

[6] Raab A G, Chukin M V, Aleshin G N and Raab G I 2014 *IOP Conf. Ser.: Mater. Sci. Eng.* **63** 012008

IOP Conf. Series: Materials Science and Engineering **672** (2019) 012054 doi:10.1088/1757-899X/672/1/012054

# Effect of the processing parameters of the hot rolling of a stainless austenitic steel on the processes of structural and phase transformation

**G I Raab**[1], **V I Semenov**[1], **P La**[2] **and S V Dobatkin**[3,4]

[1] Institute of Physics of Advanced Materials, Ufa State Aviation Technical University, 12 K. Marx st., Ufa 450008 Russia
[2] State Key Laboratory of Advanced Processing and Recycling of Nonferrous Metals, Lanzhou University of Technology, Lanzhou, China
[3] Baikov Institute of Metallurgy and Materials Science RAS, 49 Leninskiy pr., Moscow 119334 Russia Institute of Physics of Advanced Materials, Ufa State Aviation Technical University, 12 K. Marx st., Ufa 450008 Russia
[4] National University of Science and Technology "MISIS", 4 Leninskiy pr., Moscow 119049 Russia

E-mail: giraab@mail.ru

**Abstract**. We studied the mechanism of structure formation of the stainless austenitic steel with the composition 0.09%C, 17.5%Cr, 11.6%Ni, 0.5%Ti (a close analogue of the AISI 304 steel) during hot deformation by rolling at a temperature of 900 °C, strain rates of 0.5-5 $s^{-1}$ and strains of about 10-60%. It was demonstrated that with increasing strain the fraction of polygonized volumes grows, and inside them the fraction of equiaxed subgrains grows.

## 1. Introduction

In the papers [1-3] studying the structure formation processes during the hot deformation by compression of stainless austenitic steels, the authors determined a pattern of softening mechanisms depending on the temperature, strain and strain rate, and built maps of the structural states determining the regions of acting dynamic processes of polygonization and recrystallization under steady-state deformation in the coordinates «deformation temperature – strain rate».

In terms of the deformation pattern, hot compression is close to rolling, which is widely used in the industrial cycles of high-temperature thermomechanical treatment of austenitic stainless steels. Therefore, the established regularity in the change of the softening mechanisms during hot deformation by compression can also be characteristic of the processes of hot rolling. For example, to control the strengthening-softening processes the authors of the paper [4] proposed a method for the deformation-thermal treatment of austenitic stainless steels which includes plastic deformation by hot rolling to a strain of 1-2 with subsequent annealing and additional cold rolling to a true strain of 3.5-4. A decrease in the temperature of rolling [5] is accompanied by a considerable strengthening due to an increase in dislocation density and development of deformation-induced twinning. This leads to the emergence of shear bands with increasing strain. Subsequent annealing leads to the occurrence of the recovery and recrystallization processes, their kinetics depending on the preceding strain value and annealing temperature. Such approaches enable producing a set of balanced strength and ductility characteristics. Another area of research for the control of the strengthening-softening processes is to produce specified ultrafine-grained (UFG) states in austenitic stainless steels by means of severe plastic deformation processing [6-7].

It follows from the above that the study of the strengthening-softening mechanisms influencing actively the formation of a set of properties in austenitic steels is a relevant task in the physics of strength, and it is of great practical interest.

The aim of the present work is to study the features of structural transformations in the austenitic stainless steel with the composition 0.09%C, 17.5%Cr, 11.6%Ni, 0.5%Ti (a close analogue of the AISI 304 steel and the Russian steel 0.8Kh18N12T) during hot deformation by rolling depending on the temperature 900 °C, strain rate 0.5-5 $s^{-1}$ and strain in a range of 10-60%.

## 2. Material and experimental procedure

As the material for the study, we used the steel (0.09%C, 17.5%Cr, 11.6%Ni, 0.5%Ti) after cooling in water from a temperature of 1050 °C (exposure for 1 hour). Rolling was performed on a Hankook six-high strip-rolling mill for the hot rolling of sheets with strain rates of 0.5 and 5 $s^{-1}$, strains of 10, 20 and 60% (corresponding to true strains of 0.11, 0.22 and 0.92, respectively) for one pass at a temperature of 900 °C. The grain size and the fraction of recrystallized grains were determined using light microscopy, and the subgrain size, the fraction of equiaxed subgrains and the distance between elongated sub-boundaries were found using electron microscopy. The pole density of the X-ray lines $\{111\}_\gamma$, $\{200\}_\gamma$, $\{220\}_\gamma$, $\{311\}_\gamma$, $\{222\}_\gamma$ of the steel with the composition 0.09%C, 17.5%Cr, 11.6%Ni, 0.5%Ti in the process of hot deformation by rolling at $\dot{\varepsilon} = 5$ $s^{-1}$ was determined using a DRON-4 diffractometer.

## 3. Research results and their discussion

Indeed, in the process of rolling of the austenitic steel 0.09%C, 17.5%Cr, 11.6%Ni, 0.5%Ti at $T_{def}$ = 900 °C and strain rates 0.5 and 5 $s^{-1}$ (up to a strain of $\varepsilon = 0.92$) dynamic recrystallization does not take place, as it was also observed for compression [9, 10]. With increasing strain and strain rate grains become more elongated (table 1).

Study by TEM electron microscopy shows that at $T_{def}$ = 900 °C small areas are observed with a substructure typical of hot-work hardening (figure 1a), characterized by a high density of free dislocations, the presence of dislocation pile-ups and a cell substucture. In the whole volume a polygonized structure prevails, with both elongated (figure 1b) and equiaxed (figure 1c) subgrains having imperfect sub-boundaries and an increased density of free dislocations.

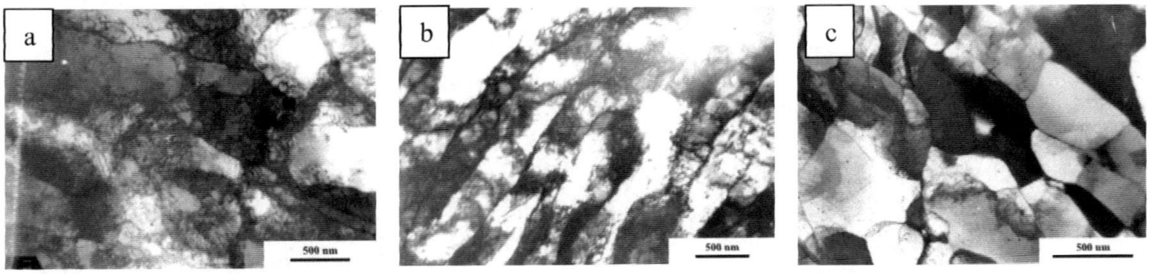


**Figure 1.** Structure (TEM) of the steel 0.09%C, 17.5%Cr, 11.6%Ni, 0.5%Ti after hot deformation by rolling at a temperature of 900 °C and a strain rate of 0.5 $s^{-1}$: a) $\varepsilon = 0.11$; b, c) $\varepsilon = 0.92$.

With increasing strain, the fraction of polygonized volumes grows, and inside them the fraction of equiaxed subgrains grows (table 1). At $T_{def}$ = 900 °C and $\varepsilon = 0.5$ $s^{-1}$ the fraction of polygonized areas increases from 29% (at $\varepsilon = 0.11$) to 42% (at $\varepsilon = 0.22$) mainly due to the formation of elongated

IOP Conf. Series: Materials Science and Engineering **672** (2019) 012054 doi:10.1088/1757-899X/672/1/012054

subgrains. When a strain of 0.92 is reached, the polygonized areas occupy about 90% of the whole volume, 60% of them being equiaxed subgrains (table 1).

**Table 1.** Structural parameters of the steel 0.09%C, 17.5%Cr, 11.6%Ni, 0.5%Ti after hot deformation by rolling.

| Material | $T_{def}$, °C | $\dot{\varepsilon}$, $s^{-1}$ | ε, % | $\bar{d}$, μm | $K_r$,% | $V_r$, % | V*, % | V**, % | Distance between sub-boundaries in elongated subgrains, μm | Size of equiaxed subgrains, μm |
|---|---|---|---|---|---|---|---|---|---|---|
| Steel 0.09%C, 17.5%Cr, 11.6%Ni, 0.5%Ti | 1060 | Quenching | | 32.1±4.5 | | | | | | |
| | 900 | 0.5 | 10 | 27.1±4.0 | - | 0 | 29 | <5 | - | - |
| | | | 20 | 26.5±3.8 | 22 | 0 | 42 | <5 | 0.35±0.02 | 0.41±0.02 |
| | | | 60 | 19.1±3.9 | 28 | 0 | 90 | 60 | 0.48±0.06 | 0.30±0.04 |
| | | 5 | 20 | 34.4±4.8 | 28 | 0 | 69 | - | 0.37±0.07 | 0.56±0.05 |
| | | | 60 | 19.5±5.2 | 36 | 0 | 90 | 59 | 0.34±0.05 | 0.60±0.05 |

where $T_{def}$ is the deformation temperature, $\dot{\varepsilon}$ is the strain rate, ε is the degree of deformation, $\bar{d}$ is the grain size, $V_r$ is the fraction of recrystallized grains, V* is the fraction of polygonized volumes, V** is the fraction of oriented subgrains.

An increase in strain rate at $T_{def}$ = 900 °C from 0.5 $s^{-1}$ to 5 $s^{-1}$ results in an increase in the fraction of polygonized volumes. The size of equiaxed subgrains and the distance between the sub-boundaries of the elongated subgrains is 0.3-0.6 μm and 0.3-0.5 μm, respectively (table 1). The fraction of equiaxed subgrains after a strain of ε =0.92 at both strain rates is 55-60%. In addition, the electron microscopy studies revealed individual recrystallized grains with a size of 2-5 μm, having a low density of free dislocations, which apparently indicates the static nature of their formation.

The changes in the texture during hot deformation by rolling and during compression have a certain similarity (figure 2) [10]. In the conditions providing the formation of a dynamically polygonized structure (at $T_{def}$ = 900 °C), with increasing strain the deformation texture becomes enhanced – the component $\{220\}_\gamma$ grows (figure 2a) [10]. Unlike high-temperature compression, the formation of a dynamically polygonized structure during hot rolling is accompanied by a smaller change in the respective texture components (figure 2) [10]. This is apparently related to the partial static recrystallization during slower cooling after rolling, as indicated by the increased pole density of the component $\{311\}_\gamma$ (figure 2) [10].

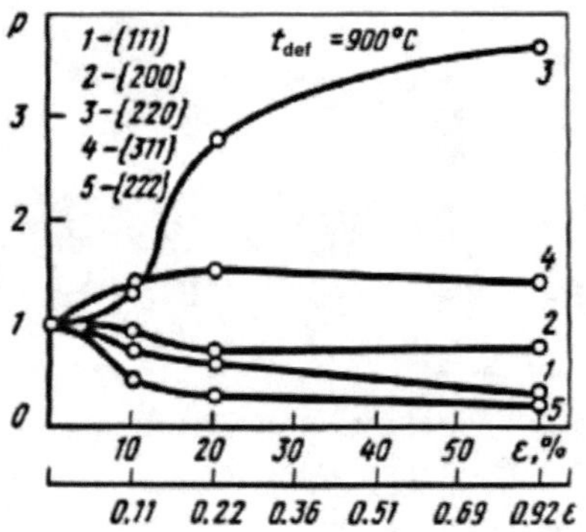


**Figure 2.** Variation in the pole density of the X-ray lines for the steel 0.09%C, 17.5%Cr, 11.6%Ni, 0.5%Ti in the process of hot deformation by rolling ($\dot{\varepsilon}$= 5 $s^{-1}$): T = 900 °C.

## 4. Conclusions

1. In the process of rolling at $T_{def}$ = 900 °C with increasing strain and strain rate grains become more elongated.
2. With increasing strain the fraction of polygonized volumes grows, and inside them the fraction of equiaxed subgrains grows, e.g. at $\varepsilon = 0.5$ $s^{-1}$ the fraction of polygonized areas increases from 29% (at $\varepsilon = 0.11$) to 42% (at $\varepsilon = 0.22$), mainly due to the formation of elongated subgrains. When a strain of 0.92 is reached, polygonized areas occupy about 90% of the whole volume, 60% of them being equiaxed subgrains. Thus, in the process of rolling of the steel 0.09%C, 17.5%Cr, 11.6%Ni, 0.5%Ti at $T_{def}$ = 900 °C with  = 0.5 and 5 $s^{-1}$ softening takes place practically only via the mechanism of dynamic polygonization.
3. It has been found that in the conditions providing the formation of a dynamically polygonized structure (at $T_{def}$ = 900 °C), with increasing strain the deformation texture becomes enhanced – the component {220}$\gamma$ grows.

**Acknowledgements**
The work was performed in the framework of a joint Russian-Chinese research project funded by the Russian Foundation for Basic Research (project No. 19-53-53022) and the National Natural Science Foundation of China (51911530119).

**References**

[1] Kaputkina L M, Dobatkin S V and Zhdanovich T K 1985 *Bull. Acad. Sci. USSR: Metals* **3** 122 [in Russian]
[2] Dobatkin S V and Kaputkina L M 2001 *Phys. Met. Metallog.* **91** 79
[3] Gorelik S S, Dobatkin S V and Kaputkina L M 2005 *Recrystallization of Metals and Alloys* (Moscow: MISIS) [in Russian]
[4] Beljakov A N, Shakhova J E and Kajbyshev R O 2013 *Russian Federation patent No. 2482197*
[5] Shakhova I, Dudko V, Belyakov A, Tsuzaki K and Kaibyshev R 2012 *Mater. Sci. Eng.* A **545** 176
[6] Dobatkin S V, Rybal'chenko O V and Raab G I 2007 *Mater. Sci. Eng.* A **463** 41
[7] Dobatkin S V, Kaputkina L M, Rybal'chenko O V and Komlev V S 2012 *Metally* **5** 28 [in Russian]
[8] Bernshtein M L, Kaputkina L M, Prokoshkin S D and Dobatkin S V 1982 *Bull. Acad. Sci. USSR: Metals* **2** 94 [in Russian]
[9] Bernshtein M L, Kaputkina L M, Prokoshkin S D and Dobatkin S V 1985 *Acta Metall.* **33** 247
[10] Bernshtein M L, Kaputkina L M, Prokoshkin S D and Dobatkin S V 1989 *Diagrams of Hot Deformation, Structure and Properties of Steels* (Moscow: Metallurgiya) [in Russian]

IOP Conf. Series: Materials Science and Engineering **672** (2019) 012055 doi:10.1088/1757-899X/672/1/012055

# High strength and high conductive copper-based alloy produced by SPD for contact wires for high speed railway lines - A short review

**B Ravisankar[1], K Sivaprasad[1], N Ramesh Babu[1], G I Raab[2] and R Z Valiev[2]**

[1] Department of Metallurgical and Materials Engineering, National Institute of Technology, Tiruchirappalli – 620 015 INDIA
[2] Institute of Physics of Advanced Materials, Ufa State Aviation Technical University, 12 K. Marx str., Ufa, 450008 Russia

E-mail: brs@nitt.edu

**Abstract**. The development of high-speed railway lines, the requirements for contact wires of catenary becomes more essential. The contact wire should be strong, wear resistant and fatigue resistant, and also possess good thermal stability along with high electrical conductivity. At present, copper wires alloyed with cadmium are used. The replacement is more of a necessity since cadmium is a material hazardous to environment, and it is already banned by several countries. Over the last several decades, growing attention has been focused on the development of copper-based alloys with high strength and good electrical conductivity. Novel severe plastic deformation (SPD) procedure, namely equal-channel angular pressing (ECAP), has been applied for obtaining ultrafine-grained (UFG) materials with high strength and good electrical conductivity. This overview mainly focuses on the suitability of ECAPed copper and its alloys for contact wires in high-speed railway lines.

## 1. Introduction

The development of high-speed railway lines, the requirements for contact wires of catenary becomes more essential. The contact wires are suspended from the catenary wires via dropper wires and are in direct contact with the pantograph, transmitting power from the rail overhead line system to the locomotive as shown in the figure 1.

The contact wire should be strong, wear resistant and fatigue resistant, and possess good thermal stability along with high electrical conductivity. The contact wires should be suitable for outdoor use in moist tropical climate and in areas subject to heavy rainfall, polluted due to industry and marine atmosphere and severe lightning.

Global practice shows that the real solution to the problem is to replace the material of contact wires with copper alloys with higher strength, high electrical conductivity and good thermal stability. At present, coper alloyed with cadmium and Cu-Mg wires are used. The replacement is more of a necessity since cadmium is a material hazardous to environment, and it is already banned by several countries. The properties of Cu-Mg alloys do not meet the requirements for high speed railway lines and, hence, replacement with another material becomes indispensable. Over the last several decades, growing attention has been focused on the development of copper-based alloys with high strength and good electrical conductivity. Strength can be increased by alloy addition, but as is well known, high strength and good electrical conductivity are often mutually exclusive in copper alloys. In order to overcome the shortcoming, a novel severe plastic deformation (SPD) process, namely equal-channel angular pressing (ECAP), has been applied for obtaining ultrafine-grained (UFG) materials with high strength and good electrical conductivity. In the ECAP process, a well lubricated billet is pressed in a die that contains two channels, equal in cross section, intersecting at an angle called the die channel angle as shown in figure 2 (a). The material is subjected to shear deformation at the intersection plane. Since the process is capable of maintaining the net dimensions of the work piece, repetitive extrusion is possible and enables for imposing high strains in the materials subjected to ECAP. Thus during the ECAP processing, significant structural refinement can be achieved. Numerous studies have demonstrated that the severe plastic deformation processing enables achieving a drastically new level of properties due to the severe working of the structure and the change in the kinetics of phase transformations in alloys. The advanced version of ECAP, i.e. ECAP-Conform (figure 2 (b)) process is capable of producing long-length wire rods required for contact wires of catenary.

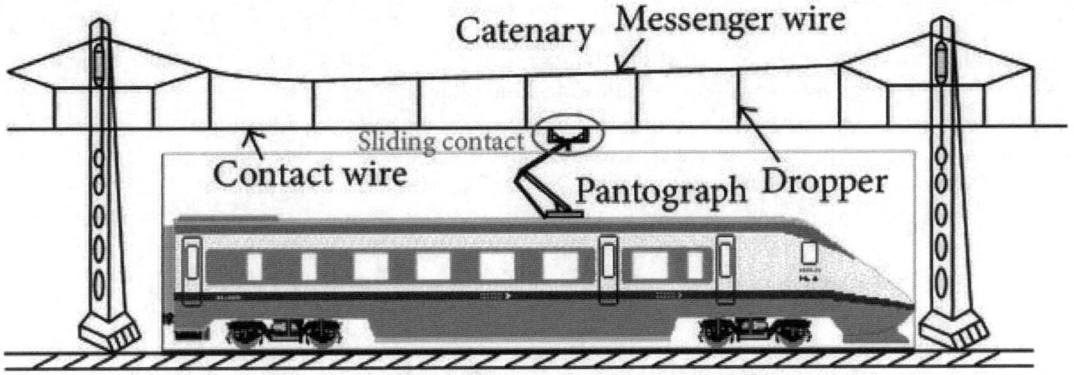


**Figure 1.** Schematic representation of catenary.

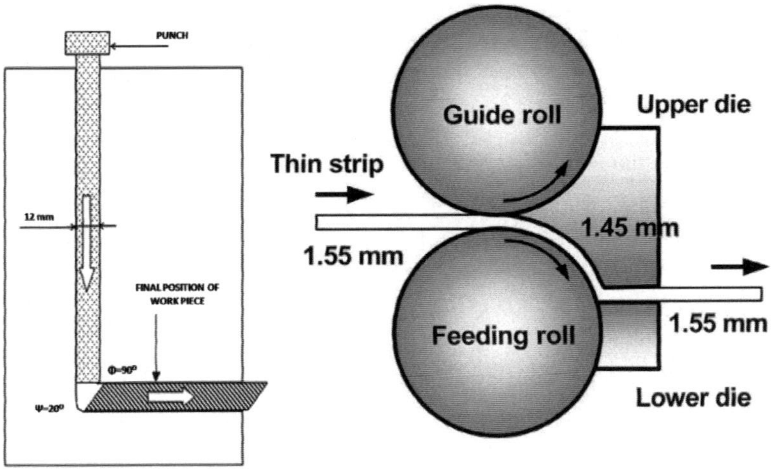


(a) (b)

**Figure 2.** Schematic representation of (a) ECAP and (b) ECAP - Conform processes.

While going for the development of high strength and high conductivity copper based alloys, the retention of properties is a major technical challenge. Therefore, in the process of development it is necessary to determine precise amount of alloying elements that can provide high strength and preserve a satisfactory level of ductility. The task of the processing by ECAP-Conform is to ensure the maximally profound solid solution decomposition at the deformation stage for strain-hardening Cu alloys and at the aging stage for precipitation-hardening Cu alloys.

This overview mainly focuses on the suitability of copper and its alloys processed by SPD techniques for contact wires in high speed railway lines.

## 2. Copper and its alloys

The other best method, apart from strengthening by alloy addition, is grain refining. The desired combination is the tensile strength of 450 MPa with electrical conductivity of 90% International Annealed Copper Standard (IACS) in case of pure copper, and 600 MPa with electrical conductivity of 83 %IACS in case of Cu-Cr-Zr alloys, for contact wires in high speed railway lines. By traditional techniques such as rolling, drawing etc., the required properties are achieved by the primary 30% thickness reduction and intermediate aging and/or annealing followed by a secondary 60% thickness reduction. The combination of tensile strength and electrical conductivity were gained for the alloy with 95% reduction by conventional techniques [1]. The reduction in dimensions and cost involved in production restricts the use of conventional techniques. The other way to attain grain refinement without dimensional changes is through severe plastic deformation techniques. Also, the SPD techniques heavily alters the microstructure, precipitation kinetics which enables for attainment of the required level of properties in copper and its alloys for high speed railway lines. Hence, the paper discusses the review of development of high strength and high conductive catenary wires especially in pure copper, copper-chromium-zirconium and copper-titanium carbide insitu composites and its processing by various severe plastic deformation processes by various scientists and researchers.

In pure copper, a tensile strength of 430±20 MPa has been achieved in ECAP with three passes [2] with a ductility of 10%. In oxygen free high conductivity copper, a tensile strength of 440 MPa is achieved with a ductility of 28% and dephosphorised copper in four passes [3]. The conductivity of the samples decreased with increasing number of ECAP passes. After 7 passes, it was estimated to be 81% International Annealed Copper Standard (IACS). The electrolytic tough pitch (ETP) copper after 12 passes of ECAP the ultimate tensile strength was about 440 MPa [4]. The ETP copper in the initial state exhibits an electrical conductivity of 100.2% IACS. Grain refinement by ECAP caused a reduction to 93.1% IACS. Though the strength properties are achieved with little sacrifice in electrical conductivity and ductility, the thermal stability of the nanostructured ECAPed copper is maintained till 400 K only, which restricts the use of pure copper in contact wires.

The research is directed towards the controlled addition of alloying elements coupled with grain refinement by severe plastic deformation. The alloying elements, which are responsible for precipitation hardening, are chosen as against solid solution straighteners, as they are reducing the electrical conductivity to a greater extent.

Cu–Cr–Zr alloys have attracted considerable interest recently because of their high strength, as well as good formability, high thermal/electrical conductivity, good fatigue resistance, excellent stability at high temperatures (473 – 673 K) and low cost. Therefore, Cu–Cr–Zr alloys are widely used in high speed railway contact wire. Cu–0.71%Cr– 0.08%Zr alloy when ECAPed in Route Bc attains a tensile strength of 450 MPa with an elongation of about 20% [5]. Based on the results of high pressure torsion (HPT) of Cu-0.1%Cr-0.06%Zr, Cu-0.5% Cr-0.08%Zr, Cu-1%Cr-0.1%Zr, and Cu-0.3%Cr-0.5%Zr alloys, it is concluded that alloying with chromium in concentration above the solubility limit does not result in additional hardening both after HPT and after subsequent aging. However, excessive alloying with zirconium allows obtaining a structure with a smaller grain size (113 nm), enhances structure stability during heat treatment and noticeably increases strength of the alloy after HPT and after additional aging. The Cu-0.3%Cr-0.5%Zr alloy has both high strength and electrical conductivity of 63% IACS [6]. The Cu-Cr-Zr alloys fetch highest strength of 720 MPa with electrical conductivity of

70% of IACS when solutionised samples are ECAPed and aged with enhanced wear resistance [7]. When the warm extruded billets of Cu-Cr-Zr alloys are processed through HPT and subsequent aging yields an electrical conductivity of 75%IACS with a tensile strength of 825 MPa. The aging temperature played crucial role in improving the electrical conductivity [8-9].

A dispersion-strengthened Cu alloy with 1 wt% TiC, having a size of 0.5–1 μm, prepared by in-situ reaction casting, has a tensile strength of 250 MPa and a conductivity of 76% IACS. With increasing the number of ECAP passes, the hardness and strength (566 MPa) of Cu-1 wt% TiC composite increased obviously, but the conductivity and elongation decreased gradually. The post-ECAP annealing at 573 K significantly changed the optical microstructure exhibits a high tensile strength of 550 MPa and a good conductivity of 82.2% IACS. The novel combined process endows the alloy the appropriate performance to serve current high-frequency electrification railway systems [10].

A number of papers investigating the kinetics of strain-induced processes in copper low-alloyed alloys report complex phase processes occurring during SPD, and dispersed particles dissolve in low-alloyed alloys, which provides an additional reserve for precipitation hardening at the aging stage. Thus, during aging, the contribution of precipitation hardening increases. [11-13]. It has been experimentally demonstrated that SPD Processing has a high potential for achieving improved electrical conductivity with high strength in copper alloys by one of the authors [14,15].

## 3. Summary

In summary, from the results obtained by the various researchers, it can be concluded that the regime of the SPD processing governs the size and volume fraction of precipitates and segregations, thereby affecting the grain size and dynamic aging, influencing the mechanical and physical properties in copper alloys. Thus, the precise control of alloying additions and parameters characterizing SPD, such as combining the operations of heat treatment, ECAP and drawing, can result in the formation of a high set of physical and mechanical properties in copper and its alloys.

## References

[1] Meng A, Nie J, Wei K, Kang H, Liu Z and Zhao Y 2019 *Vacuum* **167** 329
[2] Kommel L, Hussainova I and Volobueva O 2007 *Materials and Design* **28** 2121
[3] Goto M, Han S Z , Yakushiji T, Kim S S and Lim C Y 2008 *International Journal of Fatigue* **30** 1333
[4] Lipin´ska M, Olejnik L and Lewandowska M 2018 *J. Mater. Sci.* **53** 3862
[5] Ding R, ChengGuo and Guo S 2013 *Mater. Sci. Eng.* A **587** 320
[6] Shangina D V, Bochvar N R, Morozova A I, Belyakov A N, Kaibyshev R O and Dobatkin S 2017 *Materials Letters* **199** 46
[7] Purcek G, Yanar H, Saray O, Karaman I and Maier H J 2014 **Wear 311** 149
[8] Purcek G, Yanar H, Shangina D V, Demirtas M, Bochvar N R and Dobatkin S V 2018 *Journal of Alloys and Compounds* **742** 325
[9] Shangina D V, Bochvar N R, Gorshenkov M V, Yanar H, Purcek G and Dobatkin S V 2016 *Mater. Sci. Eng.* A **650** 63
[10] Ni S, Jiang J, Chen J, Song D and Ma A 2016 *Progress in Natural Science: Materials International* **26** 643
[11] Raab G I, Raab A G and Shibakov V G 2015 *Metalurgija* **54** 423
[12] Faizov I A, Mulyukov R R, Aksenov D A, Faizova S N, Zemlyakova N V, Cardoso K R and Yu Zeng 2018 *Letters on Materials* **8** 110
[13] Faizov I A, Raab G I, Faizova S N, Zaripov N G and Aksenov D A 2016 *Letters on Materials* **6** 132
[14] Murashkin M Yu, Sabirov I, Sauvage X and Valiev R Z 2016 *J. Mater. Sci.* **51** 33
[15] Sabirov I, Enikeev N A, Murashkin M Y and Valiev R Z 2015 *Bulk Nanostructured Materials with Multifunctional Properties* (New York: Springer International Publishing)

IOP Conf. Series: Materials Science and Engineering **672** (2019) 012056 doi:10.1088/1757-899X/672/1/012056

# Effect of cold rolling on the structure and properties of the UFG Cu-0.5wt%Cr alloy

**E A Sarkeeva[1,3], G I Raab[1], W Wei[2,3,4] and I V Alexandrov[1, 3]**

[1] Ufa State Aviation Technical University, 12 K. Marx St., 450008 Ufa, Russia
[2] School of Materials Science and Engineering, Changzhou University, 1 Gehu Road, Changzhou 213164, P.R .China
[3] Sino-Russian Joint Laboratory of Functional Nanostructured Materials, Changzhou University, Changzhou 213164, P.R. China
[4] Jiangsu Key Laboratory of Materials Surface Science and Technology, Changzhou University, Changzhou 213164, P.R. China

E-mails: sarkeeva.e@inbox.ru, benjamin.wwei@163.com

**Abstract.** The results of studying the microstructure, mechanical and electrical properties of the UFG alloy Cu-0.5wt.% Cr, obtained by equal-channel angular pressing (ECAP), followed by cold rolling and subjected to aging, are presented. It was found that, as a result, the average size of the fragmented structure is 160 nm. The results showed that the tensile strength increases with increasing degree of deformation. The maximum strength value of 605 MPa was achieved after a combination of 8 ECAP passes followed by rolling by 95%, while ductility was reduced to 13%. The number of ECAP passes and subsequent rolling practically do not affect the electrical conductivity and is 34-39% of IACS.

## 1. Introduction

An analysis of the literary published over the past decade indicates a great interest of researchers in studying different methods of increasing strength while maintaining the electrical conductivity of Cu-Cr alloys, which are actively used in various industries [1-3]. The main way to increase the strength of copper materials without compromising electrical conductivity is to grind the microstructure and increase the density of dislocations due to plastic deformation. An additional increase in strength properties can be ensured by using severe plastic deformation (SPD) as the main processing method, which leads to the formation of an ultrafine-grained (UFG) structure with a high dislocation density [4, 5]. However, industrial production requires further processing of semi-finished products to bars, tapes, sheets, plates, etc. In this regard, research aimed at analyzing the evolution of the microstructure and properties of UFG electrotechnical materials during subsequent deformation is relevant. This work is devoted to the study of the effect of cold rolling on the microstructure, mechanical properties and electrical conductivity of the UFG Cu-0.5wt.% Cr alloy.

## 2. Method of experimental research

As the material for research, we chose the industrial electrotechnical Cu-0.5wt.% Cr alloy. Samples 10×10×60 mm in size in the form of rectangular parallelepipeds with an exposure time of 0.5 hours at

1000 °C, followed by quenching in a 5% NaCl solution to form a supersaturated solid solution were used as the initial state (IS).

ECAP was carried out in a die with an internal angle ψ =90° along route $B_C$ (turning the workpiece 90° clockwise along the longitudinal axis after each pass) at room temperature at a speed of 0.2 mm/s with the number of passes 1, 2, 4 and 8. Subsequently, cold rolling (CR) was carried out at room temperature with a total reduction ratio of 95%. In one pass during rolling, the reduction in thickness was approximately 5%. The final thickness of the samples was 0.5 mm. Between the passes, billets were cooled in air to minimize the effect of deformational heating on the microstructure. The rolling direction was parallel to the direction of extrusion. The equivalent deformation in one pass of ECAP was 1.15, and after rolling by 95% - 3.5. An OLIMPUS GX51 optical microscope (OM), a JEOL JSM-6490LV scanning electron microscope (SEM) JEOL JSM-6490LV and a JEOL 2100 transmission electron microscope (TEM) JEOL 2100 were used to study the microstructure. Studies of the microstructure and properties were carried out in the rolling plane. Tensile tests were performed at a strain rate of $10^{-3}$ $s^{-1}$. Samples were cut parallel to the direction of deformation from the central part of the ECAP blanks and sheets after rolling. Uniaxial tensile tests were carried using small specimen on an Instron 8801 tester at room temperature at a speed of $10^{-3}$ $s^{-1}$. Conductivity studies were carried out at a temperature T = 23 °C by the eddy current method. The measurement error was ± 2%.

## 3. Results and discussion

The metallographic study showed the coarse-grained microstructure with an average grain size of 260 ± 11 μm in the initial state (figure 1, a). In the microstructure, inclusions of 4.3 ± 0.1μm size based on chromium are visible [6].

Deformation by the ECAP method led to a significant refinement of structural elements and transformed the microstructure of the initial state into the UFG structure. After one ECAP pass, the elongated subgrain microstructure was formed, most of the boundaries were curved and thick, which can be explained by their small misorientation. After two ECAP passes, the microstructure became more equiaxed, but most of the boundaries retained their appearance. An increase in the degree of deformation as a result of four ECAP passes led to the formation of small equiaxial fragments along with elongated ones. The boundaries became thinner and clear, possibly because of an increase in their misorientation.

a) b)

**Figure 1.** The microstructure of the alloy: a) the initial state, SEM; b) after 8 ECAP passes, TEM.

After 8 ECAP passes, a homogeneous, almost equiaxed microstructure with an average size of structural components of 250 ± 15 nm was formed (figure 1, b). An analysis of microstructure photographs after ECAP revealed the presence of particles of the secondary phases (figure 1, b). Therefore, during SPD, diffusion processes were activated, leading to the decomposition of the

IOP Conf. Series: Materials Science and Engineering **672** (2019) 012056 doi:10.1088/1757-899X/672/1/012056

supersaturated solid solution and the formation of particles with a size of about 5-10 nm. The volume fraction of precipitated particles increased with an increase in the degree of deformation.

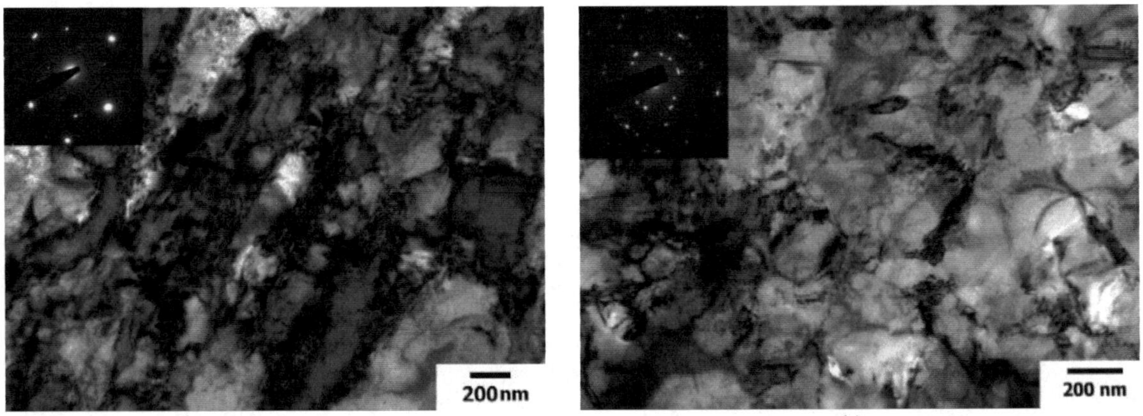


a) b)

**Figure 2.** The microstructure: a) the initial state after cold rolling; c) 8 ECAP passes and cold rolling. TEM.

As a result of rolling of the initial state, an elongated structure with nonequilibrium boundaries was formed (figure 2, a). The use of rolling after 8 ECAP passes led to a decrease in the size of structural elements to 160±10 nm (figure 2, b). The analysis of the fine structure after 8 passes of ECAP in combination with cold rolling indicates the formation of a strongly deformed UFG microstructure with a high density of dislocations that are unevenly distributed, most of the dislocations accumulate near the boundaries. The inhomogeneous diffraction contrast in the bright field image of structural elements indicates large internal stresses, the source of which is the boundaries of the fragments. Most boundaries are not clearly defined, they are uneven and curved.

**Table 1.** Mechanical characteristics and electrical conductivity.

| State | Hv, MPa | $\sigma_{UTS}$, MPa | $\sigma_{YS}$, MPa | ε, % | IACS, % |
|---|---|---|---|---|---|
| IS | 710±10 | 160±10 | 60±5 | 42±2 | 37 |
| ECAP1 | 1160±11 | 325±18 | 318±16 | 15±1 | 34 |
| ECAP2 | 1250±10 | 400±5 | 385±7 | 14±1 | 39 |
| ECAP4 | 1390±13 | 425±5 | 370±20 | 14±1 | 36 |
| ECAP8 | 1480±6 | 475±10 | 450±10 | 17±1 | 35 |
| CR | 1460±15 | 490±7 | 480±12 | 8±2 | 39 |
| ECAP1+CR | 1550±12 | 445±13 | 435±16 | 11±1 | 37 |
| ECAP2+CR | 1560±15 | 500±5 | 475±5 | 13±1 | 37 |
| ECAP4+CR | 1500±18 | 550±10 | 520±10 | 13±1 | 37 |
| ECAP8+CR | 1550±10 | 605±18 | 565±13 | 13±1 | 37 |

Table 1 shows the properties of the Cu-0.5wt.%Cr alloy in various states. Microhardness with an increase in the number of ECAP passes increases from 710 MPa to 1480 MPa. At the same time, the strength also increases and reaches saturation already after the second ECAP pass, reaching 475 MPa. Subsequent rolling increases the strength by 100-150 MPa, depending on the number of ECAP passes. The ultimate strength value is 605 MPa. The microhardness remains approximately at the same level of ~ 1500 MPa. In this case, rolling of the initial state leads to a strength value of 490 MPa. From table 1 it can be seen that the number of ECAP passes and subsequent rolling practically do not affect the electrical conductivity, which is 34-39% of IACS. The same value is characteristic of the rolled initial

state. Consequently, the contribution of the size of the fragments and dislocations to the electrical conductivity is negligible and the solid solution makes a significant contribution.
The obtained electrical conductivity values are unacceptable for industrial production. Since alloys of the Cu–Cr system are heat-hardening, it is possible to precipitate small dispersed hardening particles of the secondary phases during post-deformation heat treatment, as a result of which, on the one hand, the strength of these alloys increases, and on the other hand, the electrical conductivity is restored [1-3, 6]. It is also known that materials subjected to SPD are characterized by high internal stresses due to the high density of structural defects [1, 3]. Therefore, in the following studies, it is necessary to analyze the effect of aging on the microstructure and the formation of the optimal combination of strength and electrical conductivity of the studied alloy Cu-0.5wt.% Cr.

Thus, rolling after ECAP makes it possible to increase the strength by 100-150 MPa depending on the number of ECAP passes, while the ductility of the rolled UFG samples is greater than the ductility of the rolled initial state.

## 4. Conclusions

As a result of the study of the effect of cold rolling on the microstructure, mechanical properties and electrical conductivity of the UFG Cu-0.5wt.% Cr alloy, it was revealed that the microstructure obtained by 8 ECAP passes with subsequent rolling was characterized by the strength which is 1.2 times higher, and the ductility which is 1.6 times higher in comparison with the rolled initial state.

## Acknowledgments

Authors are grateful for the financial support of the Ministry of Education and Science within the frame work of project No. 16.1969.2017/4.6. W.Wei is grateful to be supported by the National Natural Science Foundation of China under Grant No. 51561001, the Priority Academic Program Development of Jiangsu Higher Education Institutions (PAPD), the Top-notch Academic Programs Project of Jiangsu Higher Education Institutions (TAPP), and the Science and Technology Project of Changzhou, P. R. China under grant No.CZ20180016 and CE20170028.

## References

[1] Vinogradov A, Ishida N, Kitagawa K and Kopylov V 2005 *Acta Mater.* **53** 2181
[2] Purcek G, Yanar H, Saray O, Karaman I and Maier H 2014 *Wear* **311** 149
[3] Sarkeeva E A, Abramova M M, Oreshkina R A, Sitdikov V D, Wei K X, W Wei, Musin F F and Alexandrov I V 2015 *Mater. Phys. Mech.* **24** 211
[4] Valiev R Z and Alexandrov I V 2007 *Bulk Nanostructured Metal Materials: Preparation, Structure and Properties* (Moscow: Akademkniga) [in Russian]
[5] Valiev R Z, Islamgaliev R K and Alexandrov I V 2000 *Prog. Mater. Sci.* **45** 103
[6] Sarkeeva E A, Sitdikov V D, Raab G I, Wei W and Alexandrov I V 2018 *IOP Conf. Ser.: Mater. Sci. Eng.* **447** 012065

IOP Conf. Series: Materials Science and Engineering **672** (2019) 012057 doi:10.1088/1757-899X/672/1/012057

# Computer simulation of the SPD process of a low-alloyed chromium bronze using a virtual full factorial experiment

**V I Semenov [1], N T Tontchev [2], L Sh Shuster[1], G I Raab[1] and V V Kamburov [3]**

[1] Ufa State Aviation Technical University, Ufa, Russia
[2] Todor Kableshkov Higher School of Transport, Sofia, Bulgaria
[3] Technical University of Sofia, Bulgaria

E-mail: semenov-vi@rambler.ru

**Abstract.** Using computer simulation in the DEFORM-3D software package, we performed a virtual full factorial experiment for the ECAP-Conform processing of the low-alloyed chromium bronze Cu-0.14%Cr. In the course of the experiment we evaluated the effect of independent parameters (the rotation speed of the working wheel, the friction factor at the lateral surfaces of the working wheel and the friction factor between a billet and the die). As a result of the experiment, we obtained a regression equation and determined the most important individual factors and their combinations that influence the response parameter – the deformation force.

## 1. Introduction

At present, there is interest in the studies of the strength enhancement of metals due to structure refinement to the submicrocrystalline (SMC) range through severe plastic deformation (SPD) processing [1]. Among the SPD techniques is equal-channel angular pressing (ECAP) [2, 3] and its modification – the ECAP-Conform technique [4] developed for the processing of long-length billets with a bulk SMC structure and creating prerequisites for the practical implementation of SPD processing.

However, to produce long-length semi-products, it is necessary to solve the task related to the found contradiction. To feed a billet to the deformation site, it is necessary to use the active friction forces at the lateral surfaces of the working wheel, i.e. to have the maximum friction factor $(f_1)$. In contrast, to realize the deformation process and produce a high-quality semi-product with a defect-free surface, it is necessary to provide the lowest value of the friction factor $(f_2)$ in the deformation site.

The fragmentary application of a lubricant only on those surfaces where a low friction factor is required leads to a decline in the productivity and mechanization of SPD processing. This makes the processing, which is already not cheap, even more expensive. In order to increase the efficiency of SPD processing by ECAP-Conform, it is necessary to find a compromise solution that involves one variant of billet surface preparation prior to deformation processing, enabling the feeding of the billet to the deformation site and producing semi-products of the required quality.

In the analysis of tribological systems, an important place is held by the numerical methods for the study of complex processes, including computer simulation with the use of the state-of-the-art software [5, 6]. Therefore, it seems reasonable to perform numerical simulation using a virtual full factorial experiment (FFE) [7].

The aim of the simulation is to perform virtual SPD processing by ECAP-Conform using a FFE, and to identify the rational speed regime of the processing, combined with a universal surface preparation of a billet, in the conditions of the fabrication of long-length SMC semi-products under the minimum possible deformation force.

## 2. Experimental procedure

In order to obtain the fullest possible information about the dependencies under study, the authors used an FFE in their simulation. The design of experiments is a procedure of selecting the number and conditions of experiments, necessary and sufficient to obtain a mathematical model of the process [8]. It is important to consider the following: the striving for the minimization of the number of experiments; the simultaneous variation of all variables defining the process; the selection of a clear strategy that enables making grounded decisions after each series of experiments [9].

The design of the experiment was carried out based on the simulation of the process of the fabrication of long-length semi-products from the low-alloyed chromium bronze Cu-0.14%Cr by ECAP-Conform. The principle of the facility is presented in figure 1.

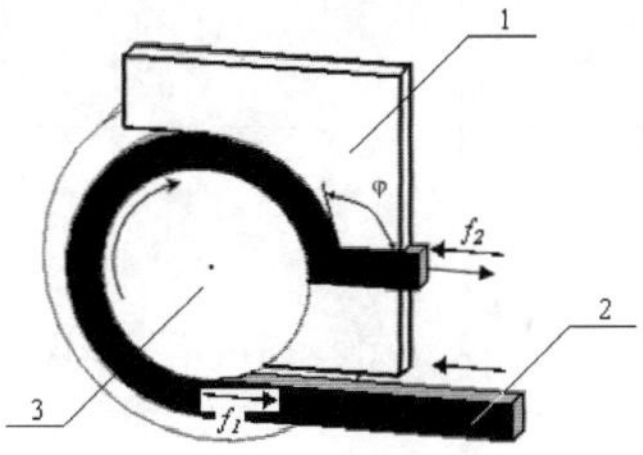

**Figure 1.** Principle of the SPD technique of ECAP-Conform: 1 – stationary die; 2 – billet; 3 – working wheel.

The object of study is the low-alloyed chromium bronze Cu-0.14%Cr, its rheological properties were taken from [10] when designing the numerical model.

To perform the numerical simulation, we used the standard software package DEFORM-3D.

With a view to perform the simulation and factorial experiment in the DEFORM-3D software package, we had previously created three-dimensional models in the Kompas-3D software.

### *2.1. Assumptions*

1) The material in the billet in the initial state is isotropic and has no initial stresses and strains;
2) The deformation temperature is taken as 200 °C;
3) The channels intersection angle is taken as 120°;
4) The tool is absolutely rigid, and the tool's geometry is taken into account automatically;
5) The material of the initial billet is taken to be ductile;
6) For the simulation we selected 100 steps taking into account the complete passage of the billet through the dies and getting a stable result;
7) The billet is divided into 43553 trapezoidal elements.

At the stage of preparation of the simulation task, we believed that the most important factors influencing the fabrication of defect-free semi-products are the factors of friction (parameters of contact) between the billet and different parts of the tool, and the deformation speed conditioned by the rotation speed of the working wheel. Therefore, we decided to perform the virtual FFE using a two-tier model with three unknowns corresponding to three variable factors, and then formalize the obtained results in the form of a regression equation and optimize the selected factors.

Thus, we selected the following as the independent variables in the process of drawing with shear, characterizing the running of the process and its efficiency in terms of the deformation force: the friction factor $f_1$ ($X_1$) at the upper and lower surfaces of the working wheel determining the efficiency of billet feeding to the deformation site, the friction factor $f_2$ ($X_2$) at the shape-forming parts of the tool, and the deformation speed (the rotation speed of the working wheel) $V$ ($X_3$). The deformation force $P$ ($Y$) was taken as the response parameter (dependent parameter).

## 3. Experiment results and discussion

It is necessary to find such values of $f_1$, $f_2$, V that will provide the minimum deformation force P. After completing the full factorial experiment, the mathematical model has the following form:

$$y=b_0+b_1x_1+b_2x_2+\ldots+b_{12}x_1x_2+b_{13}x_1x_3+ b_{23}x_2x_3\ldots+b_{123}x_1x_2x_3, \quad (1)$$

where $b_i$ is the regression coefficient.

To calculate the coefficients, in this model we used an extended matrix for the experiment design and results (table 1).

**Table 1.** Extended matrix of the experiment design $2^3$ and results.

| Experiment No. | $x_0$ | $x_1$ | $x_2$ | $x_3$ | $x_1x_2$ | $x_1x_3$ | $x_2x_3$ | $x_1x_2x_3$ | $Y(kN)$ |
|---|---|---|---|---|---|---|---|---|---|
| 1 | + | + | + | + | + | + | + | + | 13.3 |
| 2 | + | - | + | + | - | - | + | - | 16.5 |
| 3 | + | + | - | + | - | + | - | - | 10.7 |
| 4 | + | - | - | + | + | - | - | + | 12.7 |
| 5 | + | + | + | - | + | - | - | - | 23.2 |
| 6 | + | - | + | - | - | + | - | + | 19.8 |
| 7 | + | + | - | - | - | - | + | + | 13.5 |
| 8 | + | - | - | - | + | + | + | - | 12.9 |

Figure 2 shows the solution of the task of the numerical simulation of the ECAP-Conform process. As a result, the minimum deformation force was obtained.

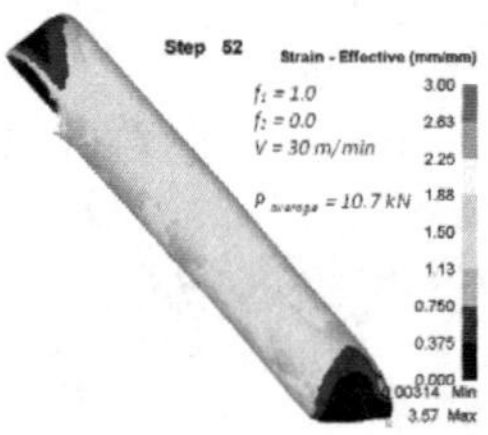


**Figure 2.** Result of the simulation of the ECAP-Conform process – fields of accumulated strain distribution. The deformation force $P_{average}$ = 10.7 kN.

On the basis of the calculations, we obtained the following general form of a linear regression equation:

$$y=15.33X_0-0.15X_1+2.88X_2-2.03X_3+0.20X_1X_2-1.15X_1X-1.28X_2X_3-0.50X_1X_2X_3. \quad (2)$$

Equation (2) demonstrates that it is the friction factor $f_2$ ($X_2$) in the sliding contact between the billet and the tool, and the deformation speed $V$ ($X_3$) that have the greatest influence on the deformation force. It can be seen from the coefficients of the regression equation that the deformation force will decrease as both of these factors increase. The factor $f_1$ ($X_1$) of active friction at the upper and lower surfaces of the working wheel, which provides the feeding of the billet to the deformation site, has a much smaller effect on the deformation force. Since the greatest and unidirectional influence is exerted by the factors $X_2$ and $X_3$, this makes it possible to select a variant of universal preparation of the billet surface. It should be noted that two-way and three-way mutual interactions have an ambiguous interpretation, and therefore complex interactions should be analyzed separately and with a reference to the specific operation conditions of a multi-component system.

A priori, it can be stated that in the conditions under consideration the minimum value of deformation force can be obtained under the optimum combination of independent parameters taken in this study.

The design of the experiment showed that in the considered conditions the deformation force will be the lowest when there is high friction at the upper and lower surfaces of the working wheel ($f_1 \approx 1.0$), the friction at the shape-forming parts of the tool tends to have the minimum values ($f_2 \approx 0.00$), and the deformation speed is $V \approx 25$ m/min. When the above-mentioned values of independent parameters are ensured, it is possible to provide the deformation force $P \approx 10.7$ kN (figure 2).

However, the task of this study was to provide the SPD processing by ECAP-Conform with the minimum possible deformation force under the condition of a universal preparation of the billet surface.

By solving the reverse task, we managed to select a variant of universal surface preparation and the deformation speed that provide a deformation force of $P \approx 12.5$ kN. For this purpose, it is necessary to ensure $f_1 = f_2 = 0.3$ and a deformation speed of $V \approx 30$ m/min.

## 4. Conclusions

1. As a result of a virtual full factorial experiment, it has been found that it is the friction factor $f_2$ ($X_2$) in the sliding contact between the billet and the tool, and the deformation speed V ($X_3$) that have the greatest effect on the deformation force.
2. For the practical implementation of the ECAP-Conform processing of the low-alloyed chromium bronze Cu-0.14%Cr, a variant of billet surface preparation can be proposed that combines the application of an under-lubricant coating and a processing lubricant.

## References

[1] Valiev R Z and Alexandrov I V 2007 *Bulk Nanostructured Metallic Materials: Production, Structure and Properties* (Moscow: Akademkniga) [in Russian]
[2] Segal V M, Reznikov V I and Kopylov V I 1994 *Processes of Plastic Structure Formation in Metals* (Minsk: Science and Technology) [in Russian]
[3] Segal V M 2004 *Mater. Sci. Eng.* A **386** 269
[4] Raab G I and Valiev R Z 2000 *Izvestiya Vuzov. Tsvetnaya Metallurgiya* **5** 50 [in Russian]
[5] Abdullah O I and Schlattmann J 2012 *Tribology in Industry* **34** 206
[6] Petrović Savić S, Adamović D, Devedžić G, Ristić B and Matić A 2014 *Tribology in Industry* **36** 354
[7] Belhocine A, Abu Bakar A R and Bouchetara M 2014 *Tribology in Industry* **36** 49
[8] Novick F S and Arsov Y B 1980 *Optimization of Technology Processes for Metals by the Methods of Experiment Design* (Moscow: Mashinostroenie) [in Russian]
[9] Adler Yu P 1978 *Experiment Design* (Moscow: Znanie) [in Russian]
[10] Adler Yu P 1969 *Introduction to Design of Experiments* (Moscow: Metallurgiya) [in Russian]
[11] Frolova N Yu, Zeldovich V I, Khomskaya I V, Kheifets A E and Shorokhov E V 2015 *Diagnostics, Resource and Mechanics of Materials and Structures* **5** 99

IOP Conf. Series: Materials Science and Engineering **672** (2019) 012058 doi:10.1088/1757-899X/672/1/012058

# Evaluation of the tribological properties of a magnesium alloy of the Mg-Ca system depending on the structure of the material under study

**V I Semenov**[1], **H-C Lin**[2], **L Sh Shuster**[1] **and C Chiu**[3]

[1] Ufa State Aviation Technical University, Ufa, Russia
[2] National Taiwan University, Taipei, Taiwan
[3] National Taiwan University of Science and Technology, Taipei, Taiwan

E-mail: semenov-vi@rambler.ru

**Abstract**. The paper presents the results of a study of the tribological characteristics of the contact of the tool steel of composition Fe-18W-4Cr-1.2V with a magnesium alloy of the *Mg-Ca* system, strengthened by SPD using the method of ECAP. It has been established that the SPD processing of a magnesium alloy, due to an increase in strength as a result of the grain structure refinement, makes it possible to significantly reduce the integral value of the friction coefficient in the moving friction contact. In addition, the SPD processing of the material under study leads to a decrease in the molecular component of the friction coefficient.

## 1. Introduction

In modern mechanical engineering, in particular in friction units, one of the ways to reduce the weight of vehicles is to apply light alloys with enhanced mechanical properties. Magnesium is interesting for use as a structural material due to the attractive strength-to-weight ratio that exceeds the one for aluminum and other light metals and alloys [1-3]. The high damping capacity of magnesium alloys allows using them effectively to manufacture automobile and aircraft wheels, various components for motor-and-tractor and aerospace engineering, rollers for cargo conveyors [4, 5] etc. Severe plastic deformation (SPD) can effectively increase the strength of bulk metals due to the fabrication of an ultrafine-grained structure in them [6, 7]. SPD techniques can be considered as alternative techniques to dispersion strengthening of composite materials. However, the tribological behavior of ultrafine grain magnesium alloys is studied insufficiently for application in friction units [8, 9].

This study presents the results of determining the integral value of the coefficient of friction and its molecular component. In addition, a comparative assessment of the strength of adhesive bonds and the molecular component of the friction coefficient of a magnesium alloy of the Mg-1%Ca composition, hardened as a result of SPD by the equal channel angular pressing (ECAP) method in sliding tribological contact with tool steel of composition Fe-18W-4Cr-1.2V is given. The material of the counter-body - tool steel - was selected because of its high wear resistance to maximize the elimination of errors during tribological tests.

## 2. Experimental procedure and materials

In comparative tests, the composition of the magnesium Mg-1% Ca alloy was used in the initial (coarse-grained) state and after two passes of ECAP.

Tribological tests were carried out on the friction machine "Timken". Figure 1 presents the machine and the processing scheme.

Testing of the initial and dispersion-strengthened materials was conducted under a normal load of *10 N* and *50 N* and a disk rotation speed of $250\ min^{-1}$ and $1000\ min^{-1}$, the slip distance was 1650 m in all the tests.

The friction coefficient *f* was calculated according to the formula:

$$f = \frac{F}{P}, \tag{1}$$

where *F* – the friction force, *N*; *P* – the normal loading force, *N*.

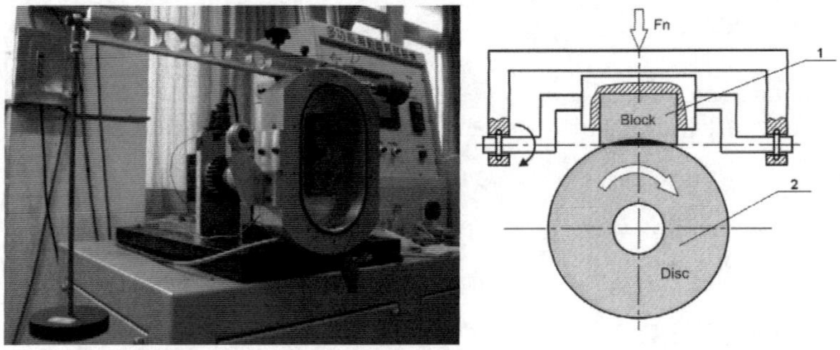


a b

**Figure 1.** Friction machine "Timken" (*a*) and testing scheme (*b*): 1 – tested sample (magnesium alloy of the composition Mg-1% Ca); 2 – rotating steel disk (tool steel of the composition Fe-18W-4Cr-1.2V).

Figure 2 presents the ECAP scheme that was chosen for strain hardening of the initial material [10].

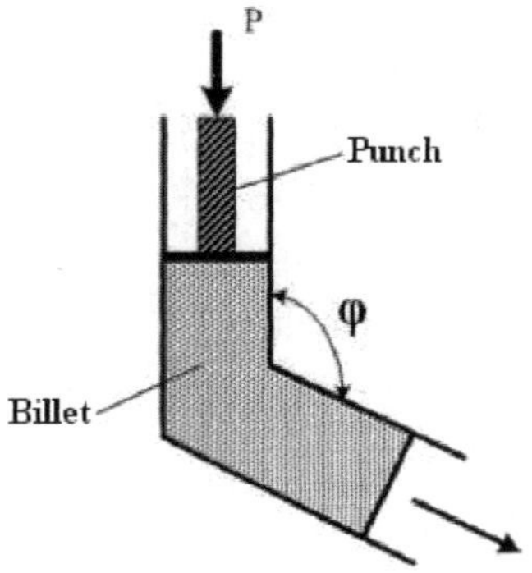


**Figure 2.** ECAP scheme.

The employed scheme allows achieving high degrees of accumulated strain as a result of shear in the conjugating channels. In this case, the angle between the conjugating channels $\varphi$ was 120°.

Studies on the evaluation of the adhesive bond shear strength and the molecular component of the friction coefficient were carried out on the one-ball friction machine at temperatures of 20, 150 and 300°C according to the scheme shown in figure 3 [11].

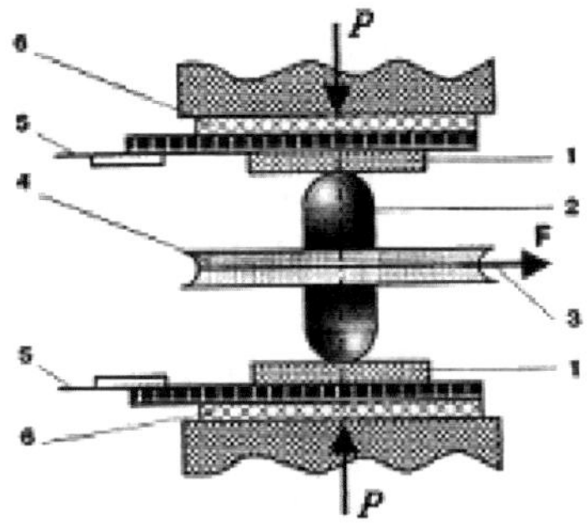


**Figure 3.** Scheme for testing the molecular component of the friction coefficient: 1 – plane-parallel tested samples (magnesium alloy of the composition Mg-1%Ca); 2 – spherical indenter (tool steel of the composition Fe-18W-4Cr-1.2V); 3 – pulling rope; 4 – disk with a groove for the rope; 5 – line for contact heating of the tested samples; 6 – electroinsulating pads.

Adhesive bond shear strength ($\tau_n$) and adhesive (molecular) component of the friction coefficient ($f_m$) were determined by the method presented in [11].

## 3. Results of experiments and their discussion

Figure 4 presents graphs with the results of tribological tests according to the "block-disk" scheme (figure 1).

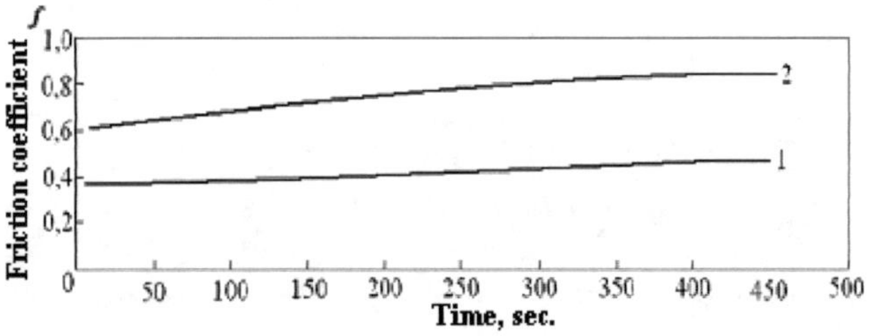


**Figure 4.** Change in the integral value of the coefficient of friction depending on the microstructure of a magnesium alloy of the composition Mg-1%Ca: 1 - fine-grained after SPD treatment by ECAP; 2 - coarse-grained after annealing.

It can be seen from the graph that the friction coefficient is higher for the studied material in the coarse-grained state after annealing than for the fine-grained after two cycles of SPD processing by ECAP throughout the entire test period. Moreover, the coefficient of friction of a material with a fine-grained structure varies slightly depending on the test time. In order to explain these results, let us consider the data received with the help of the one-ball friction machine.

Figure 5 demonstrates that the dependence of the adhesive bond shear strength $\tau_n$ on the normal pressure $p_r$ is described by the binomial dependence:

$$\tau_n = \tau_o + \beta p_r \tag{2}$$

where $\tau_o$ – adhesive bond shear strength without normal loading force; $\beta$ – piezocoefficient.

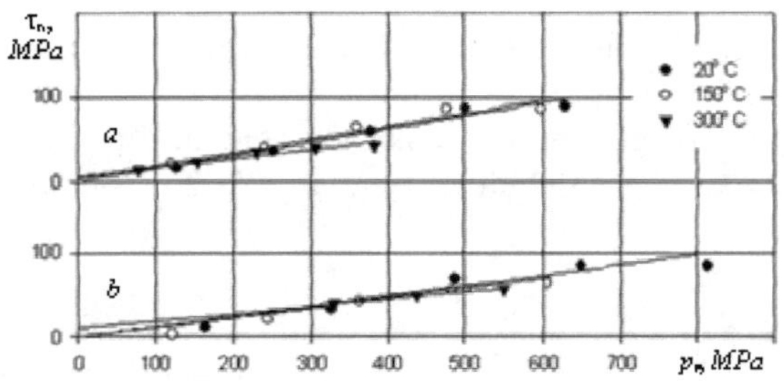


**Figure 5.** The dependence of the adhesive bond shear strength on the normal pressure in the contact with tool steel of the composition Fe-18W-4Cr-1.2V: (*a*) - Mg-1%Ca in the initial state; (*b*) Mg-1%Ca after two ECAP passes.

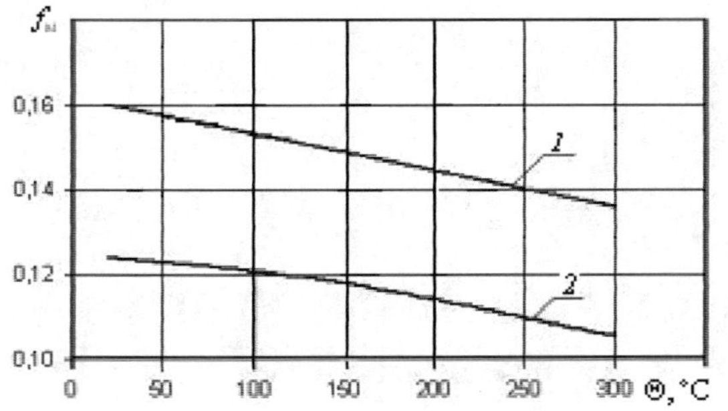


**Figure 6.** The dependence of the molecular constituents of the friction coefficient on the temperature for friction couples: 1) Mg-1%Ca – Fe-18W-4Cr-1.2V; 2) Mg-1%Ca after two ECAP passes – Fe-18W-4Cr-1.2V.

The molecular constituent of the friction coefficient $f_M$ can be defined as:

$$f_M = \frac{\tau_n}{p_r} = \frac{\tau_o}{p_r} + \beta \tag{3}$$

It is seen from formula (3) that the molecular component of the friction coefficient increases when the normal pressure $p_r$ decreases.

A decrease in the slip rate reduces the temperature of the friction contact, which, according to the data in figure 6, enhances the molecular component $f_m$ and the total friction coefficient $f$.

The experimental data given in figures 5 and 6 testify to the fact that the deformation treatment of the initial magnesium alloy provides a lower value of the adhesive constituent of the friction coefficient with a high bearing capacity of the frictional contact [12].

## 4. Conclusion

On the basis of the conducted research, it was established that the magnesium alloy of the composition Mg-1%Ca can be effectively strengthened by severe plastic deformation using the equal-channel angular pressing technique. The deformed material significantly improves the tribological properties in the moving friction contact and there is a decrease in both the integral value of the friction coefficient and its molecular component. In this case, an increase in the bearing capacity of the frictional contact is observed. Thus, the use of a magnesium alloy after ECAP in friction units is quite promising.

## References

[1] Harrington W C 1994 Metal matrix composites applications *Mechanical properties of Metallic Composites* ed S Ochiai (New York: Mercel-Dekker) pp 759–73
[2] Kaczmar J W, Pietrzak K, Wlosinski W 2000 *J. Mater. Process. Technol.* **106** 58
[3] Yeong Y M 2000 *Influence of Types of Reinforcements on Properties of Magnesium Alloy* (Eng. Thesis) (National University of Singapore)
[4] Galdin N M 1989 *Color Casting: Reference Book* (Moscow: Mashinostroenie) [in Russian]
[5] Reshetov D N 1995 *Mechanical Engineering. Vehicle Components. Structural Strength. Friction, Wear, Lubrication* Vol IV-1 (Moscow: Mashinostroenie) [in Russian]
[6] Valiev R Z and Alexandrov I V 2007 *Bulk Nanostructured Metallic Materials: Processing, Structure and Properties* (Moscow: Akademkniga) [in Russian]
[7] Valiev R Z, Estrin Y, Horita Z, Langdon T G, Zehetbauer M J and Zhu Y T 2006 *JOM* **58(4)** 33
[8] Huang S-J and Liu H-W 2003 *The Chinese Journal of Mechanics* **19(2)** 57
[9] Sinha S K, Reddy S U and Gupta M 2006 *Tribology International* **39** 184
[10] Valiev R Z 2006 *Rossiyskiye Nanotechnologii* **1(1-2)** 208 [in Russian]
[11] Shuster L Sh 1999 *Adhesive Interaction of Metallic Solids* (Ufa: Gilem) [in Russian]
[12] Semenov V I, Shuster L Sh, Chertovskikh S V and Raab G I 2005 *Trenie i Iznos* **26** 74 [in Russian]

# The effect of combined thermomechanical treatment on the structure, phases and hardness of a superelastic Ti-Zr-Nb bar stock

**V Sheremetyev[1], A Kudryashova[1], M Derkach[1], K Lukashevich[1], V Andreev[2], S Galkin[1], S Prokoshkin[1] and V Brailovski[3]**

[1] National University of Science and Technology "MISiS", Moscow, Russia
[2] Baikov Institute of Metallurgy and Materials Science, RAS, Moscow, Russia
[3] Ecole de technologie superieure, Montreal, QC, Canada

E-mail: sheremetyev@misis.ru

**Abstract**. The superelastic Ti-19Zr-14Nb (at.%) alloy was subjected to a novel sequence of thermomechanical treatment, including radial shear rolling, rotary forging and post-deformation annealing. The features of the grain structure, phase composition and hardness of a long-length bar stock for fabrication of bone implants were analyzed. It was shown that after the aforementioned combined thermomechanical treatment followed by annealing in the 550-750 °C temperature range, the produced bar stock demonstrates uniformly distributed grains of $\beta$-phase and uniform hardness across its section.

## 1. Introduction

An increasing interest in metastable $\beta$-Ti alloys as promising materials for load-bearing implants, is caused by two main features: superelasticity and the absence of toxic elements in their composition [1, 2]. The superelasticity of these alloys is due to stress-induced reversible martensitic transformation from body center cubic $\beta$-phase to orthorhombic $\alpha''$-phase [1]. These alloys exhibit high strength, good corrosion resistance, a relatively low Young's modulus, and excellent biocompatibility [1-3]. Ti-Zr-Nb alloys have been particularly investigated due to their large (up to 7%) transformation strain and, therefore, outstanding superelasticity [4].

It is known that thermomechanical treatment (TMT) is an effective technique to improve the mechanical and functional properties of shape memory alloys via the control of their microstructure and phase composition [5]. For example, traditional TMT including cold rolling ($e$=0.3) and post-deformation annealing (PDA) led to the formation in Ti-Zr-Nb alloys of a polygonized dislocation substructure of $\beta$-phase with nano- to submicrometer-sized subgrains and to stable functional fatigue behavior with a Young's modulus as low as 30-40 GPa [6]. It was also shown that combining hot radial shear rolling (RSR) and hot and cold rotary forging (RF) allows manufacturing bar stocks with a dynamically polygonized substructure of $\beta$-phase, homogeneous across-the-section grain size distribution and superior functional fatigue behavior [7].Thus, the consistent application of the RSR, RF, and PDA operations to Ti-Zr-Nb alloys looks like a promising way to produce long-length bar stock with an optimal combination of phase composition, microstructure and functional properties. In this study, the effect of such a combined TMT on the phase composition, structure and mechanical properties of a Ti-19Zr-14Nb bar stock is investigated.

## 2. Experimental procedure

A 20 kg-weight, 55 mm-diameter Ti-19Zr-14Nb (at.%) ingot was produced by vacuum arc remelting and isostatically-pressed (900 °C, 100 MPa, 2 h). The ingot was cut and subjected to hot radial shear rolling at 900 °C using "14-40" and "10-30" rolling mills to form a 11.5 mm-diameter bar stock (an accumulated true strain of $e$=3.2). The bar was then subjected to hot rotary forging at 600 °C using an RKM-2 forging machine to form a 7.8 mm-diameter bar ($e$=0.8). After both the RSR and RF operations, the bars were air-cooled. The hot-forged bar was then subjected to room temperature RF to form a 6.6 mm-diameter bar stock ($e$=0.3). Next, the bar was straightened and grinded to obtain a 6.0 mm-diameter bar stock. Samples for structural characterization and hardness measurements were cut from the bar stock and annealed for 0.5 h in argon atmosphere in the 500 to 750 °C temperature range, and then water-quenched. The cross-section surfaces of the RF and RF+PDA samples were prepared using an ATM Saphir 560 grinding and polishing machine. Specimens for optical microscopy and X-ray diffractometry were mechanically polished and then etched in $1HF:3HNO_3:6H_2O$ solution to remove a damaged surface layer.

The microstructure of the alloy and its phase composition before and after TMT were studied using a *Union Versamet-2 Union* optical microscope equipped with a *Nikon D90* camera and a *PANalytical X'pert Pro* X-ray diffractometer. The lattice parameters of the studied phases were calculated from the angular coordinates of the $\beta$ and $\alpha''$-phase XRD peaks. The maximum $\beta\rightleftarrows\alpha''$ transformation lattice strain ($\varepsilon_{max}$), which corresponds to a theoretical recovery strain limit, was calculated from the lattice parameters of the $\beta$ and $\alpha''$-phases, using the methodology described in [7]. Hardness was measured using a *Metkon microhardness tester* with an applied load of 1 kg and a loading time of 10 sec.

## 3. Results and discussion

The metallography images obtained after different RF+PDA combinations and the corresponding average grain size values $d$ are shown in figure 1. It can be seen that immediately after RSR+RF, microstructure formed in the bar stock is heterogeneous across the section. It is well known, that during RSR, metal flows helicoidally, thus forming a very peculiar deformed state with a characteristic inhomogeneity [7-9]. In this case, the microstructure of the peripheral zone is composed of grains with an average size of $d\approx 6$ μm (figure 1a). When moving to the center, in the intermediate zone, the average grain size increases to $d\approx 9$ μm (figure 1b), and finally, in the central zone, to $d\approx 13$ μm (figure 1c). After PDA at 550 °C, a significant amount of small statically recrystallized grains can be observed on the local scale, but the initially heterogeneous microstructure is preserved (figure 1d-f). With an increase of the PDA temperature to 600 °C, the recrystallization process extends over the entire cross-section of the bar stock (figure 1g-i). PDA at 750 °C leads to an intensive grain growth, and the grain distribution across the section line becomes uniform (figure 1j-l).

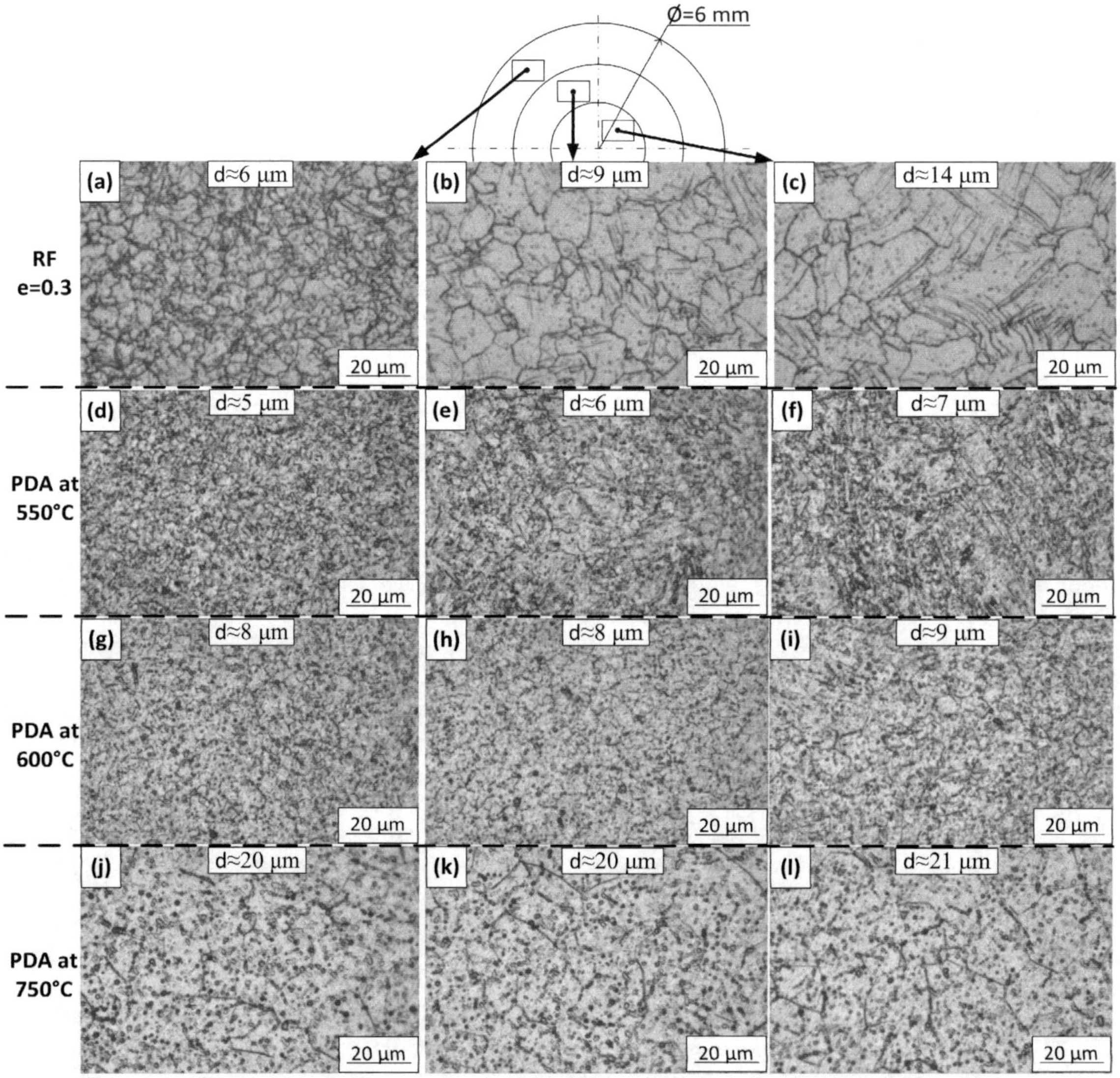


**Figure 1**. Images of the microstructure of Ti-Zr-Nb alloy after RF (e=0.3) (a, b, c), RF+PDA at 550 °C (d, e, f), RF+PDA at 600 °C (g, h, 1), and RF+PDA at 750 °C (j, k, l).

The X-ray diffractograms obtained directly after RF, and then after RF+PDA at different temperatures, are shown in figure 2a. The main phase constituent in all the cases is BCC $\beta$-phase. After PDA in the 500-525 °C range, a certain amount of cooling- or annealing-induced $\alpha$-phase is also observed. After cold RF and PDA in the 525-600 °C range, $\alpha''$-martensite lines are observed. The lattice parameters of the $\beta$- and $\alpha''$-phases, as well as the crystallographic limit of the recovery strain, are presented in table 1. The recovery strain value is limited to ~5.9% irrespective of the TMT route.
The evolution of the level of lattice defects after various treatments was evaluated from the changes in the $\beta$-phase half-width X-ray lines (figure 2b). Immediately after cold RF, the $\beta$-phase X-ray diffraction lines are strongly broadened, thus reflecting significant lattice defectness and substructural hardening of the material. With an increase of the PDA temperature from 500 to 750 °C, the width of the $\beta$-phase lines first decreased rapidly from RT to 525 °C, and then monotonically, from 550 to 750 °C.

These observations can be interpreted as the results of polygonization (500-550 °C), and recrystallization (550-750 °C) processes occurring in Ti-Nb-Zr SMA during post deformation annealing after cold rotary forging [6], which correlate well with the results of optical microscopy. Note that directly after RF, the $\{200\}_{\beta}$ X-ray line is not narrower than the $\{211\}_{\beta}$ X-ray line, whereas after PDA at 500 °C, the width of the former drops far below the width of the latter. These observations correlate with a transition from a highly dislocated work-hardened substructure to a polygonized (subgrained) dislocation substructure of $\beta$-phase during the polygonization process, like it was observed in Ti-Ni SMA earlier [9].

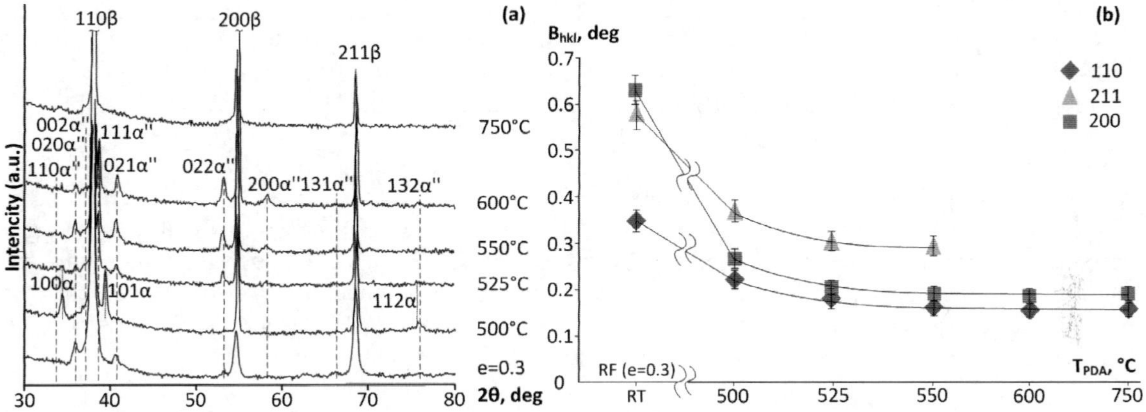


**Figure 2.** Results of X-ray diffraction analysis of Ti-Zr-Nb alloy after RF and RF (e=0.3)+PDA: X-ray diffractograms (a), $\beta$-phase X-ray line width *vs* PDA temperature (b).

**Table 1.** Lattice parameters of $\beta$- and $\alpha''$-phases, and crystallographic (theoretical) recovery strain limit of Ti-Zr-Nb alloy after RF+PDA at different temperatures.

| | 500 °C | 525 °C | 550 °C | 600 °C | 750 °C |
|---|---|---|---|---|---|
| $a_{\beta}$, Å | 3.344±0.002 | 3.345±0.002 | 3.347±0.002 | 3.339±0.002 | 3.346±0.002 |
| $a_{\alpha''}$, Å | - | 3.172±0.003 | 3.164±0.003 | 3.158±0.003 | - |
| $b_{\alpha''}$, Å | - | 5.001±0.009 | 5.011±0.011 | 5.002±0.010 | - |
| $c_{\alpha''}$, Å | - | 4.747±0.007 | 4.749±0.004 | 4.733±0.005 | - |
| $\varepsilon_{max}$, % | - | 5.88±0.18 | 5.87±0.23 | 5.94±0.20 | - |

The results of hardness measurements are presented in figure 3. After RF, the alloy exhibits the most hardened state. Additional PDA in the 500-550 °C range leads to rapid alloy softening and then in the 600-750 °C range, to stabilization of HV values (figure 3a). These hardness behavior correlates well with changes in the $\beta$-phase line widths related to the polygonization and recrystallization processes. Figure 3b demonstrate heterogeneous distribution of hardness across the section of bar stock after RF and RF+PDA (500-525 °C), which can be explained by different structure states at different points of the cross-section line (see figure 1a,b,c). After PDA at 550-750 °C, the formation of a homogeneous grain structure due to the process of recrystallization results in the leveling of the mechanical properties along the cross-section (figure 3b).

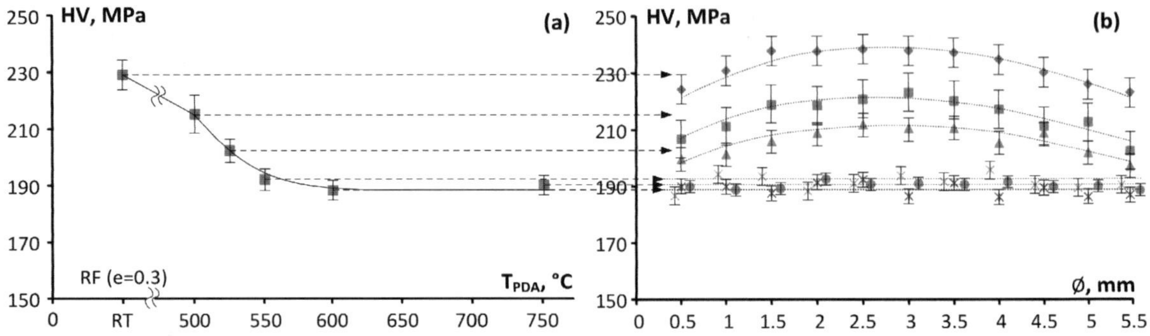


**Figure 3.** Results of hardness measurements of Ti-Zr-Nb alloy after RF and RF+PDA: HV *vs* PDA temperature (a), HV measured at different radial positions along the bar stock cross-section (b).

## 4. Summary

As a result of applying combined thermomechanical treatment, which comprises hot radial shear rolling and hot and cold rotary forging operations, to the superelastic Ti-19Zr-14Nb alloy, a strongly non-uniform microstructure was formed across the bar stock section with a subsequent non-uniform distribution of the mechanical properties. After post deformation annealing of the deformed alloy in the 550-750 °C temperature range, the cross-section $\beta$-phase grain and hardness distributions become uniform, and the bar stock semi-product becomes acceptable for implant manufacturing.

**Acknowledgments**

The present work has been carried out under financial support of the Russian Foundation for Basic Research (project №18-79-00247).

**References**

[1] Miyazaki S 2017 *Shape Memory and Superelasticity* **3** 279
[2] Konopatsky A S, Dubinskiy S M, Zhukova Y S, Sheremetyev V, Brailovski V, Prokoshkin S D and Filonov M R 2017 *Mater. Sci. Eng.* A **702** 301
[3] Geetha M, Singh A K, Asokamani R and Gogia A K 2009 *Prog. Mater. Sci.* **54(3)** 397
[4] Kim H Y, Fu J, Tobe H, Kim J I and Miyazaki S 2015 *Shape Memory and Superelasticity* **1** 107
[5] Prokoshkin S, Brailovski V, Dubinsky S, Zhukova Y, Sheremetyev V, Konopatsky A and Inaekyan K 2016 *Shape Memory and Superelasticity* **2** 130
[6] Sheremetyev V, Kudryashova A, Dubinskiy S, Galkin S, Prokoshkin S and Brailovski V 2018 *Journal of Alloys and Compounds* **737** 678
[7] Sheremetyev V, Kudryashova A, Cheverikin V, Korotitskiy A, Galkin S, Prokoshkin S and Brailovski V 2019 *Journal of Alloys and Compounds* **800** 320
[8] Akopyan T K, Belov N A, Aleshchenko A S, Galkin S P, Gamin Y V, Gorshenkov M V, Cheverikin V V and Shurkin P K 2019 *Mater. Sci. Eng.* A **746** 134
[9] Prokoshkin S D, Turenne S, Khmelevskaya I Yu, Brailovski V and Trochu F 2000 *Canadian Metallurgical Quarterly* **39(2)** 225

IOP Conf. Series: Materials Science and Engineering **672** (2019) 012060 doi:10.1088/1757-899X/672/1/012060

# Mechanical behavior at elevated temperatures of the ultrafine-grained titanium alloy VT8M-1 processed by rotary swaging

**A G Stotskiy[1], T V Yakovleva[1], Yu M Modina[1], A V Polyakov[1], I A Ramazanov[2] and I P Semenova[1]**

[1] Institute of Physics of Advanced Materials, Ufa State Aviation Technical University, Ufa 450008 Russia

[2] Research and Educational Center "Nanostructured Materials and High Technologies", Ufa 450008 Russia

E-mail: stockii_andrei@mail.ru

**Abstract.** This paper investigates the mechanical properties at elevated operation temperatures of the Ti alloy VT8M-1(Ti-5.3Al-4.0Mo-1.2Zr-1.3Sn-0.2Si) in the coarse-grained (CG) and ultrafine-grained (UFG) states. The duplex UFG structure was produced by rotary swaging at a temperature of 750 °C. The UFG state was studied in terms of thermal stability during annealing at temperatures 500, 550, 600, 700 °C for a period of 1 hour. It was demonstrated that the recrystallization processes in the UFG structure start after annealing at 550 °C and above. We performed tensile mechanical tests at such temperatures. In a range of 300-450 °C, the ultimate tensile strength of the UFG alloy was stably above that of the CG alloy. Mechanical tests at 500 and 550 °C revealed a more significant decrease in the ultimate tensile strength of the UFG alloy as compared to the CG alloy, conditioned by the intensification of the recovery and recrystallization processes at these temperatures.

## 1. Introduction

Titanium alloys are applied in aviation and engine building primarily due to their high strength and small weight. In particular, the two-phase ($\alpha+\beta$) Ti alloy VT8M-1 (Ti-5.3Al-4.0Mo-1.2Zr-1.3Sn-0.2Si) is used for producing compressor blades of gas-turbine engines (GTE) and is operated at temperatures up to 450 °C. Currently, the demands to modern engines are such that require a material that not only has high strength characteristics at elevated temperatures, but also does not degrade in long-term temperature-stressed conditions. In order to increase the hot-strength characteristics of Ti alloys, alloying with rare-earth elements and heat treatments are used, which have largely exhausted their potential [1].

One of the ways to improve the physical and mechanical properties of metals and alloys is the formation of ultrafine-grained (UFG) structures by means of severe plastic deformation (SPD) processing [2]. However, structures refined by large strains are usually stable at room temperature and at close temperatures, which is conditioned by a high accumulated internal energy leading to fast relaxation and a decrease in the recrystallization onset temperature [3]. Correspondingly, a question arises about the applicability of these methods for engineering materials operating at elevated temperatures.

The possibility in principle to produce a UFG structure in the VT8M-1 alloy by rotary swaging was demonstrated earlier [4]. The deformation by rotary swaging is implemented by a gradual slight swaging of a billet with a very high frequency, as a result of which a very high strain is introduced in a material and a rather uniform deformation takes place [5]. As a result of such a processing, a UFG structure with a grain size of 0.3 μm was produced in a bulk billet of the VT8M-1 alloy [4]. This method enables producing long-length billets, which is an undoubted advantage for industrial application. In this connection, the aim of the present work is to study the stability of the mechanical properties of the VT8M-1 processed by rotary swaging, at temperatures close to the operation ones.

## 2. Material and experimental methods

As the material for the study, we used the VT8M-1 (Ti-5.3Al-4.0Mo-1.2Zr-1.3Sn-0.2Si) alloy in the form of hot-rolled rods with a diameter of 70 mm (produced by VSMPO-AVISMA Corporation, Verkhnaya Salda, Russian Federation).

Rotary swaging (RS) was conducted at a temperature of 750 °C with a gradual reduction in the diameter from 70 mm to 32 mm. This allowed to introduce in the material a true strain of 1.56. The strain was calculated according to the relation $e = \ln((S_0 - S_1)/S_1)$, where $S_0$ and $S_1$ are the cross-sectional areas of the billet before and after deformation, respectively.

Tensile mechanical tests were performed using an Instron universal testing machine at elevated temperatures (300, 400, 450, 500, 550 °C) with a strain rate of $1\times10^{-3}$ $s^{-1}$ in compliance with the Russian standard GOST 9651-84. For the tests, we used cylindrical samples cut out in the longitudinal direction of the billet.

The microstructure of the samples was studied in the cross section using a JEOL JSM 6390 scanning electron microscope.

## 3. Results and discussion

The microstructure of the VT8M-1 alloy in the initial condition represented a duplex structure typical for the hot-rolled condition. The average grain size of the primary α-phase was 3 μm, and its volume fraction was $\alpha_{glob} \approx 65\%$ [6]. The VT8M-1 alloy after RS retains a duplex structure, the grains of the primary α-phase are elongated and twisted, and the lamellar mixture of the (α+β) phases is divided into fragments of a globular shape (figure 1a). The average size of the grains/subgrains of the secondary α- and β-phases is 0.3 μm [6].

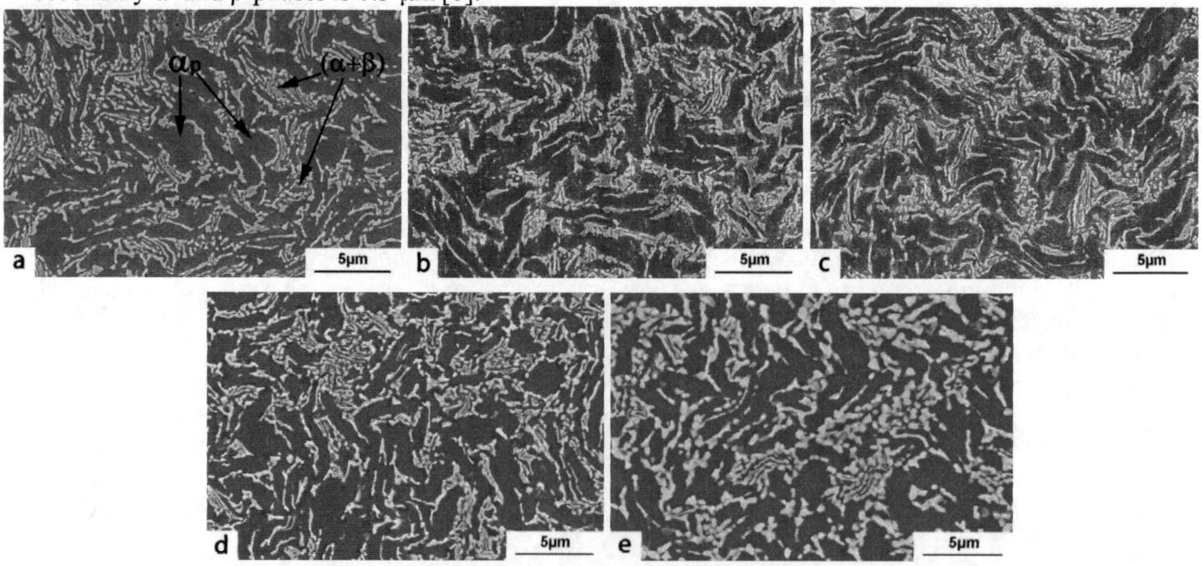


**Figure 1.** Microstructure of the VT8M-1 alloy after rotary swaging (a) and further annealings for 1 hour: (b) – 500 °C, (c) – 550 °C, (d) – 600 °C, (e) – 700 °C.

In order to evaluate the thermal stability of the alloy's UFG structure, we performed a series of short-term annealings for 1 hour at 500, 550, 600, 700 °C [7]. Based on the results of the one-hour annealings, it can be seen that the structure after annealings at 500 and 550 °C does not exhibit any changes, which indicates the stability of the UFG structure at these temperatures. After annealing at 600 °C, we observed in the UFG structure growth of the sizes of secondary phases, resulting from the start of recrystallization (figure 1d). At the annealing temperature of 700 °C these processes became more noticeable when the size of the secondary phases reached 0.5 μm (figure 1e). It should be noted that the volume fraction of the primary α-phase practically does not change and remains at a level of 65%, whereas the sizes of the secondary α- and β-phases retain their ultrafine range.

Taking into account the high thermal stability of the UFG structure during static heating at temperatures of 500 and 550 °C, the tensile mechanical tests were performed in a range of 300-550 °C. The test results are shown in figure 2. It can be seen that the UFG alloy VT8M-1 exhibits higher, almost by 200 MPa, values of ultimate tensile strength (UTS) and yield stress (YS) in comparison with the initial CG state up to a temperature of 450 °C. Worth noting are two characteristic slopes of the UTS and YS curves for the UFG state. In a temperature range of 300-450 °C the slope of the curve for the UFG alloy is parallel to the slope of the curve for the CG state (figure 2).

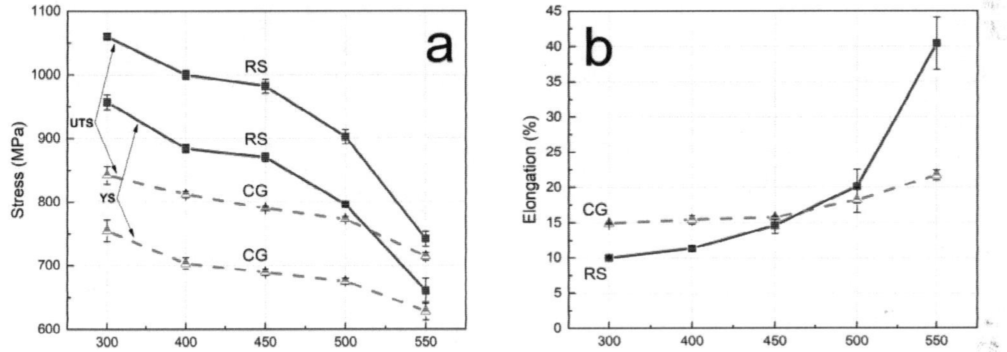


**Figure 2.** Mechanical tests of the VT8M-1 alloy in the UFG and CG states at temperatures of 300-550 °C: (a) curves showing the variation of UTS and YS with temperature; (b) curves showing the variation of ductility with temperature.

At temperatures of 450-550 °C the slope of the curve for the UFG alloy is much more pronounced than the one for the CG state, which indicates a more intensive softening of the UFG alloy, in spite of the fact that static heating at these temperatures does not lead to a considerable growth of the α- and β-phases (figure 1c). Apparently, the generation of stresses during tensile deformation can promote the development of dynamic recrystallization and a more intensive coarsening of structural elements [8]. After testing at a temperature of 550 °C the UTS and YS values for the UFG and CG states are close, and elongation is visibly higher as compared to the CG state (40 and 18%, respectively) (figure 2b). Such a mechanical behavior of the UFG alloy is an undoubted advantage from the perspective of further processing and shape-forming of items at lower deformation temperatures.

Thus, grain refinement in the structure of the VT8M-1 alloy by means of high-rate rotary swaging leads to the strength enhancement of the VT8M-1 alloy at temperatures of 300-450 °C, i.e. in the temperature range of operation of GTE components [9]. The obtained research results provide grounds to predict an enhanced strength of the UFG alloy under long-term applied stresses at operation temperatures (creep rupture strength), and this will be the focus of our further studies.

## 4. Conclusions

Based on the results of the study into the stability of the microstructure and mechanical properties at elevated temperatures of the VT8M-1 alloy samples with a UFG structure formed by rotary swaging, the following conclusions can be made:

1. The VT8M-1 alloy produced by rotary swaging is thermally stable during one-hour annealings at temperatures up to 550 °C. Some growth of the secondary α- and β-phases with increasing annealing temperature to 600 and 700 °C is conditioned by the development of recrystallization, while the structure's ultrafine range is retained.
2. The mechanical behavior of the UFG alloy is characterized by the fact that its UTS and YS values are stably higher by 200 MPa in comparison with the CG alloy up to a temperature of 450 °C, which creates prospects of its successful application for producing GTE components.

**Acknowledgments**

This work was supported by the Ministry of Education and Science of the Russian Federation in the scope of the State Assignment.

The investigation was performed to the Center for Common Use of the Research and Educational Center "Nanostructured Materials and High Technologies» of Ufa State Aviation Technical University".

## References

[1] Il'enko V M and Shalin R E 1995 *Titan* **1–2(5–6)** 25 [in Russian]
[2] Valiev R Z 2004 *Nature Materials* **3** 511
[3] Estrin Y and Vinogradov A 2013 *Acta Mater.* **61** 782
[4] Modina Iu M, Polyakov A V, Dyakonov G S, Yakovleva T V, Raab A G and Semenova I P 2019 *IOP Conf. Ser.: Mater. Sci. Eng.* **461** 012056
[5] Semiatin S L 2005 *ASM Handbook. Metalworking: Bulk Forming* vol 14A, pp 179–182
[6] Stotskiy A G, Polyakov A V, Dyakonov G S and Semenova I P 2019 Thermal stability of titanium alloy VT8M-1 with ultrafine-grained structure *The 14th World Conf. on Titanium (Nantes)*
[7] Pavlenko D V, Kovalenko T A and Ovchinnikov A V 2016 *Technological Systems* **1(74)** 60 [in Russian]
[8] Novikov I 1986 *Theory of the Heat Treatment of Metals* (Publishing house "Metallurgy") pp 62–66
[9] Pavlova T V, Kashapov O S and Nochovnay N A 2012 *All Materials. Encyclopedic Handbook* 5, pp 8-14 [in Russian]

IOP Conf. Series: Materials Science and Engineering **672** (2019) 012061 doi:10.1088/1757-899X/672/1/012061

# Microstructure and properties of permanent joints of ultrafine-grained titanium alloys produced by linear friction welding

**G R Suleymanova[2], A Y Medvedev[1], G S Dyakonov[1], T V Yakovleva[1], N F Izmailova[1], G I Raab[1] and I P Semenova[1]**

[1] Institute of Physics of Advanced Materials, Ufa State Aviation Technical University, 12 K. Marx St., Ufa, 450008, Russian Federation
[2] PJSC «UEC-Ufa Engine Industrial Association», 2 Ferin St., Ufa, 450039, Russian Federation

E-mail: guzel.suleymanova@bk.ru

**Abstract**. This paper presents the results of a study of the microstructure and mechanical properties of welded joints obtained by linear friction welding of samples from heat-resistant titanium alloys VT8M-1 (Ti-5.4Al-1Sn-1Zr-4Mo-0.15Si) and VT25U(Ti-6.5Al-1.8Sn-4Zr-4Mo-1W-0.2Si). In the VT8M-1 alloy, an ultrafine-grained (UFG) structure was formed prior to welding. The features of the phase and structural transformations in the areas of contact and thermomechanical effect of welding were shown, as well as the microhardness distribution across the width of the welded joints. The obtained values of strength characteristics of a welded joint during short-term and long-term tests at 400 °C indicate their correspondence to the level of the monolithic VT8M-1 alloy with a UFG structure.

## 1. Introduction

Titanium alloys are widely applied as structural materials in aircraft building owing to their small specific weight and corrosion resistance [1]. In particular, the wheel blades of gas-turbine engines (GTEs) operate in complex operating conditions of high static and dynamic loads, erosive and corrosive action, temperature exposure. Therefore, a whole range of technical requirements are imposed on such products in terms of their static strength, fatigue resistance, long-term strength at operational temperatures, etc. [2].

In today's GTE compressors, such wheels are used where the blades and the disc do not have a mechanical mounting, but represent a single unit. The only method for producing such joints that can be used in industry is linear friction welding (LFW) [3]. The transition from dismountable wheels to welded monowheels provides an opportunity to produce the blade and disc workpieces separately. For example, to produce blades one can use ultrafine-grained (UFG) alloys with enhanced strength before welding to a disc. In the present work, to produce a solid-phase joint by LFW we used the heat-resistance Ti alloys VT8M-1 and VT25U. The VT8M-1 alloy is widely applied in blades, and the VT25U alloy is intended for the production of large-sized GTE rotor discs [2]. In the VT8M-1 alloy, a UFG structure was formed by high-rate rotary swaging (RS) in order to produce enhanced strength properties [4,5]. The aim of this work is to study the microstructure and mechanical properties of the

IOP Conf. Series: Materials Science and Engineering **672** (2019) 012061 doi:10.1088/1757-899X/672/1/012061

solid-phase joint between the VT8M-1 alloy having a UFG structure and the VT25U alloy, imitating the materials of a GTE compressor blade and disc.

## 2. Material and experimental procedure

For the investigation, we used rods produced by OJSC VSMPO-AVISMA Corporation (Verkhnyaya Salda, Russia) with the following chemical compositions: VT8M-1 (Ti–5.7Al–3.8Mo–1.2Zr–1.3Sn–0.16Fe, wt.%) and VT25U (Ti–6.7Al–3.9Mo–4.2Zr–1.8Sn– 1.3W, wt.%)

The rotary swaging of the rods from the VT8M-1 alloy was implemented with diameter reduction from 70 to 32 mm at a temperature of 750 °C. The total strain was ε~1.7, the strain rate was higher than 300 mm $s^{-1}$. For implementing LFW, samples were produced from the VT25U alloy rods in the as-received state and the VT8M-1 alloy rod with a UFG structure. The samples had the shape of a parallelepiped with a height of 35 mm and a section of 13×26 mm. LFW was performed by means of reciprocating motion of the welded parts with a frequency of about 50 Hz and an amplitude of up to 2 mm, compressed for a tight contact.

Tensile tests (short-term strength) at an operational temperature of 400 °C were conducted in accordance with the Russian standard GOST 9651-84. Long-term strength was estimated in accordance with the Russian standard GOST 10145-81. For this purpose, samples were produced with a gage portion of d = 6 mm and a gage length of L = 35 mm, having threaded heads.

To reveal the microstructure, the sample surface was etched with a solution of hydrofluoric and nitric acids in water. Microstructural studies were performed using an Olympus GX-51optical microscope and a TescanMira 3 LMH scanning electron microscope. Microhardness was tested using a PMT-3 microhardness tester with a step of 0.5 mm under a load of 2 N applied for 10 s.

## 3. Results and their discussion

### *3.1. Microstructure of the VT8M-1 alloy before and after rotary swaging*

The microstructure of the alloyin the as-received state had a duplex character and consisted of the primary α-phase grains with an average size of 3 μm and a volume fraction of about 60%, in the (α+β)- regions the thickness of the α-phase lamellae was on average 0.2 μm (figure 1a).

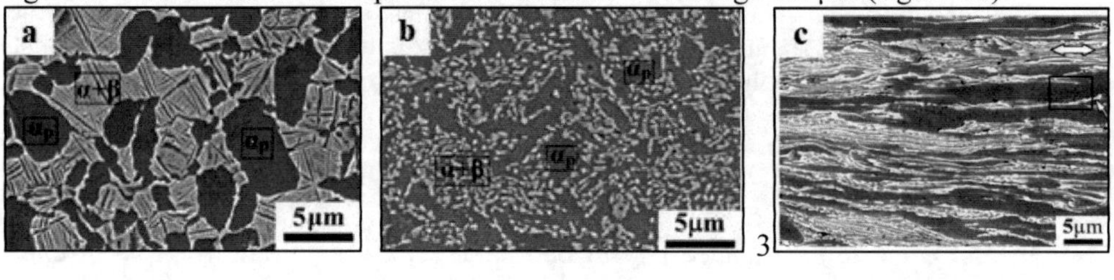


**Figure 1.** Microstructure of VT8M-1 alloy: (a) as-received state; (b) after RS in the transverse section and (c) in the longitudinal section.

After deformation by RS, in the transverse section of the rod we observed the spheroidization of the lamellar (α+β)-constituent of microstructure (figure 1b), and in the longitudinal section a typical for RS metallographic texture formed, characterized by the elongation of the primary α-phase along the deformation direction (figure 1c). The average size of the recrystallized grains of the secondary α-phase was about 0.3 μm (figure 1b).

In the metallographic images from a section of the welded joint (figure 2) one can distinguish four characteristic zones. Zones 1 and 4 correspond to the structure of the samples, respectively, from the UFG VT8M-1 alloy (1) and from the VT25U alloy (4) with a coarse-grained (CG) duplex structure, which did not undergo any changes in the course of welding (figure 2 a,b,d). Regions 2 and 3 are the thermo-mechanical effected zones, zone 2 refers to the UFG VT8M-1 alloy, and zone 3 refers to the VT25U alloy (figure 2a). In Fig. 2d distinctly visible is a pronounced metallographic texture oriented in the direction of the deformation flow of the material, which is apparently related to the thermomechanical effect of friction welding. In zone 2, from the side of the UFG alloy, this type of texture is weakly expressed (figure 2b). The images of the microstructure in the area of contact between zones 2 and 3 are characterized by recrystallized grains of the β-phase with a martensitic structure and sizes ranging from 5 to 15 μm (figure 2a and c), which is related to local heating to temperatures above the β-transus temperature of the welded alloys and subsequent rapid cooling of the sample.

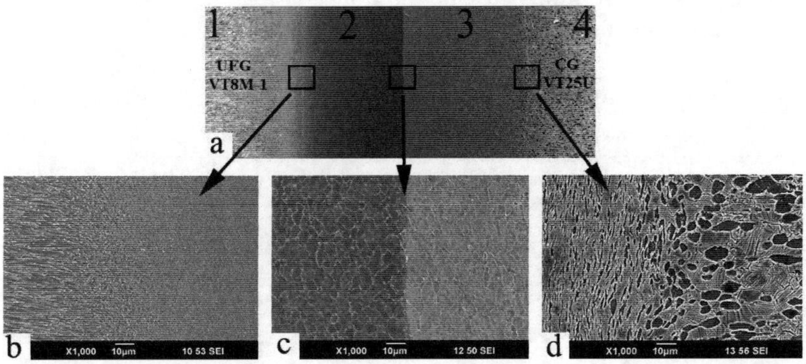


**Figure 2.** Microstructure of the solid-phase joint between the UFG VT8M-1 alloy and the VT25U alloy: (a) general view; (b) zones 1 and 2 – UFG VT8M-1; (c) zones 2 and 3 – welded seam; (d) zones 3 and 4 – VT25U.

Figure 3 shows the HV microhardness distribution across the section of the welded seam. On the whole, these results agree well with the microstructural zones revealed earlier, shown in figure 2.

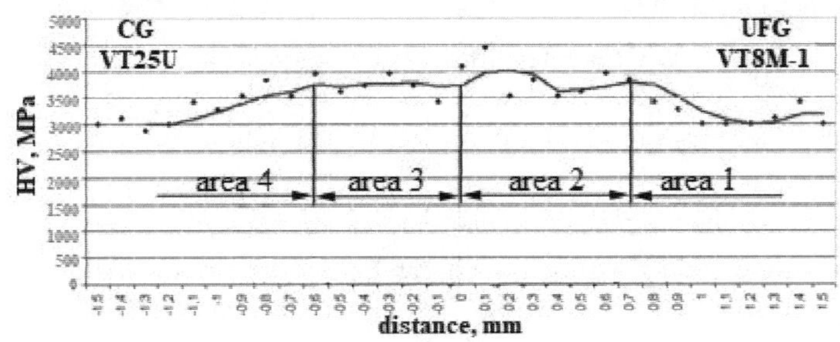


**Figure 3.** Microhardness distribution across the section of the solid-phase joint between VT8M-1/VT25U.

In particular, in the contact area (zones 2 and 3) the formation of a martensitic structure resulted in higher microhardness values, up to 4500 MPa on average, as compared to zones 1 and 4 with the initial (before welding) structure. We should also note the small difference in the microhardness of VT25U (HV 3200 MPa) and VT8M-1 with a UFG structure (HV 3500 MPa). The increased microhardness values are characteristic for a rather extensive region with a length of 1200 μm (areas 2 and 3). It should be noted that in [6] a more narrow interval of increased hardness (about 500 μm) was observed in the samples of the CG alloy Ti-6Al-4V, as compared to the joint of UFG samples [6]. Apparently, the smoother distribution of microhardness across the width of the welded seam of the UFG samples was promoted by the presence of the residual α-phase in the martensitic matrix [6].

In the present work, to produce a solid-phase joint we used different alloys, where the more heavily-alloyed VT25U alloy is noticeably stronger than the VT8M-1 alloy in the initial CG state [2]. The UFG structure formation in the VT8M-1 alloy enabled, first, increasing its microhardness and equalizing it with the strength of VT25U, and second, forming a practically full-strength joint without local "jumps" in microhardness in the contact area (figure 3).

### *3.2. Short-term and long-term strength*

To evaluate the strength of the obtained solid-phase joint, we performed the tensile mechanical testing of the samples at a temperature of 400 °C, corresponding to the operational temperature. It can be seen from table 1 that the short-term strength of the VT8M-1 alloy in the case of UFG structure formation increases from 800 to 1100 MPa, while the short-term strength of the VT25U alloy at this temperature is ~ 860 MPa [2]. For testing long-term strength, specimens were cut out from the solid-phase joint so that their welded seam was in the middle. The 100-hour strength at 400 °C was 850 MPa, and the fracture took place far from the junction, in the area of the VT25U.

**Table 1.** Short-term and long-term strength at a temperature of 400 °C of the investigated materials and their welded joint.

| Alloy/state | Short-term strength at T=400 °C, $\sigma^{400}$, MPa | 100-hour long-term strength at T=400 °C, $\sigma^{400}$, MPa |
|---|---|---|
| VT25U CG | 860* | 850* |
| VT8M-1 CG | 800 | 720 |
| VT8M-1 UFG | 1100 | 850 |
| UFG VT8M-1/VT25U welded joint | 900 | 850 |

* according to the data given in [2].

The obtained values of the strength characteristics of the welded joint under short-term and long-term tests indicate that they correspond to the level of the monolithic VT8M-1 alloy having a UFG structure, which is higher than the strength properties of the alloys in the CG state.

## 4. Conclusions

The following conclusions can be drawn from the results of the conducted studies:

1. The main characteristic zones have been revealed in the section of the welded joint of samples differing in terms of composition and microstructure, namely the VT8M-1 alloy with a UFG structure and the VT25U alloy. It has been shown that the contact area of the alloys (welded seam) is characterized by recrystallized grains of the β-phase with a martensitic structure, which is related to local heating to temperatures above the β-transus temperature of the welded alloys and subsequent rapid cooling after LFW.
2. Microhardness across the width of the welded joint varies uniformly without any sharp jumps, which characterizes the full-strength joint. This is also confirmed by the mechanical test results of the welded joint between the UFG VT8M-1 alloy and the VT25U alloy at a temperature of 400 °C, which revealed a 100-hour long-term strength of 850 MPa, corresponding to the level of the monolithic material with a UFG structure. This allows us to talk about the possibility of using this approach in production of high strength welded joints.

## Acknowledgments

This work was supported by the Ministry of Education and Science of the Russian Federation in the scope of the project part of State Assignment No. 11.1235.2017.

## References

[1] Boyer R, Welsch G and Collings E W 1998 *Materials Properties Handbook: Titanium Alloys* (ASM International, USA)
[2] Il'enko V M and Shalin R E 1995 *Titan* **1–2(5–6)** 25 [in Russian]
[3] Bhamji I, Preuss M, Threadgill Ph L and Addison A C 2010 *Mater. Sci. Technol.* **27(1)** 2
[4] Valiev R Z 2004 *Nature Mater.* **3** 511
[5] Langdon T G 2013 *Acta Mater.* **61** 7035
[6] Polyakova V, Gabitova S, Semenova I, Yakushina E, and Rosochowski A 2012 *Proc. of the 2nd Int. Conf. on Recent Trends in Structural Materials (Pilsen)* pp 1–6

# Effect of precipitation under long-term aging at 650 °C on the tensile strength of advanced 10%Cr heat-resistant steel

**N Dudova, R Mishnev, M Tikhonova and R Kaibyshev**

Belgorod State National Research University, Pobeda, 85, Belgorod, 308015, Russia

E-mail: dudova@bsu.edu.ru

**Abstract**. The effect of long-term aging for $10^3$ to $3.9 \cdot 10^4$ h at 650 °C on the microstructure, dispersion of secondary phases, and tensile strength properties was studied in a low-nitrogen 10%Cr martensitic steel with 3%Co and 0.008%B additives. Tensile tests of small specimens cut from grip portions of creep tested specimens were carried out at ambient and elevated (650 °C) temperatures. An increase in the yield stress and ultimate tensile strength after 10,000 h of aging was revealed to be associated with the precipitation of V-rich MX carbonitrides, which compensated for the coarsening of the boundary particles and depletion of W and Mo solutes from the ferritic matrix. The effect of changes of the microstructure and dispersion of secondary phases on strengthening of the steel is discussed.

## 1. Introduction

High-Cr ferritic/martensitic steels remain the major part of the materials used for components of fossil power plants operable at ultra-supercritical steam parameters [1]. This fact determines further improvement of their creep strength. The main factor providing high creep resistant of these steels is the stability of the tempered martensite lath structure (TMLS), which consists of prior austenite grains, packets, blocks and laths with a high dislocation density in the lath interiors [2]. The dispersion strengthening of these steels is ensured mainly, by nanoscale boundary $M_{23}C_6$-type carbides and MX carbonitrides homogeneously distributed in the laths.

The 9-11%Cr martensitic steels recently developed in accordance with the new approach by the increasing the B content to approximately 0.01wt% and decreasing the N content to approximately thousandths of a percent, demonstrate high creep resistant at an elevated temperature of 650 °C [3,4]. The role of precipitates of secondary phases ($M_{23}C_6$, MX, Laves phase) in the high stability of TMLS under creep condition is the subject of research interest.

The aim of this paper is to examine the effect of long-term aging at 650 °C on the tensile strength of advanced 10% Cr steel with low N and high B contents. Specific attention will be paid to the effect of changes in the microstructure and dispersion of the secondary phases on the strength.

## 2. Experimental

A 10% Cr steel with the following chemical composition (in wt.%) 0.1C, 0.06Si, 0.1Mn, 10.0Cr, 0.17Ni, 0.7Mo, 0.05Nb, 0.2V, 0.003N, 0.008B, 2.0W, 3.0Co, 0.002Ti, 0.006Cu, 0.01Al and Fe-balance was examined. The vacuum induction-melted steel was subsequently hot-forged by Lasmet, Chelyabinsk. The steel was subjected to standard heat treatment: normalization at 1060 °C/30 min and tempering at 770 °C/3 h. Small specimens for tensile tests were cut from the grip portions of creep

IOP Conf. Series: Materials Science and Engineering **672** (2019) 012062 doi:10.1088/1757-899X/672/1/012062

tested specimens at 650 °C, i.e. they were subjected only to long-term thermal aging for 1000, 10000, 28286 and 39437 hours [4]. Tensile tests were carried out using flat specimens with a gauge length of 4 mm and cross-sectional dimensions of 1 mm x 1 mm using an Instron 5882 testing machine at ambient temperature and a temperature of 650 °C and a strain rate of $\sim 10^{-3}$ $s^{-1}$.

The structural characterization was performed using a Jeol JEM-2100 transmission electron microscope (TEM) with an INCA energy dispersive X-ray spectrometer. The transverse lath/subgrain sizes were measured on TEM micrographs by the linear intercept method, counting all clear visible (sub)boundaries. The dislocation densities were estimated by counting individual dislocations in the (sub)grain/lath interiors per unit area on at least six arbitrarily selected typical TEM images for each data point.

## 3. Results and discussion

### *3.1. Effect of long-term aging on the tensile properties*

Figure 1 shows the effect of long-term aging at 650 °C on the tensile properties of the 10%Cr steel: the yield stress (YS) and ultimate tensile strength (UTS) at 20 and 650 °C. It has been revealed that aging for a long time (up to $\sim 3 \cdot 10^4$ h) does not lead to a decrease in YS and UTS. Moreover, the slight increase in the YS and UTS values occurs. The maximal increase in the strength of +3...6% is observed after 10,000 hours of aging. The subsequent aging (39,437 hours) results in an insignificant decrease in UTS by 4...6% as compared with tempered condition, while the YS values return to the level of the tempered condition. Therefore, the 10%Cr steel obviously demonstrates higher tensile strength characteristics both at ambient and elevated temperatures after aging at 650 °C for 10,000 h.

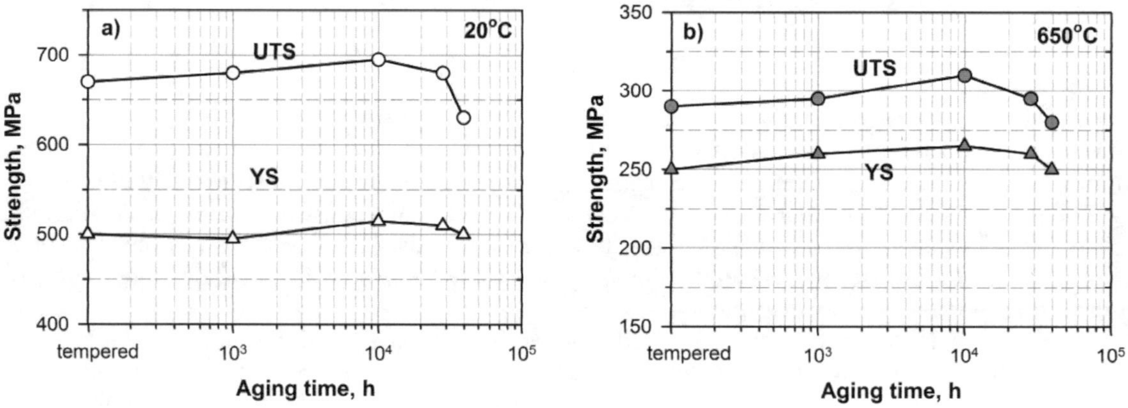


**Figure 1.** Effect of long-term aging at 650 °C on the yield stress (YS) and ultimate tensile strength (UTS) at 20 °C (a) and 650 °C (b).

### *3.2. Evolution of microstructure and dispersion of the secondary phases during long-term aging*

The initial microstructure of the 10%Cr steel in the tempered condition is a typical tempered martensite lath structure with the parameters presented in table 1 [4]. The studied steel is characterized by a smaller size of $M_{23}C_6$ carbides (70 nm) and their higher volume fraction (2.05%) in contrast to conventional high-chromium steels with standard N and B contents. Also, a peculiarity of the tempered 10%Cr steel is the absence of V-rich MX carbonitrides in contrast to conventional steels.

The lath structure of the steel remains almost unchanged during long-term aging. A 60% increase in the lath width and a 4-fold decrease in the dislocation density is observed after ~40,000 hours of aging (table 1).

**Table 1.** Parameters of the structure of the 10%Cr steel before and after long-term aging at 650 °C.

| Structural parameters | tempered | After aging at 650 °C | | | |
|---|---|---|---|---|---|
| | | 1000 h | 10,000 h | 28,286 h | 39,437 h |
| Lath width (nm) | 380 | 409 | 507 | 566 | 614 |
| Dislocation density, x$10^{14}$ ($m^{-2}$) | 1.7 | 1.23 | 1.06 | 0.61 | 0.43 |
| Mean size of particles: | | | | | |
| $M_{23}C_6$ (nm) | 70 | 72 | 81.9 | 83.6 | 96 |
| Nb-rich MX (nm) | 35 | 35 | 31.3 | 31.6 | 35 |
| V-rich MX (nm) | - | - | 26.5 | 40 | 58 |
| Laves phase (nm) | - | 145.6 | 197.6 | 298.8 | 319 |
| W content in the solid solution (wt.%) | 2 | 1.1 | 0.95 | 0.85 | 0.8 |

The precipitation processes occurring in the steel during long-term aging are as follows:
1) an insignificant coarsening of $M_{23}C_6$ carbides from 70 nm to 96 nm after ~ 40,000 hours. $M_{23}C_6$ carbides demonstrate a high resistance to coarsening under aging condition (figure 2a);
2) the precipitation of the Laves phase particles ($Fe_2(W,Mo)$) and their growth, which leads to the depletion of W and Mo from the solid solution. The Laves phase particles precipitate at the lath boundaries and high-angle boundaries of the blocks, packets, and PAGs (figure 2b). This process is accompanied by the depletion of W and Mo from the ferritic matrix [4]. The onset of the Laves phase coarsening appears after $10^3$ h and accelerates after $10^4$ h;
3) as it was revealed [4], during aging for the time between 1000 and 10,000 hours, fine V-rich MX carbonitrides precipitate in the lath interiors (figure 2a). The mean size of particles is 26.5 nm. The following aging for 28,286 and 39,437 hours results in an increase in the mean size to 40 and 58 nm, respectively. Their volume fraction is small and comprises 0.00862% (as calculated by Thermo-Calc). Aging for ~40,000 hours does not lead to MX→Z-phase (CrVN) transformation due to the low content of N.

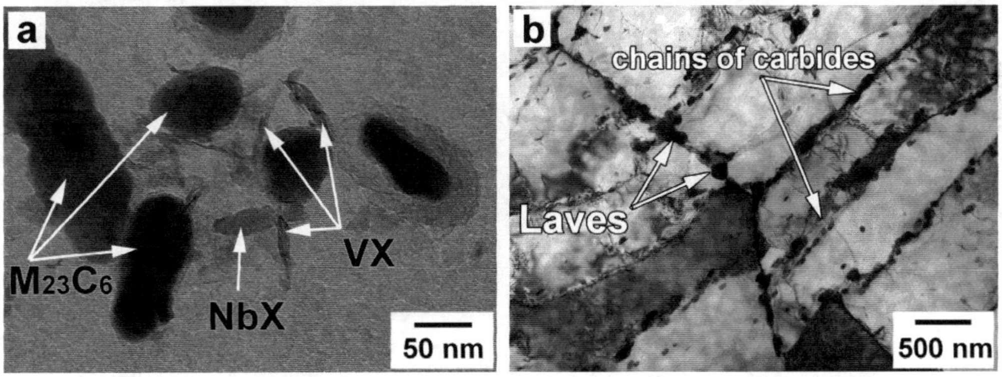


**Figure 2.** TEM micrographs of the 10%Cr steel after long-term aging: a) precipitation of V-rich MX particles after aging for 10,000 hours (replica); b) retained lath structure after aging for 39,437 hours (foil).

*3.3. Effect of precipitation on the tensile strength*

Usually, a recovery of non-equilibrium tempered martensite lath structure occurs during long-term aging at high temperature that results in the softening of high-Cr martensitic/ferritic steels [5,6]. In contrast, in the studied 10%Cr steel, an obvious stabilization of the tensile strength after long-term aging and a slight strengthening after 10,000 hours of aging is observed.

Recovery of the lath structure of the 10%Cr steel under aging condition leads to the lath widening, a decrease in the dislocation density and the depletion of W and Mo from the solid solution. These processes reduce the dislocation, substructure and solid solution strengthening of the steel. On the other hand, precipitation hardening is affected by precipitation of the Laves phase particles and their coarsening, precipitation of fine V-rich MX carbonitrides, and retention of stable against the coarsening $M_{23}C_6$ carbides. The strength of the aged 10%Cr steel retaining at a high level corresponding to that for the tempered condition is indicative of the fact that dispersion strengthening compensated for the lath widening, decrease in the dislocation density and the depletion of the ferritic matrix.

The strengthening after 10,000 hours of exposure seems to be associated with the precipitation of fine V-rich MX carbonitrides, since the precipitated Laves phase particles enlarged by ~35% during 1000…10,000 hours of aging and did not increase the dispersion strengthening. The strengthening due to V-rich MX phase can explain why the creep rate significantly decreases to ~$10^{-11}$ $s^{-1}$ at 120 MPa providing a high rupture time of ~40,000 hours for the 10%Cr steel [4] (figure 3).

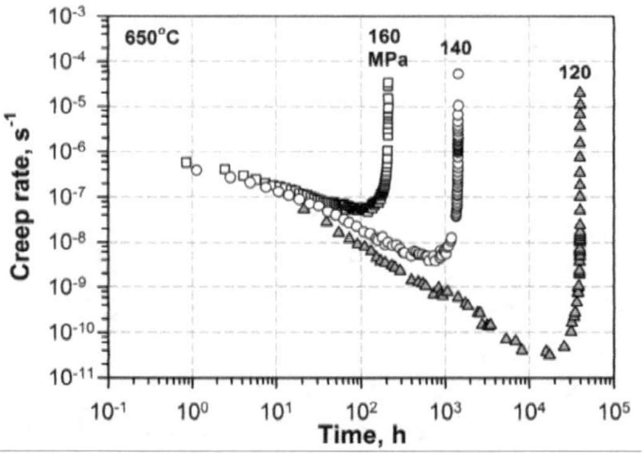


**Figure 3.** Creep rate vs time curves of the 10%Cr steel at 650 °C.

## 4. Conclusion

The effect of long-term aging at 650 °C on the tensile strength and the changes in the microstructure and precipitations of secondary phases were studied in advanced 10% Cr martensitic steel with 0.008% B and 0.003% N. The main results can be summarized as follows:

1. Aging for 1000…~30,000 hours leads to a slight increase in the yield stress and ultimate tensile strength, reaching the maximum values (+3…6%) at 10,000 hours of aging, whereas subsequent aging for ~40,000 hours results in the 4…6% lower ultimate tensile strength as compared with the tempered condition.
2. Strengthening of the steel after 10,000 hours of exposure is associated with the precipitation of fine V-rich MX carbonitrides in the lath interiors.

## Acknowledgments

One of the authors (N.D.) greatly acknowledges the joint support by DAAD (Deutscher Akademischer Austausch Dienst) under Grant No. 57447934 and the Ministry of Education and Science of Russian Federation under Grant No. 11.13447.2019/13.2 in the Michael Lomonosov Programm. The authors are grateful to the staff of the Joint Research Center, "Technology and Materials", Belgorod State University, for providing the equipment for instrumental analysis.

## References

[1] Abe F, Kern T-U and Viswanathan R 2008 *Creep-Resistant Steels* (Cambridge: Woodhead Publishing)
[2] Hald J 2008 *Int. J. Press. Vess. Piping* **85** 30
[3] Abe F 2011 *Procedia Engineering* **10** 94
[4] Mishnev R, Dudova N and Kaibyshev R 2018 *Mater. Sci. Eng.* A **713** 161
[5] Sahoo K C, Vanaja J, Parameswaran P, Vijayanand V D and Laha K 2017 *Mater. Sci. Eng.* A **686** 54
[6] Xu Y, Nie Y, Wang M, Li W and Jin X 2017 *Acta Mater.* **131** 110

IOP Conf. Series: Materials Science and Engineering **672** (2019) 012063 doi:10.1088/1757-899X/672/1/012063

# Study of the dynamic strength of the ultrafine-grained titanium VT 1-0

**A D Evstifeev[1] and R R Valiev[1,2]**

[1] St. Petersburg State University, 198504, Universitetskiy pr., D. 28, Old Peterhof, St. Petersburg, Russia
[2] Ufa State Aviation Technical University, Institute of Physics of Advanced Materials, 450008, Karl Marx St., 12, Ufa, Russia

E-mail: Rovaliev@gmail.com

**Abstract.** We report the results from the testing of the strength properties of the ultrafine-grained Ti alloy VT 1-0 under dynamic and static loading. A comparative analysis was made of the ultimate static and dynamic strength of the small-sized samples of the material subjected to severe plastic deformation. The importance of studying the dynamic strength of Ti alloys is closely related to their practical application in structures and parts operating under high dynamic loads.

## 1. Introduction

In the past two decades, there has been active development of ultrafine-grained (UFG) metallic materials produced by severe plastic deformation (SPD) for practical applications [1]. At present, there are many publications on the effect of the UFG structure on the mechanical and fatigue properties of Ti alloys [1-3]. However, there have been practically no studies into the dynamic strength of UFG alloys. It is known that such characteristics as the ultimate tensile strength and yield stress of ultrafine-grained alloys under dynamic loads differ significantly from the ones found during static tests. It is also known that the deformation of high-strength Ti alloys by SPD often leads to a considerable decline in their ductility. This was demonstrated through the example of the UFG titanium alloy Grade 5 (VT6 rus) and other titanium materials [3]. In its turn, the decline in the ductile characteristics of the material affects its performance properties: fatigue strength decreases, the characteristics of crack resistance and impact toughness decline, and erosion resistance also decreases [4,5]. The ductile characteristics of titanium materials normally depend a lot on the SPD processing regimes [3], which is the subject of the present study. This paper presents the data on the mechanical characteristics under dynamic and static loading of commercially pure (CP) Ti VT 1-0 processed by equal-channel angular pressing (ECAP) via different regimes, including additional thermomechanical treatment by extrusion.

## 2. Material and research methods

As the material for the study, we used CP Ti Grade 4 (VT 1-0 rus) in the form of hot-rolled rods with a diameter of 20 mm. To improve the internal structure, the initial rod with a diameter of 20 mm was held in a furnace at a temperature of 680 °C for 1 hour and cooled in air. As a result of this treatment, a homogeneous structure with an average grain size of 20-25 μm was formed. With a view to produce various UFG structures in the samples, we performed ECAP processing via three different regimes: 4

passes at 400 °C, 8 passes at 400 °C and 8 passes at 400 °C with subsequent extrusion at 300 °C. Recent studies [3] indicate that these ECAP regimes enable producing in Ti a UFG structure with various grain shapes and sizes. The ECAP processing was carried out via route Bc with a channels intersection angle of 120°. The structure of the produced samples was studied in a JEOL JEM-2100 transmission electron microscope.

For static and dynamic tests, we cut out small samples of reduced sizes from the rods. The gauge length of such samples was 5 mm, the width was 2 mm, and the thickness was 0.7 mm. Static tests were conducted using the method of uniaxial tension on a Shimadzu AG-50kNX universal testing machine with a rate of $10^{-3}$ $s^{-1}$. Dynamic tests were conducted using an Instron CEAST 9350 drop tower impact system with an accelerator. Previously, in the paper [6], using the SPS (Sign-Perturbed Sums) method [7] the applicability of samples of such a geometry was demonstrated for the testing of materials in a wide variation range of loading parameters.

## 3. Results and discussion

The TEM studies of the structure after the deformation processing show that after 4 passes a UFG structure is formed with an average grain size of 450 nm. Increasing the number of passes to 8 promoted a stronger structure refinement to an average grain size of about 400 nm. The structure of the billet after ECAP processing for 8 passes and extrusion represents elongated grains with an average grain size of 300 nm.

Figure 1 shows the characteristic “stress-strain” curves for the tension of the small-sized samples in all the structural conditions. The test results indicate that the strongest condition is the UFG Ti produced by ECAP for 8 passes and extrusion. Its ultimate tensile strength (UTS) is about 920 MPa, which is much higher than the UTS after ECAP for 4 passes and 8 passes. However, its elongation to failure is lower and amounts to about 17%, whereas after ECAP the elongation to failure is over 20%. Thus, while strength grows in UFG Ti and the grain size decreases, its ductility declines, as it was observed in earlier studies [3]. The material after ECAP processing for 4 and 8 passes practically has no differences in the total strain to failure, but its static strength differs by 20 %.

Therefore, it can be noted that the alloy after 8 ECAP passes has the optimum strength characteristics. When a lower strain is produced in the material, its strength declines; when a higher strain is produced in the material, its ductility declines considerably. Since the enhanced strength characteristics in the region of quasi-static loading do not guarantee good dynamic properties of the material in the region of dynamic loads, and materials from Ti alloys normally operate under dynamic loads, we continued the study of the strength characteristics.

Figure 2 presents the results of the dynamic tests of UFG Ti produced via different processing regimes. The diagram shows the experimental data and calculated curves in the framework of the structure-time approach [8,9].

$$\frac{1}{\tau}\int_{t-\tau}^{t}\frac{\sigma(s)}{\sigma_c}ds \le 1, \tag{1}$$

where $t$ is time, $\sigma$ is the dependence of rupture stress on time, $\sigma_c$ is the ultimate tensile strength under quasi-static loading conditions, $\tau$ is the incubation time of fracture, which is a measure of strength in the dynamic range of the parameters of external action. As a model load, we accepted one linearly growing with time, which is closest to the experimental conditions. The parameter $\tau$ in this set up is determined numerically by selecting the optimum correspondence of the calculated curves to the experimental curves.

The numerical values of the material’s parameters are listed in table 1. The slope of the curves in figure 2 is determined by the material’s parameter τ.

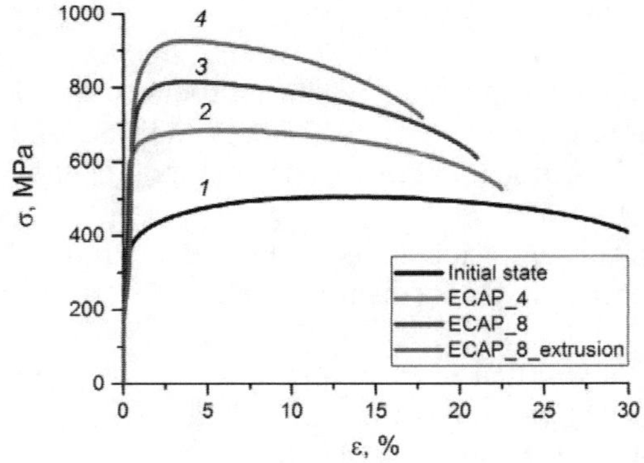


**Figure 1.** Tensile curves of the samples in the initial condition and after ECAP via different regimes.

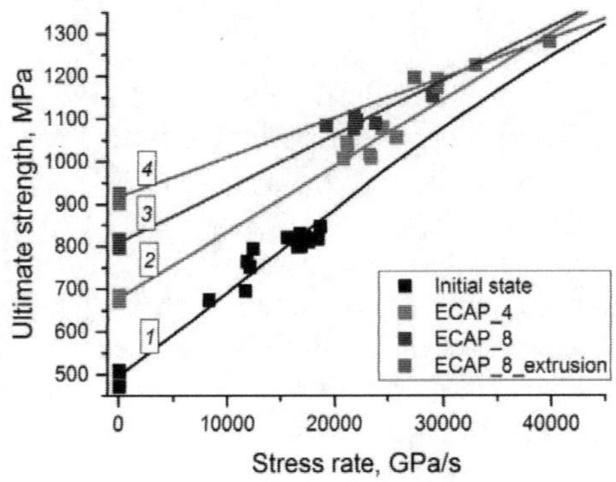


**Figure 2.** Results of the dynamic tests of the initial Ti and after ECAP via different regimes.

**Table 1**. Parameters and mechanical characteristics of the material in different structural conditions, where δ is the elongation to failure, $d_{av}$ is the average grain size.

| Condition | $\sigma_c$, MPa | δ, % | $d_{av}$, μm | $\tau$, μs |
|---|---|---|---|---|
| CG | 495 | 30 | 25 | 39 |
| ECAP_4 | 678 | 22 | 0.45 | 31 |
| ECAP_8 | 808 | 21 | 0.4 | 25 |
| ECAP_8_Extrusion | 917 | 17 | 0.3 | 19 |

The results of the static and dynamic tension experiments of VT 1-0 in different structural conditions show an indirect relationship between the structure and ductile characteristics, and the parameter of the material's dynamic strength, τ. Introducing into our analysis a material parameter responsible for dynamic strength has revealed the need to produce a set of strength and ductile properties in a material in order to attain high strength values under extreme loads. It has been visibly demonstrated that increasing strength in the static range of loads is not enough to achieve high deformational characteristics in the region of dynamic loads. For example, a slight decrease in the ductility parameters of the Ti alloy after additional extrusion significantly reduced the material's dynamic strength, which is expressed in the intersection of the dynamic curves in the diagram $\sigma - \dot{\sigma}$ (figure 2). In the region of loading rates $\dot{\sigma} > 40000\ GPa/s$ the material after ECAP for 8 passes and extrusion has a lower strength than the alloy in the initial condition, although in the region of static loads the initial alloy's strength is almost two times lower. The material after ECAP for 8 passes and additional extrusion is inferior already after $\dot{\sigma} > 30000\ GPa/s$, which considerably reduces its application areas. The obtained results from the tension of Ti samples in the ranges of static and dynamic loads demonstrate that it is not enough to increase the material's static strength to ensure the material's safe operation in the region of dynamic loading. With increasing strain rate, the role of the balance of strength and ductility grows, and the high strength of the material stops being dominant in this pair.

## 4. Conclusions

1. Using severe plastic deformation by equal-channel angular pressing, we have produced in VT 1-0 Ti an ultrafine-grained structure with an average grain size ranging from 450 nm in the case of 4 ECAP passes to 300 nm in the case of 8 ECAP passes and additional thermomechanical treatment by extrusion.
2. The results of the mechanical tests of the small-sized samples demonstrate that the strongest condition (917 MPa) is the one after 8 ECAP passes and extrusion, but its ductility is much lower than that of the samples after 4-8 ECAP passes.
3. Using the structure-time approach, we have shown that the optimum set of strength characteristics for the use in the static and dynamic ranges of loading parameters is featured by the material after 8 ECAP passes having a combination of enhanced strength and ductility.

## Acknowledgements

The authors gratefully acknowledge the financial support from Saint Petersburg State University in the framework of Call 3 project (id 26130576).

## References

[1] Valiev R Z and Alexandrov I V 2000 *Nanostructured Materials Produced by Severe Plastic Deformation* (Moscow: Logos) [in Russian]
[2] Semenova I P, Raab G I and Valiev R Z 2014 *Nanotechnologies in Russia* **9(5-6)** 311
[3] Valiev R Z, Zhilyaev A P and Langdon T G 2014 *Bulk Nanostructured Materials: Fundamentals and Applications* (Hoboken: John Wiley & Sons, Inc.)
[4] Klevtsov G V, Valiev R Z, Semenova I P, Klevtsova N A, Matchin A A, Kashapov M R and Klassen N A 2012 *Vestnik of Yaroslav the Wise Novgorod State University* **67** 19 [in Russian]
[5] Evstifeev A D, Kazarinov N A, Petrov Y V, Atroshenko S A and Valiev R R 2017 *Physics of the Solid State* **59** 1794
[6] Evstifeev A D, Volkov G A, Chevrychkina A A and Petrov Yu 2019 *Technical Physics* **64** 620
[7] Csaji B C, Campi M C and Weyer E 2015 *IEEE Transactions on Signal Processing* **63(1)** 169
[8] Petrov Y V and Utkin A A 1989 *Sov. Mater. Sci.* **25** 153
[9] Petrov Y V and Morozov N F 1994 *ASME J. Appl. Mech.* **61** 710

IOP Conf. Series: Materials Science and Engineering **672** (2019) 012064 doi:10.1088/1757-899X/672/1/012064

# Corrosion resistance of ultrafine-grained pure titanium in simulated sea water

**X Li[1,4], Y Dong[1,2], I V Alexandrov[3], Z Dan[1,2] and H Chang[1]**

[1] College of Materials Science and Engineering/Tech Institute for Advanced Materials, Nanjing Tech University, Nanjing 211816, China
[2] State Key Laboratory of Metal Material for Marine Equipment and Application, Anshan 114000, China
[3] The Department of Physics, Ufa State Aviation Technical University, Ufa 450008, Russia
[4] Jiangsu Collaborative Innovation Center for Advanced Inorganic Function Composites, Nanjing Tech University, Nanjing 211816, China

E-mails: xinge@njtech.edu.cn, dongyuecheng@njtech.edu.cn

**Abstract**. Corrosion resistance of ultrafine-grained Ti processed by equal channel angular pressing with 3 passes was investigated. Results indicated that corrosion rate of ultrafine-grained Ti significantly reduced in comparison with coarse grain Ti. Furthermore, corrosion resistance of 3 passes ECAP sample had an obvious distinguish for different extraction planes; normal plane has shown a best corrosion resistance than transversal and extrusion plane, which could be attributed to higher fraction of low angle grain boundary.

## 1. Introduction

Ti and Ti alloy are widely used in the field of petrochemical, biomedical and marine engineering due to their high specific strength and excellent corrosion resistance [1]. Severe plastic deformation (SPD) could achieve unique mechanical and service properties due to the grain refinement of materials, which has received extensive attention in recent years [2]. Among them, ECAP is one of most popular methods to obtain bulk ultrafine-grained (UFG) materials with a uniform structure by the introduction of large plastic deformation without changing the material size [3].

In comparison with the extensive researches on strength, fatigue and biocompatibility behaviors of UFG-Ti alloy fabricated by SPD, the study on corrosion resistance is relatively limited and controversial. Balyanov et.al [4] studied the corrosion resistance of UFG-Ti processed by ECAP in different concentrations of HCl and $H_2SO_4$ solutions, the results indicated that UFG-Ti has better corrosion resistance than coarse-grained Ti (CG-Ti), which is mainly because of the rapid growth of a passivation film in UFG-Ti. The same conclusions have been also achieved in other studies [5, 6] in various corrosive environments. As we all know, the grain boundary characteristics distribution (GBCD) plays a key role in the formation of the properties of UFG materials, as well as corrosion resistance. However, in the related literature on the corrosion of UFG materials, most studies were focused on the variation of grain size, and the role of grain boundary in corrosion behavior were less related. Low angle grain boundaries (LAGBs) have very good corrosion behavior due to the low interfacial energy and impurity segregation [7]. Since a large volume fraction of high angle grain boundaries (HAGBs) could be generated by ECAP, it is interesting to investigate the relationship between the grain size and the corrosion resistance with the analysis of GBCD.

## 2. Materials and methods

Pure Ti was used in the present studies. 3 passes of ECAP at 400 °C in route C were used to obtain UFG-Ti. Different extraction planes of the sample after 3 ECAP passes were selected for the investigation; figure 1 shows a schematic diagram of different extraction planes.

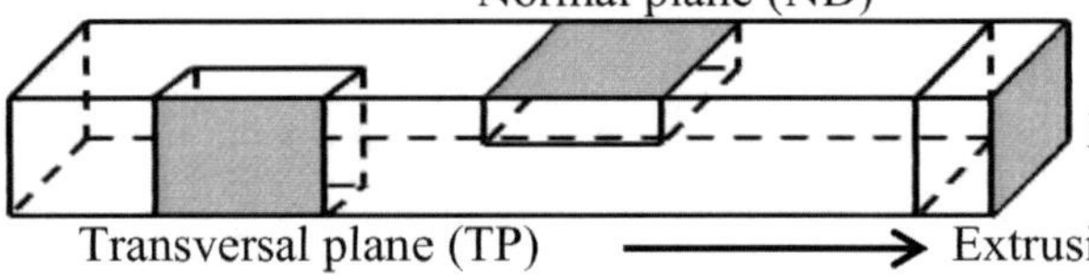


**Figure 1.** Schematic diagram of different extraction planes.

Microstructure of the samples were observed and quantitatively characterized by EBSD. Potentiodynamic polarization and electrochemical impedance spectroscopy (EIS) test were used to characterize the corrosion behavior of the samples in 3.5 wt.% NaCl by CHI660E electrochemical workstation. The details can be found in the other published paper of authors [8].

## 3. Results and discussion

### *3.1. Microstructural characterization*

Fig. 2 shows EBSD figures of the samples with different extraction planes. As shown in figure 2a, as-received Ti exhibits an equiaxed microstructure with an average grain size of ~ 143 μm. After ECAP, the grain size decreased to 620 nm, 653 nm, and 683 nm for the NP, EP, and TP, respectively.

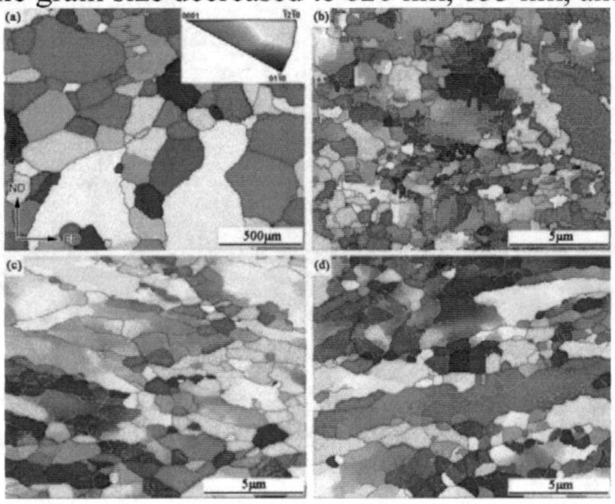


**Figure 2.** EBSD figures of as-received (a) and 3 passes ECAP samples of NP (b), EP (c), TP (d), respectively.

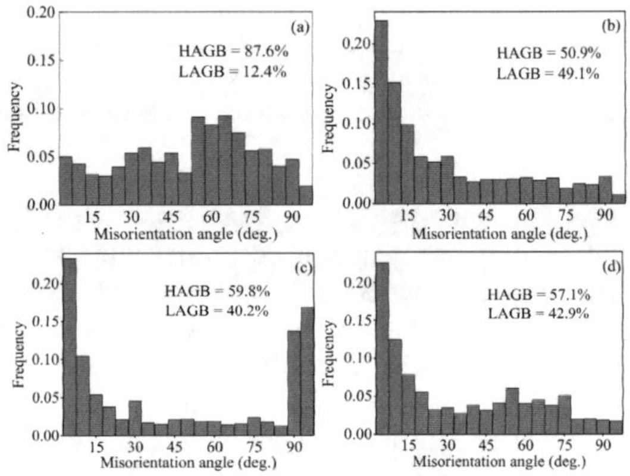


**Figure 3.** Misorientation distributions (a, b, c, d) of as-received (a) and 3 passes ECAP samples with NP (b), EP (c) and TP (d) plane.

Figure 3 displays the misorientation distribution of CG-Ti and UFG-Ti. The results indicate that GBCDs change obviously after ECAP, accompanied by refining of the grain size. It shows that a fraction of HAGBs is high to 87.6% for CG-Ti, after ECAP process, the fraction of HAGBs decreased dramatically. The fraction of HAGBs is distinguished from each other for different extraction planes of the sample after 3 passes ECAP, the lowest fraction of HAGBs is NP, which could get 50.9%. Meanwhile, the fraction of HAGBs increased to 59.8% for EP. Furthermore, the fractions of HAGBs exceeding 90° reached to 30% for EP, which is differential from the others.

### 3.2. Corrosion behavior

Polarization curves of the samples in 3.5 wt.% NaCl are shown in figure 4a. All polarization curves of the tested samples are analogous, which indicate that grain size refining cannot influence the corrosive mechanism. High $Cl^-$ in simulated seawater could penetrate the passive film to substrate of Ti, which led to pitting corrosion. The corrosion current density ($i_{corr}$) and corrosion potential ($E_{corr}$) can be obtained by linear fitting of Tafel region. The fitting results are shown in table 1.

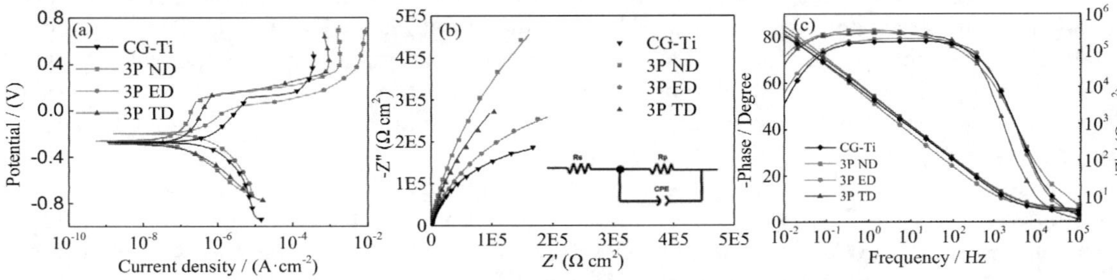


**Figure 4.** Polarization curves (a) and EIS (b, c) of CG-Ti and UFG-Ti in 3.5 wt.% NaCl.

**Table 1.** Electrochemical properties and equivalent circuit fitting parameters of CG-Ti and UFG-Ti.

| Sample | $E_{corr}$ V | $E_{pit}$ V | $i_{pass}$ μA·cm$^{-2}$ | $i_{corr}$ μA/cm$^{-2}$ | $R_s$ Ω·cm$^2$ | *CPE* $10^{-5}$ S·s$^n$·cm$^2$ | $n$ | $R_p$ $10^5$ Ω·cm$^2$ | Chi-squared |
|---|---|---|---|---|---|---|---|---|---|
| CG-Ti | -0.271 | 0.006 | 0.898 | 0.8987 | 4.787 | 3.575 | 0.8721 | 5.219 | 0.00053 |
| 3P NP | -0.259 | 0.112 | 0.100 | 0.0941 | 5.154 | 2.577 | 0.8932 | 42.520 | 0.00915 |
| 3P EP | -0.195 | 0.033 | 0.667 | 0.4746 | 4.936 | 2.919 | 0.8824 | 7.256 | 0.00511 |
| 3P TP | -0.286 | 0.135 | 0.158 | 0.1358 | 4.976 | 2.173 | 0.9094 | 13.940 | 0.00874 |

As shown in table 1, $E_{corr}$ did not change obviously. However, it can be found that $E_{pit}$ of UFG-Ti is generally higher than CG-Ti, and $i_{pass}$ is much lower than CG-Ti, which indicates that the pitting resistance was enhanced for UFG-Ti. The corrosion resistance of the materials can be directly reflected by $i_{corr}$. It shows that the corrosion resistance from strong to weak is NP > TP > EP > CG-Ti.

The Nyquist diagram composed of semicircular arcs is illustrated in figure 4b, which was obtained from the test under a stable OCP. Generally speaking, the corrosion resistance increases as the radius of the arc increases [8, 9]. In order to get the corrosion behavior of the material, the experimental data were fitted by the ZSimDemo software. The used equivalent circuit is shown in figue 4b. Among them, $R_s$ is the solution resistance; CPE is the capacitance; $R_p$ is the polarization resistance, which can reflect the corrosion resistance of materials intuitively, i.e. the larger $R_p$ value, the better corrosion resistance. The fitting results are listed in table 1. Obviously, $R_p$ value of the samples is NP > TP > EP > CG-Ti. A little difference of $n$ values indicates that uniformity of the passivation film on the surface of all samples is not much different. Figure 4c shows the bode plot of CG-Ti and UFG-Ti in 3.5 wt.% NaCl solution. It indicates that the frequency is lower than $10^3$ Hz, the spectra displays a linear slope of about −1, which shows the characteristic response of a capacitive behavior of the passive film [10]. UFG-Ti exhibits a higher impedance and phase angle than the CG-Ti at low frequencies. It indicates that the corrosion resistance of the samples is NP > TP > EP > CG-Ti.

As we all know, Ti and Ti alloys surfaces are easy to form a dense oxide film, it can hinder the direct contact between the substrate of metal and the corrosive medium, then improves corrosion resistance of the materials effectively. Exactly speaking, the corrosion behavior of Ti is more concerned with the state of the surface oxide film.

As shown in table 1, $E_{pit}$ of UFG-Ti is higher than CG-Ti, which indicates that the occurrence of the pitting corrosion is difficult due to a higher potential. On the other hand, $i_{pass}$ of UFG-Ti is lower than CG-Ti, which indicates that UFG-Ti is easier to form passivation on the surface. In comparison with CG-Ti, the grain size decreased with the process of ECAP and high stored energy can be formed in the dislocations and grain boundaries of the material, which is beneficial to the formation of the passivation film and increase self-repairing ability of the passivation film. On the other side, not only grain refinement, but the improvement activity of electrons near the grain boundaries also can promote the formation of passivation film of metals [11]. It is worthy to note that the ECAP processing can introduce a large number of dislocations at grain boundaries and grain interior simultaneously, which is benefits to the uniformity of nucleation of the passivation film and improve the density of the passivation film, then can hinder the penetration of chloride ions via the passivation film to the substrate effectively.

As mentioned above, the corrosion resistance of the sample after 3 ECAP passes with a different extraction plane is significantly different, although the grain size seems similar. GBCD plays an important role in the corrosion resistance of UFG materials, and LAGBs have a very good corrosion resistance due to the low interfacial energy and the impurity segregation. In the present study, the fraction of LAGBs of NP reached to 49.1%, which is higher than TP (equal to 42.9%) and EP (equal to 40.2%). From the perspective of grain boundary engineering, if corrosion occurred at a high-angle segment of a grain boundary and extended, it could be encountered by arrested LAGBs, then the intergranular corrosion was arrested [12]. The EP has the lowest fraction of LAGBs, as well as the concentration of HAGB exceeds 90°, which is probably the reason to cause the worst corrosion resistance of all the deformed planes.

## 4. Conclusions

The corrosion resistance of UFG-Ti processed by ECAP was investigated by potentiodynamic polarization and EIS in 3.5 wt.% NaCl solution. Electrochemical results indicated that the corrosion resistance of UFG-Ti was improved obviously in comparison with CG-Ti Ti. Furthermore, NP of the sample after 3 ECAP passes has shown a better corrosion resistance than TP and EP, which could be attributed to the higher fraction of LAGBs.

## Acknowledgements

This work was supported by State Key Laboratory Open Source for Metal Materials and Applications for Marine Equipment (No. SKLMEA-K201807), the financial assistance from the China Postdoctoral Science Foundation (No. 2017M623392), Postgraduate Research & Practice Innovation Program of Jiangsu Province (No. SJCX19_0324), the Priority Academic Program Development of Jiangsu higher education institutions (PAPD).

## References

[1] Wang X *et al.* 2017 *Journal of Alloys and Compounds* **728** 709
[2] Hoseini M, Shahryari A, Omanovic S and Szpunara J A 2009 *Corrosion Science* **51** 3064
[3] Valiev R Z and Langdon T G 2006 *Progress in Materials Science* **51** 881
[4] Balyanov A *et al.* 2004 *Scripta Materialia* **51** 225
[5] Gurao N P *et al.* 2013 *Metallurgical & Materials Transactions* A **44** 5602
[6] Chuvil'deev V N *et al.* 2017 *Journal of Alloys and Compounds* **723** 354
[7] Zhao Y, Cheng I C, Kassner M E and Hodge A M 2014 Acta Materialia 67 181-188
[8] Li X *et al.* 2019 *Acta Metallurgica Sinica* **55** 967
[9] Yang D S *et al.* 2018 *Materials and Corrosion* **69** 1455
[10] Hodgson A W E *et al.* 2002 *Electrochimica Acta* **47** 1913
[11] Nikfahm A, Danaee I, Ashrafi A and Toroghinejad M R 2013 **16** 1379
[12] Hu C, Xia S, Li H, Liu T, Zhou B, Chen W and Wang N 2011 *Corrosion Science* **53** 1880

IOP Conf. Series: Materials Science and Engineering **672** (2019) 012065 doi:10.1088/1757-899X/672/1/012065

# Microstructure and mechanical properties of workpieces of the ultrafine-grained two-phase Ti alloy produced by die forging

**T V Yakovleva, G S Dyakonov, A G Stotskiy and I P Semenova**

Institute of Physics of Advanced Materials, Ufa State Aviation Technical University, 12 K. Marx street, Ufa, 450008, Russian Federation

E-mail: yakovleva_tatyana93@mail.ru

**Abstract**. The paper presents the results of studies of the microstructure and mechanical properties of experimental workpieces of the ultrafine-grained (UFG) VT8M-1 (Ti–5.7Al–3.8Mo–1.2Zr–1.3Sn in wt%) alloy produced by isothermal die forging (IDF). The UFG structure in the VT8M-1 alloy was formed by a high-performance method, namely rotary swaging (RS). The structural parameters (the shape and size of the α- and β-phases) and the homogeneity of the UFG structure were analyzed in different sections of forgings. It was shown that the fraction of the primary α- phase decreases, and the size of the secondary equiaxed grains of the α- and β-phases increases from 0.3 to 0.9 μm as a result of recrystallization during IDF. The relationship between the microstructure and mechanical properties of the forged workpieces of the UFG VT8M-1 alloy is discussed.

## 1. Introduction

Among structural materials, titanium alloys are widely applied in aviation and engine building due to their high specific strength, corrosion resistance and specific hot strength [1]. In particular, the two-phase Ti alloy VT8M-1 (Ti–5.7Al–3.8Mo–1.2Zr–1.3Sn in wt%) is applied in modern engine building for the production of compressor blades, which operate under the severe conditions of high static and dynamic loads, corrosive attack and temperature exposure. In this connection, a higher set of requirements is imposed on them, such as a combination of high strength, ductility, fracture toughness and fatigue resistance.

The conventional technology for the fabrication of blade workpieces from Ti alloys, having a rather complex surface configuration, is die forging [2]. It is known that the temperature of isothermal forging for coarse-grained Ti-based materials is about 950 °C, and the use of UFG billets in the shape-forming of blades creates an opportunity to implement isothermal forging at lower temperatures of 750-780 °C. The decrease in the forging temperature is related to the manifestation of superplasticity in UFG materials at lower temperatures and higher strain rates [3]. Such a temperature range makes it possible to preserve a homogeneous UFG structure, which ensures a higher level of properties in an experimental item. In this connection, the aim of the present paper is to study the effect of isothermal forging at lower temperatures on the microstructure of the UFG VT8M-1 alloy, and also to evaluate the mechanical properties of the produced workpiece.

IOP Conf. Series: Materials Science and Engineering **672** (2019) 012065 doi:10.1088/1757-899X/672/1/012065

## 2. Material and experimental procedure

For the investigation, we used the VT8M-1 alloy (VSMPO-AVISMA Corporation, Russia) with the following composition: Ti–5.7Al–3.8Mo–1.2Zr–1.3Sn–0.16Fe (wt.%). The β-transus temperature (TPT) in the alloy is 980 ± 5 °C. The ultrafine-grained structure of the VT8M-1 alloy was formed by rotary swaging (RS) at 750 °C with a gradual reduction along the diameter of the billets from 70 to 32 mm (true strain e ~ 1.56, the strain rate was higher than 300 mm $s^{-1}$ ).

An experimental forging simulating a compressor blade was produced by IDF at a temperature of 780±10 °C. Specimens for metallographic studies and testing of mechanical properties were cut out from the airfoil and root of the experimental forging (figure 1).

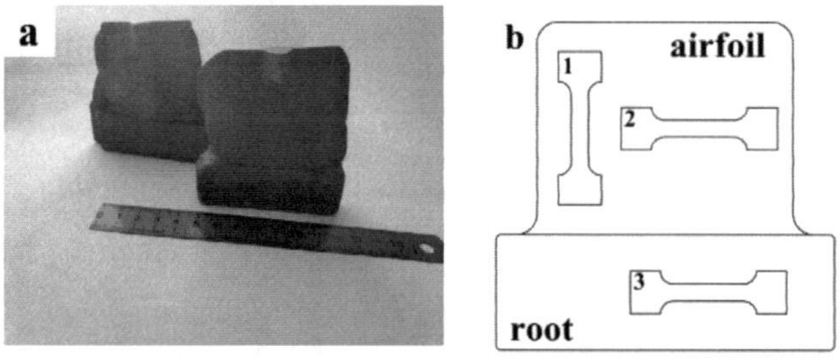


**Figure 1.** (a) Experimental forging produced from the UFG alloy VT8M-1; (b) scheme of cutting out of the specimens from the experimental forging.

The microstructure was characterized by scanning electron microscopy (SEM) using a JSM 6390, and by transmission electron microscopy (TEM) using a JEM-2100. Tensile mechanical tests of flat specimens with a gage length of 16 mm and a width of 3 mm, cut out from the root and airfoil of the experimental forging, were performed using an Instron universal testing machine at room temperature with a strain rate of $1\times10^{-3}$ $s^{-1}$ in compliance with ASTM E8/E8M.

## 3. Results and discussion

### *3.1. Microstructure and mechanical properties of the VT8M-1 alloy billet produced by RS*

The initial structure of the VT8M-1 alloy rod (figure 2a) was a duplex microstructure. The average size of the primary α-phase is 3 μm, and its fraction in the alloy is 60%, the thickness of the lamellae is 0.2 μm. Deformation by RS led to the spheroidization of the lamellar (α+β)-constituent (figure 2b) and the elongation of the primary α-phase along the draw direction (figure 2c) [4]. In the cross section, the primary α-phase was bent in accordance with the flow direction during rotary swaging, its average width decreased from 3 μm to 1.5 μm (figure 2b). The electron-microscopic images show that after the deformation the (α+β)-structure consists of fragments of severely distorted plates, as well as the small globular grains of the α- and β-phases with an average size of 0.3 μm (figure 2b). In the interiors of the primary α-phase grains, an increase in dislocation density with the formation of a developed substructure is observed (figure 2c).

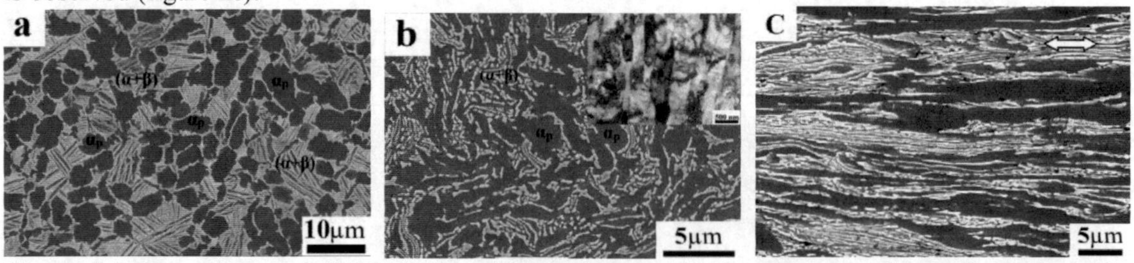


**Figure 2.** Microstructure of VT8M-1 alloy: (a) initial state; (b) after rotary swaging in the cross section and (c) in the longitudinal section. The arrow indicates the draw direction.

IOP Conf. Series: Materials Science and Engineering **672** (2019) 012065 doi:10.1088/1757-899X/672/1/012065

In the as-received state, the VT8M-1 alloy exhibited the following mechanical characteristics: ultimate tensile strength (UTS) ~1056 MPa and elongation δ=13% (table 1). As a result of the UFG structure formation, the alloy's strength increased by 22% to UTS = 1290 MPa and elongation decreased to δ=9%, which is typical of the majority of metals and alloys subjected to SPD [5,6,7].

*3.2. Microstructure and mechanical properties of the experimental forgings produced from the UFG alloy by IDF*

The microstructure of the forged workpieces produced by IDF (T=780 °C) is shown in figure 3. Due to a complex geometrical shape of the experimental forgings, different strains are observed in the airfoil and in the root part. The estimation of strain showed that the root part of the experimental forging experienced a lower strain (60%) in comparison to the forging's airfoil (75%). The SEM study of the microstructure of the forgings revealed a similar character of microstructure in the airfoil part and in the root part. The primary α-phase grains are elongated along the deformation direction and their thickness is 2-3 μm (figure 3a,c). As it was noted above, such an elongated shape formed at the stage of processing by rotary swaging. It was found that forging led to a reduction in the volume fraction of the globular constituent of the α-phase due to a decrease in grain length, as compared to the condition after RS (from 65% to 29% in the airfoil zone and to 35% in the root part) (figure 3a,c). TEM study at a larger magnification revealed (figure 3b,d) that the microstructure of the (α+β)-region is represented by equiaxed recrystallized and deformed grains, the average grain size in the root part being larger than that in the airfoil and amounting to 0.9 and 0.7 μm, respectively. This is related to the fact that in the more massive root part, cooling after forging is slower than in the airfoil, which promotes more intensive recrystallization processes.

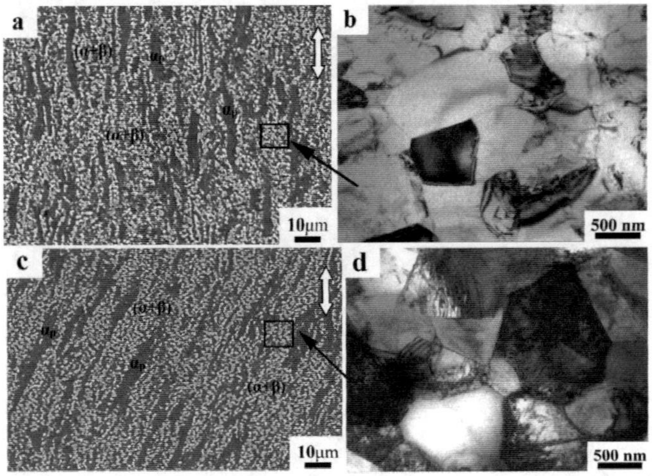


**Figure 3.** Microstructure of the experimental forging: (a,b) the airfoil of the forging; (c,d) the root of the forging; (a,c) SEM; (b,d) TEM. The arrow indicates the draw direction.

The mechanical properties of the specimens cut out from the root and airfoil zones of the experimental forging are listed in table 1. In the longitudinal direction of the experimental forging's airfoil (specimen 1 in figure 1b) strength declines to 1100 MPa as compared to the condition after RS (UTS = 1290 MPa), but at the same time ductility increases visibly, in particular, the uniform and percentage elongations grow to 7 and 15%, respectively (table 1). Such a decline in strength in conditioned by a decrease in dislocation density, the propagation of recrystallization and an increase in the sizes of small globular grains of the α- and β-phases (from 0.3 to 0.7 μm) (figure 3). The mechanical properties of the specimens in the longitudinal (specimen 1) and cross sections (specimen 2 in figure 1b) of the airfoil zone have some differences. In particular, the ultimate tensile strength (UTS) in the cross section of the airfoil is slightly higher (1118 MPa), while the ductility, percentage and uniform elongations are lower. This is apparently conditioned by the presence of the inherited after RS metallographic and crystallographic texture in the forged workpiece. The ultimate tensile strength in the root part of the experimental forging amounts to UTS=1109 MPa with an elongation of $\delta$=10%, and it is close to the level of properties of the alloy in the cross section of the airfoil (specimen 3 in figure 1b and table 1). Thus, the results of this investigation demonstrate a possibility in principle of preserving a UFG structure during the shape-forming of an item by isothermal forging, and correspondingly, of preserving enhanced mechanical properties.

**Table 1.** Mechanical properties of the Ti alloy VT8M-1 in different conditions.

| Condition | UTS, MPa | YS 0.2, MPa | Elong., % | Uniform elongation, % |
|---|---|---|---|---|
| As-received | 1056±10 | 967±5 | 13±0.5 | 4.7±0.5 |
| RS | 1290±5 | 1195±10 | 9±1 | 3.4±0.5 |
| IDF 780 °C specimen 1 airfoil | 1100±30 | 947±25 | 15±1 | 7.0±0.5 |
| IDF 780 °C specimen 2 airfoil | 1118±10 | 1048±20 | 10±0.6 | 0.7±0.2 |
| IDF 780 °C specimen 3 root | 1109±15 | 1105±15 | 10±0.5 | 0.2±0.1 |

## 4. Conclusions

1. As a result of processing by rotary swaging, a UFG structure is formed in the VT8M-1 alloy, characterized by the primary α-phase grains elongated in the draw direction with a thickness of ~1.5 μm, and ultrafine α- and β-grains ~0.3 μm in size in the two-phase regions. This type of UFG structure leads to an increase in the ultimate tensile strength to UTS = 1290 MPa and some decline in elongation to $\delta$=9%, as compared to the initial condition.
2. It is shown that as a result of isothermal forging at a temperature of 780 °C, the workpiece retains the ultrafine grain size which increased to 0.7 μm in the airfoil and 0.9 μm in the root due to the processes of recovery and recrystallization.
3. The differences in the mechanical properties of the root and airfoil zones are conditioned by different strains accumulated during the forging and the different cooling rates of the massive root and a thin airfoil part.

**Acknowledgments**
This work was supported by the Ministry of Education and Science of the Russian Federation in the scope of the project part of State Assignment No. 11.1235.2017.

**References**

[1] Leyens C and Peters M 2003 *Titanium and Titanium Alloys. Fundamentals and Applications* (Weinheim: Wiley-VCH Verlag GmbH)
[2] Ermachenko A G, Lutfullin R Ya and Mulyukov R R 2011 *Rev. Adv. Mater. Sci.* **29** 68
[3] Salischev G A, Imaev R M, Imaev V M and Gabdulin N K 1993 *Mater. Sci. Forum* **113** 613
[4] Modina Iu M, Polyakov A V, Dyakonov G S, Yakovleva T V, Raab A G and Semenova I P 2018 *IOP Conf. Ser.: Mater. Sci. Eng.* **461** 012056
[5] Meyers M A and Mishra A and Benson D J 2006 *Prog. Mater. Sci.* **51** 427
[6] Valiev R Z 2004 *Nature Mater.* **3** 511
[7] Semenova I P, Dyakonov G S, Raab G I, Grishina Y F, Huang Y and Langdon T G 2018 *Adv. Eng. Mater.* **20** 1700813

IOP Conf. Series: Materials Science and Engineering **672** (2019) 012066 doi:10.1088/1757-899X/672/1/012066

# Optimisation of laser assisted forming conditions for improved formability of aerospace materials

**E Yakushina[1], M McArthur[2], M Canavan[1]**

[1] Advanced Forming Research Centre (AFRC), University of Strathclyde, United Kingdom

[2] Department of Design, Manufacture and Engineering Management, University of Strathclyde, United Kingdom

E-mail: Evgenia.yakushina@strath.ac.uk

**Abstract.** ISF technology was gradually considered as a manufacturing technology for fabricating sheet metal parts for both commercial and military aircraft. However, many challenges remain with respect to improving the consistency of the process, such as part thinning and springback. This is especially critical for two phase titanium alloy Ti-6AL-4V which has limited formability at room temperature. Therefore, understanding the correlation of laser system parameters and material properties is a key aspect for successful ISF of non-fractured Ti-6AL-4V. The laser heating is modelled to find the relationships between laser power, beam radius, scan speed, and the resultant temperature. The analysis shows that the temperature from laser heating increases with an increase in power, decreases significantly with an increase in beam radius, and decreases slightly with an increase in scan speed.

## 1. Introduction

Single point incremental sheet forming (SPIF) is a promising manufacturing technique in which sheet metal is formed without a dedicated die. By removing the need for a die, SPIF significantly reduces lead times on parts and can reduce costs for prototyping and small batch production [1]. Another major advantage of SPIF is the increase in the forming limit of the metal due to localised stresses generated during the forming process [1].

However, numerous deformation mechanisms during the SPIF process have significant impact on the geometrical accuracy of the finished part [2]. Additionally, hard to deform materials such as Ti-6Al-4V still have formability issues in SPIF at room temperature. One potential solution to both of these problems is laser assisted single point incremental sheet forming (LASPIF), in which a laser preheats the area being formed.

The localised heating supplied by the laser alleviates formability issues and improves the geometrical accuracy of the final product [3]. Duflou et al. introduced LASPIF using a laser on the opposite side of the sheet to the tool with a small forward offset [3]. It was found that laser assistance reduced the forming forces, improved part accuracy in 65Cr2 blanks, and increased the formability of Ti-6Al-4V sheets.

Gottmann et al. developed their own LASPIF set up in which the laser was applied to the same side of the sheet as the tool at a forward offset of 45 mm [4]. His work focussed on titanium alloys with the aim of improving formability and accuracy to make LASPIF a suitable process for aerospace. The maximum forming depth of a Ti-6Al-4V sheet was significantly increased. In other papers a method for measuring and controlling temperature during LASPIF was to embed a thermocouple in the tool [4] and the deformation mechanisms involved in LASPIF of Ti-6Al-4V were investigated [5].

It has been demonstrated several times that LASPIF can improve the accuracy of parts made with low formability alloys such as Ti-6Al-4V, but the process isn't fully understood. The optimal laser-tool offset is unknown as different papers have used different offsets [3] [4] [6] without any conclusive investigation into which offset gives the most accurate results. Additionally, the laser parameters for each experiment's desired temperature are often found using trial and error or thermal models.

Thermal models are developed in this work reflecting the relationships between the laser parameters and the resultant temperature in the simplest case of a straight path. Due to the specifics of the Hybrid Laser Metal Deposition (LMD) machine used in the current work, the laser system and the SPIF tool (stylus) cannot be used simultaneously. Therefore, laser setups were designed to produce a preheated area on a Ti-6Al-4V sheet and the developed model is used to predict what temperature drop could be expected during tool change between the laser head and the forming tool. Gottmann et al. found that coupled setup of the laser and forming tool induced overheating of the part and tool and leads to significant tool and component surface degradation and therefore negatively influence the materials' formability [7]. It is expected that the machine set up modelled in the current work and used for subsequent validation will avoid the aforementioned issues.

## 2. Methods and experimental set-up

### *2.1. Modelling*

The models were run on Abaqus 6.14.4. To simulate Ti-6Al-4V, the model used temperature dependent values for density, thermal conductivity, and specific heat, [8] an emissivity of 0.25, and an absorptivity of 0.34 for Ti-6Al-4V [9]. A convective heat transfer coefficient of 50 $W/m^2K$ was used. To model the temperatures resulting from a straight laser path, only half of the heated volume was modelled as symmetry across the laser path can be expected. The laser was modelled as a Super-Gaussian surface heat flux of order 10 using the DFLUX subroutine. Laser parameters ranging from 2mm to 5mm beam radius, 20 mm/s to 50 mm/s scan speed, and 200 W to 1400 W were modelled, with a sheet thickness of 1.6 mm. In order to determine peak temperatures under the laser, the laser path was three times as long as the laser beam diameter. To allow comparison between different setups when each model contains thousands of data points, the highest temperature at any point under the laser was recorded as peak temperature and these peak temperatures were compared.

### *2.2. Path modelling*

To model the temperatures resulting from a laser path designed for the AFRC's setup, a 200 mm by 120 mm region of 1.6 mm thick Ti-6Al-4V sheet was modelled with a 5mm mesh across the entire surface. No symmetry could be exploited. The tool path was designed to preheat a region of sheet before a 25 second tool change and is shown in figure 1a. The path was coded in Fortran and applied using DFLUX. An 85 second air cooling step was considered, to predict the temperatures after the 25 seconds required for tool change and for up to 60 seconds of forming cycle. The laser model followed the path in figure 1a with a scan speed of 50 mm/s and a beam radius of 10 mm. The laser repeated the loop until the sheet lost heat at the same rate as the laser provided heat and the temperature distribution became steady. This required 60 loops. Different laser powers from 675 W to 1200 W were tested to find a range of temperatures.

## 3. Results and discussion

The temperature range of interest for LASPIF of Ti-6Al-4V is 400 °C to 700 °C as optimal material formability was observed [5]. The peak temperatures from selected thermal models are listed in table 1. The peak temperature was found to increase as laser power increased, decrease significantly as beam radius increased, and decrease slightly as scan speed increased.

In general, the choice of beam radius will depend on the width of heating required by the process. The smallest radii that was modelled can produce high temperatures with low powers. To produce temperatures in the range of 400-700 °C using a larger beam radius, high powers and slow scan speeds are required. The scan speed is limited by the range of appropriate tool feed rates when the forming tool follows the laser.

**Table 1.** Prediction of peak temperatures for a range of powers, radii, and speeds.

| Beam Radius (mm) | Scan Speed (mm/s) | Peak Temp at 200W | Peak Temp at 600W | Peak Temp at 1000W | Peak Temp at 1400W |
|---|---|---|---|---|---|
| 2 | 50 | 314 | 762 | 1130 | - |
| 3 | 30 | 245 | 595 | 866 | - |
| 3 | 50 | 195 | 487 | 723 | 919 |
| 5 | 20 | 161 | 400 | 601 | 782 |
| 5 | 30 | 134 | 332 | 501 | 653 |
| 5 | 50 | 107 | 266 | 406 | 529 |

The temperature distribution produced by the laser path designed for the AFRC's machine setup is shown in figures 1b and 1c. When the laser heating ends there are steep temperature gradient with temperature ranging from 900 °C to 100 °C. After 25 seconds of cooling the temperature in the central part dropped to 340 $^{O}$C and the lower range of the temperature gradient remained almost unchanged.

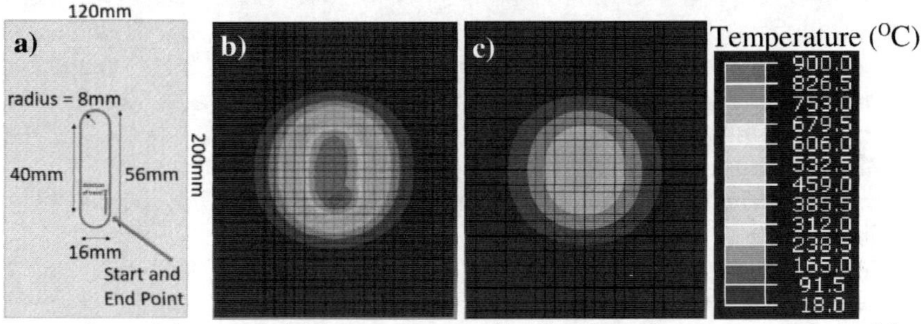


**Figure 1.** a) laser path used for model; b) temperature distribution after laser heating with power of 1200W after 60 loops; c) temperature distribution after 25 seconds of cooling.

Figure 2 shows the temperature in the centre of the upper surface sheet during 85 seconds of cooling. The peak temperature occurred directly under the laser, not in the centre of the sheet, and so the peak temperature for each model is included in the legend.

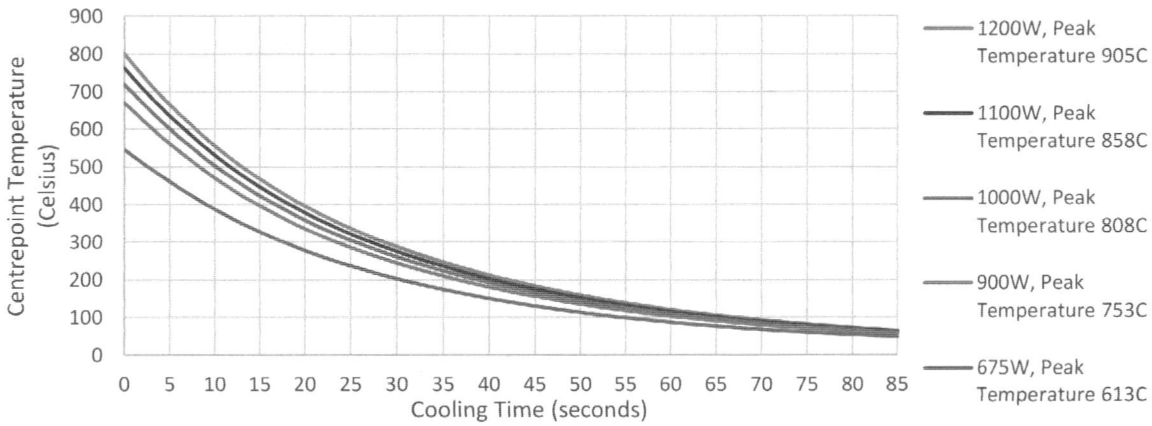


**Figure 2.** Temperature distribution in the centre of the modelled sheet.

IOP Conf. Series: Materials Science and Engineering **672** (2019) 012066 doi:10.1088/1757-899X/672/1/012066

The temperatures after 25 seconds ranged from 200–340 °C which is lower than the target temperature range of 400–700 °C. To achieve temperatures above 300 °C after 25 seconds of cooling, the sheet must be heated above 700 ºC. After 85 seconds, the temperature drops to around 60 °C. The resultant temperatures could be increased by using a thicker sheet, which may require a more powerful laser. Some improvements could also be made by increasing the length of the laser path and increasing the laser power, as these changes would decrease the peak temperature in the sheet without reducing the rate of heat input.

## 4. Summary and conclusions

Thermal models representing laser heating of a sheet of Ti-6Al-4V were produced to investigate the effect of laser parameters on the temperature distribution. The thermal models found that:

- The peak temperature due to laser heating increases as laser power increases, decreases significantly as beam radius increases, and decreases slightly as scan speed increases.
- Laser heating designed to account for a 25 second tool change has a significantly less localised temperature distribution than is normally found in LASPIF.
- The temperature range of 200-340 °C after a 25 second tool change is achievable for a 1.6 mm Ti-6Al-4V sheet heated by a 1200 W laser.

## References

[1] Adams D and Jeswiet J 2014 *Journal of Engineering Manufacture* **229** 754
[2] Jackson K and Allwood J 2009 *Journal of Materials Processing Technology* **209** 1158
[3] Duflou J, Callebaut B, Verbert J and Baerdemaeker H D 2007 *CIRP Annals* **56** 273
[4] Gottmann A, Bailly D, Bergweiler G, Bambach M, Stollenwerk J, Hirt G and Loosen P 2013 *The International Journal of Advanced Manufacturing Technology* **67** 22195
[5] Mosecker L, Gottmann A, Saeed-Akbari A, Bleck W, Bambach M and Hirt G 2013 *Key Engineering Materials* **549** 372
[6] Mohammadi A, Vanhove H, Bael A V and Duflou J R 2016 *International Journal of Material Forming* **9** 339
[7] Gottmann A, Diettrich J, Bergweiler G, Bambach M, Hirt G, Loosen P and Poprawe R 2011 *Production Engineering* **5** 263
[8] Mills K C 2002 *Recommended Values of Thermophysical Properties for Selected Commercial Alloys* (Wiltshire: Woodhead Publishing Limited)
[9] Yang J, Sun S, Brandt M and Yan W 2010 *Journal of Materials Processing Technology* **210** 2215

IOP Conf. Series: Materials Science and Engineering **672** (2019) 012067 doi:10.1088/1757-899X/672/1/012067

# Microstructure and mechanical properties of nanostructured Cu-Zn alloys by ECAP and rolling

**L I Zaynullina[1,3], I V Alexandrov[1,3] and W Wei[2,3]**

[1] Ufa State Aviation Technical University, Ufa,450008, Russia
[2] School of Materials Science and Engineering, Changzhou University, Changzhou, Jiangsu 213164, P R China
[3] Sino-Russian Joint Laboratory of Functional Nanostructured Materials, Changzhou University, Changzhou 213164, P R China

E-mails: lele4ka.06@mail.ru, benjamin.wwei@163.com

**Abstract**. The results of investigation of the Cu–Zn system alloys with 10 wt.% and 30 wt.% Zn, the stacking fault energy (SFE) of which differs by a factor of 2.5, are presented. ECAP and subsequent flat rolling resulted in the refinement of the microstructure. Flat rolling with the reduction degree 95% resulted in the decrease of the concentration of nanotwins formed during ECAP. Flat rolling of ultra-fine grained (UFG) alloys leads to the strength increase and ductility decrease explained by the refinement of the microstructure and decrease in the concentration of deformation twins.

## 1. Introduction

Since the 90s of the last century, bulk nanostructured metallic materials obtained by the severe plastic deformation (SPD) methods and characterized by attractive properties cause great interest of researchers [1]. Nanostructured billets obtained by SPD in the manufacture of industrial products must take the required dimensions and shape. The method of flat cold rolling is widely used to produce sheet blanks from coarse grained (CG) metal blanks on an industrial scale. At the same time, investigations of influence of flat rolling on the microstructure and properties of bulk nanostructured materials obtained by SPD are relevant.

Alloys of the Cu-Zn system belong to brass and have different SFE values. As a result, they are characterized by different microstructures, deformation mechanisms and level of properties [2]. In reports [3, 4], the influence of SFE on the microstructural changes of materials subjected to SPD was investigated. In some reports, the effect of SPD methods such as equal channel angular pressing (ECAP) [5] and high-pressure torsion (HPT) [6–8] on the microstructure and properties of Cu–Zn alloys was investigated. In addition, the influence of flat rolling at room or cryotemperatures on the formation of UFG structure in these alloys was studied [9–11]. However, the effect of using a combined deformation treatment scheme, including ECAP with subsequent flat rolling, remains little investigated.

In this paper the results of investigation of the alloys of Cu–Zn system with a content of 10 wt.% and 30 wt.% Zn are presented. The SFE of these alloys differs by a factor of 2.5. The alloys were subjected to ECAP and subsequent flat rolling in order to investigate their influence on the microstructure and mechanical properties.

## 2. Experimental

The materials used in this work belong to the Cu-Zn system and contain 10 and 30 wt.% Zn. The SFE of these alloys is equal approximately 35 $mJ \cdot m^{-2}$ and 14 $mJ \cdot m^{-2}$, correspondingly. The rods of 60 mm in length and the edge of the square cross of 8 mm were subjected to 2 ECAP passes using route $B_c$ (90° clockwise rotation around the billet axis between the consecutive passes) at a temperature of 150 °C for Cu-10% Zn alloy and 300 °C for Cu-30% Zn alloy correspondingly. The pressing rate was 37 $mm \cdot min^{-1}$. The inner and outer die channel intersection angles were selected as $\varphi = 90°$ and $\psi = 0°$ correspondingly. After ECAP, the samples were rolled at the same temperatures into thin ribbons with different reduction degrees from 30% to 95%. The rolling direction was parallel to the longitudinal axis of the rods. Also, the initial rods of the alloys were rolled under the conditions described above.

Thin foils to prepare the samples for transmission electron microscope (TEM) investigations were cut from the planes of the ribbons. The thin foils were examined using a TEM JEM-2100 in dark and bright fields at an accelerating voltage of 200 kV. Final thinning of the foils was performed by the ink-jet electrolytic polishing method in a Tenupol-5 special unit using the electrolyte of the following composition: 25% phosphoric acid, 25% ethanol and 50% deionized water with a current of 50-80 mA.

The tensile specimens were cut by electronic spark-cutting technique from the planes of the ribbons along the longitudinal axis and had dog-bone shape with 12 mm in gauge length and the initial working part of 4 mm. The tensile tests were carried out on an Instron 8801 tensile testing machine at room temperature with a strain rate of 0.24 mm/min. Microhardness by Vickers was determined on a Struers Duramin equipment, for which a load of 100 grams was applied for 10 seconds.

## 3. Results and analysis

### *3.1. Microstructure*

The initial microstructure of the rods, obtained by homogenizing annealing, is characterized by equiaxed grains with an average size of 117 ± 5 μm for the Cu–10 wt.% Zn alloy and 93 ± 3 μm for the Cu–30 wt.% Zn alloy. Figure 1 a,b presents dark-field images of the microstructures of Cu-10 wt.% Zn and Cu-30 wt.% Zn alloys after two ECAP passes. The calculation the subgrains/grains sizes was produced by the standard secant method on dark-field images. After two ECAP passes, the size of structural elements for the Cu-10 wt.% Zn alloy was 800 ± 66 nm, for the Cu-30 wt. Zn alloy - 290 ± 55 nm.

The presence of deformation twins was observed in the structure after ECAP of both alloys (white arrows indicated in figure 1 a,b). Double diffraction points on the captured electron diffraction patterns for these states confirm the presence of twins in the microstructure (inserts in the lower right corner of figure 1 a,b). However, in the Cu-30 wt.% Zn alloy after two ECAP passes, twins of different orientations are observed, which is probably associated with different SFEs. In the Cu-10 wt.% Zn alloy, the average twin length is 1000 ± 100 nm and their thickness is 40 ± 4 nm. The calculated twin concentration (as the ratio of the twin area to the area of the observed photo) is $\rho_{tw} = 3.3\%$. In the Cu-30 wt.% Zn alloy, the average length of twins is 290 ± 38 nm, the thickness is 20 ± 3 nm, and twin concentration $\rho_{tw} = 5.3\%$.

Thus, plastic deformation leads to the nanoscale grain refinement and formation of deformation twins. It should be noted that in the alloy with low SFE value, the grain size is smaller, and twin concentration is higher. This can be explained by the fact that in the nanoscale range, the twinning occurs via the emission of partial dislocations by grain boundaries, which differs from the twinning mechanism in the CG state [12]. Low value of SFE makes easier for the dissociation of a dislocation into two partial dislocations, as a result a wider stacking fault ribbon is formed between them. It acts as a barrier for full dislocation, and the dislocation either cross slip or climb when it encounters a barrier. This promotes the formation of narrower twins [9].

Further flat rolling of the obtained UFG states for different reduction degrees leads to further changes in the microstructure. Figure 1 c,d shows the images of the investigated alloys subjected to 2 ECAP passes and subsequent flat rolling with a reduction of 95%.

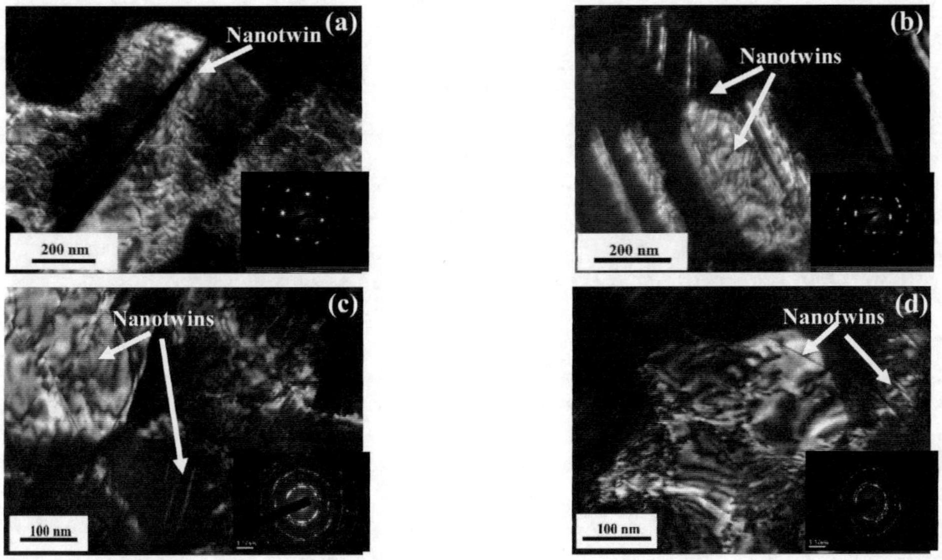


**Figure 1.** Microstructure of the Cu-10 wt.% Zn alloy (a) and Cu-30 wt.% Zn alloy (b) after ECAP. (c) Cu-10 wt.% Zn alloy and (d) Cu-30 wt.% Zn alloy after ECAP and flat rolling with a reduction of 95% with nanotwins.

In both alloys, in the body of some grains, there is a clear alignment of the dislocation walls in low-angle boundaries and the formation of a subgrain structure. The average size of the structural elements is 220 ± 30 nm and 130 ± 10 nm for alloys with a content of 10 wt.% and 30 wt.% Zn, respectively. As can be seen, subgrains and grains contain nanoscale deformation twins. In the Cu-10 wt.% Zn alloy, the average twin length is 150 ± 10 nm, the thickness is 4.0 ± 0.4 nm, and the twin concentration $\rho_{tw} = 0.8\%$. In the Cu–30 wt.% Zn alloy, the average twin length is 100 ± 8 nm, the thickness is 3.0 ± 0.3 nm, and the twin concentration $\rho_{tw} = 1.3\%$. Thus, further flat rolling leads to the microstructure refinement and formation of the nanotwins. The concentration of twins is reduced compared with ECAP, but remains higher in the alloy with a lower SFE value. It can be assumed that as reduction degrees increases, the dislocation density increases, a significant number of dislocations pile up at the twin boundaries, since they act as a barrier for dislocation slip. As a result of the interaction, the original atomically-flat coherent twin boundaries are bent and transformed into incoherent ones. Further straining transforms the incoherent boundaries into high-angle grain boundaries. This leads to the detwinning [13].

### *3.2. Tensile properties*

The mechanical tensile test results, including Vickers microhardness measurements, are presented in table 1 for conditions after ECAP and ECAP with subsequent flat rolling with reduction 95%. As can be seen from table 1, the values of microhardness after rolling increase for two alloys compared to ECAP.

For an alloy with a lower SFE value, the ultimate strength is 1.2 times higher after flat rolling. This can be explained by the fact that at low SFE value, the dislocations cross slip is limited due to a wider stacking fault ribbon between partial dislocations. As a result, plastic deformation occurs with a large contribution of twinning, which is confirmed by the results of TEM investigations described above.

The ultimate elongation of the alloys decreases as a result of rolling. This is explained by the decrease in the twins concentration, as well as by the structure refinement.

**Table 1.** Mechanical properties of Cu-Zn alloys after ECAP and ECAP with subsequent flat rolling.

| | | ECAP-2p | ECAP-2p+roll 95% |
|---|---|---|---|
| | Hv | 145±9 | 216±6 |
| Cu-10 wt.% Zn | $\sigma_{uts}$, MPa | 413±78 | 670±120 |
| | ε, % | 18±5 | 9±3 |
| | Hv | 222±7 | 250±6 |
| Cu-30 wt.% Zn | $\sigma_{uts}$, MPa | 590±89 | 812±130 |
| | ε, % | 12±3 | 8±2 |

In the Cu-10 wt.% Zn and Cu-30 wt.% Zn alloys, after flat rolling, the concentration of twins decreases by 4 times, the structural elements are refined by 2-4 times. At the same time, the ultimate elongation for the Cu-10 wt.% Zn alloy is halved to 9 ± 3%. For the Cu-30 wt.% Zn alloy, the elongation decreases to a lesser extent from 12 ± 3% to 8 ± 2%.

## 4. Conclusions

Flat rolling of UFG Cu-10 wt.% Zn and Cu-30 wt.% Zn alloys leads to the refinement of the microstructure and the formation of nanotwins, the concentration of which decreases as the reduction degree increases. However, the concentration of nanotwins remains higher in the alloy with a lower SFE value. In the future, such microstructural changes contribute to an increase in strength of the material and, however, they reduce its ductility.

## Acknowledgments

Authors are grateful to be supported by the financial support of the Ministry of education and science within the framework of the project No. 16.1969.2017/4.6. W. Wei is grateful to be supported by the National Natural Science Foundation of China under Grant No. 51561001 and the Science and Technology Project of Changzhou, P. R. China under grant No.CZ20180016 and CE20170028.

## References

[1] Valiev R, Islamgaliev R and Alexandrov I 2000 *Prog. Mater. Sci.* **45** 103
[2] Carter C and Ray I 1977 *Philos. Mag.* **35** 189
[3] Wang Z, Wang Y, Liao X, Zhao Y, Lavernia E, Zhu Y, Horita Z and Langdon T 2009 *Scr. Mat.* **60** 52
[4] Zhao Y, Horita Z, Langdon T and Zhu Y 2008 *Mat. Sci. Eng.* A **474** 342
[5] Zhang Z, Duan Q, An X, Wu S, Yang G and Zhang Z 2011 *Mat. Sci. Eng.* A **528** 4259
[6] An X, Lin Q, Wu S, Zhang Z, Figueiredo R, Gao N and Langdon T 2011 *Scr. Mat.* **64** 249
[7] Ungar T, Balogh L, Zhu Y, Horita Z, Xu C and Langdon T 2007 *Mat. Sci. Eng. A* **444** 153
[8] Bahmanpour H, Youssef K, Horky J, Setman D, Atwater M, Zehetbauer M, Scattergood R, Koch C 2012 *Acta Mat.* **60** 3340
[9] Roy B, Kumar N, Gopalakrishnan Nambissan P and Das J 2014 *AIP Advances* **4** 067101
[10] Konkova T, Mironova S, Korznikova A, Korznikova G, Myshlyaev M and Semiatin S 2015 *Mat. Charact.* **101** 173
[11] Jamaati R and Toroghinejad M 2014 *Mat. Sci. Eng.* A **598** 263
[12] Li Y, Zhao Y, Liu W, Xu C, Horita Z, Liao X, Zhu Y, Langdon T and Lavernia E 2010 *Mat. Sci. Eng.* A **527** 3942
[13] Wang Y, Liao X, Zhao Y, Lavernia E, Ringer S, Horita Z, Langdon T and Zhu Y 2010 *Mat. Sci. Eng.* A **527** 4959

IOP Conf. Series: Materials Science and Engineering **672** (2019) 012068 doi:10.1088/1757-899X/672/1/012068

# Study on the extrusion load during equal channel angular pressing

**D T Zhang [1], Y X Tong [1], L Li [1] and Y F Zheng [2]**

[1] Institute of Materials Processing and Intelligent Manufacturing, College of Materials Science and Chemical Engineering, Harbin Engineering University, HRB 150001, China

[2] Department of Materials Science and Engineering, College of Engineering, Peking University, BJ 100871, China

E-mail: zhangdiantao@hrbeu.edu.cn

**Abstract**. The equal channel angular pressing or extrusion technique, ECAP/ECAE, is a severe plastic deformation process employed to produce bulk ultra-fine grained materials. In this work, the extrusion load for performing the ECAP process is studied. The resistance includes deformation resistance and frictional resistance during ECAP processing. Base on the slip line method, upper bound solution and finite element analysis, the formula of extrusion load is proposed. The calculated results of extrusion load are in a good agreement with the experimentally measured curves. According to the formula, the influencing factors on the extrusion load are the length of the samples, friction coefficient, geometric parameters of die, yield strength and work hardening coefficient of the materials.

## 1. Introduction

Ultra-fine grained (UFG) materials with outstanding mechanical properties and special physical property has attracted much attention of researchers in the field of materials science in recent years [1]. The equal channel angular pressing or extrusion technique, ECAP/ECAE, is a severe plastic deformation process employed to produce bulk ultra-fine grained materials [2]. During ECAP processing, a well-lubricated billet is forced to pass through two intersecting channels with a constant cross-sectional area. The billet undergoes a large amount of plastic strain by simple shear within the deformation zone located at the intersection of the die channels [3].

The extrusion load used to make billet plastic deformation is one of the most important parameters in the ECAP process. At present, the main methods to study the extrusion load include the slip line field theory [4], the upper-bound analysis [5] and FEM-based simulations [6]. However, the frictional resistance caused by the lateral expansion of the billet during the ECAP process was ignored in most of the studies. Undoubtedly, the frictional resistance is the main reason why it is difficult to produce specimens with a high length to diameter ratio. In the present paper, formulas of the extrusion load were derived by force analysis. The influence of the main factors was considered comprehensively, such as the length-diameter ratio of the billet, friction coefficient, geometric parameters of die, yield strength and work hardening coefficient of the materials.

## 2. Analysis

In this study, analytical modeling of the ECAP die with a circular cross-section and an external angle $\psi=\pi-\Phi$ was established, as shown in figure 1. As demonstrated by Zhang et al. [7], there were three zones during the ECAP process, namely, a small deformation zone, a main deformation zone and a non-deformation zone. The ECAP system reaches a steady force-balance when the plunger descends with a constant speed, the solution for calculating the pressing force $F$ is given by equation (1):

$$F = f_{\mathrm{I}} + f_{\mathrm{II}} + f_{\mathrm{III}} + F_{\tau} \tag{1}$$

where $F_\tau$ denotes the resistance of shear deformation and $f_I$, $f_{II}$, $f_{III}$, denote the frictional resistance in a different zone.

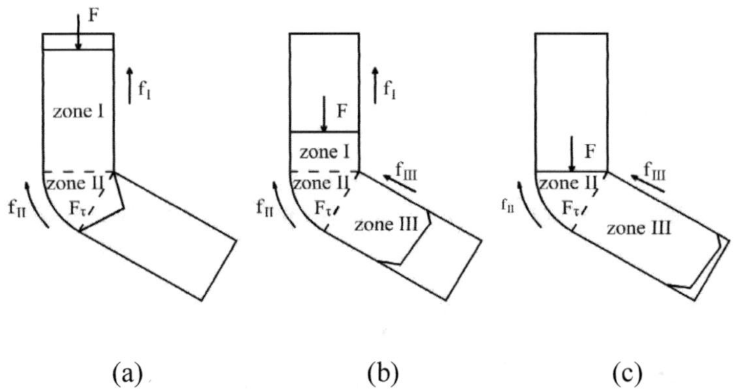


**Figure 1.** Force analysis of ECAP process at a different stage.

According to Coulomb law of friction and the von Mises criteria, the frictional resistance is given by equation (2):

$$f = \mu F_N = \frac{\mu A \sigma_s}{\sqrt{3}} \tag{2}$$

where $\mu$ denotes the friction coefficient, $A$ denotes the contact area and $\sigma_s$ denotes the flow stress. Considering the geometry size of die and the movement of the sample, the frictional resistance in a different zone is given by equation (3):

$$f_I = \frac{\mu\pi D l_I \sigma_{sI}}{\sqrt{3}} \qquad f_{II} = \mu(\pi D)^2 \left(\frac{180-\phi}{360}\right)\sigma_{sII} \Big/ \sqrt{3} \qquad f_{III} = \frac{\mu\pi D l_{III}\sigma_{sIII}}{\sqrt{3}} \tag{3}$$

where $D$ denotes the diameter of channel cross-section, $\Phi$ denotes the internal angle, $l_I$, $l_{II}$, $l_{III}$, denote the length of the sample in a different zone, $\sigma_{sI}$, $\sigma_{sII}$, $\sigma_{sIII}$, denote the flow stress in a different zone.

Base on the study of Segal et al. [8], the ECAE force is calculated using one shear plane and without consideration of the friction as follows:

$$F = kA\gamma \tag{4}$$

where $k$ denotes the shear stress, $A$ denotes the cross-section area of the channel and $\gamma$ denotes the shear strain.

According to Spuskanyuk et al. [4], the shear stress can be calculated from a stress-strain relationship. From the analysis of Iwahashi et al. [9], the shear strain $\gamma=\psi$ is attained when the external angles $\psi=\pi-\Phi$, therefore the deformation resistance is given by equation (5):

$$F_\tau = \frac{\sigma_s}{\sqrt{3}}\frac{\pi D^2}{4}\psi = \frac{\sigma_s \pi D^2 \psi}{4\sqrt{3}} \tag{5}$$

From the upper-bound theorem, Luri et al. [10] analyzed the pressing force in the ECAP process for a die with circular cross-section, friction conditions. The pressing force is given by equation (6):

$$F = \frac{\sigma_s}{\sqrt{3}}\pi\left(\frac{D}{2}\right)^2 \frac{(\pi-\phi)}{\sin(\frac{\phi}{2}+\frac{\psi}{2})} + \mu\frac{\sigma_s}{\sqrt{3}}\pi D L_{init} \tag{6}$$

IOP Conf. Series: Materials Science and Engineering **672** (2019) 012068 doi:10.1088/1757-899X/672/1/012068

The same equation of the deformation resistance can be obtained considering die structure and frictionless.

The commercial finite element software Deform v10.2 was used to carry out the simulations. Three dimensional finite element simulations of single pass ECAP were carried out with different internal angles Φ=90°, 100°, 110°, 120°, external angles ψ=π-Φ and coefficient of friction μ=0, without considering friction resistance. These simulations were carried out with the constant yield strength $\sigma_s$=100 MPa without considering work hardening. As shown in figure 2, cylindrical billet with diameter d=12 mm and length l=60 mm was considered for simulations. All simulations were performed with a speed of 2 mm/s.

Figure 3 shows the simulated pressing force versus ram time step during the ECAP process. It should be noted that three stages of the ECAP force can be distinguished in this particular case. At the first and second stages, the bent front part of the billet does not entirely fill the channel, therefore the force of the third stage represents the deformation resistance.

It can be stated that the error between the analytical and FEM results is lower than 10%, which means that a high level of agreement exists between both analyses, as shown in table 1. The difference between the results could be due to the gap between the billet and channel [11] and inhomogeneous deformation of the billet [12].

**Figure 2.** 3D FEM model.

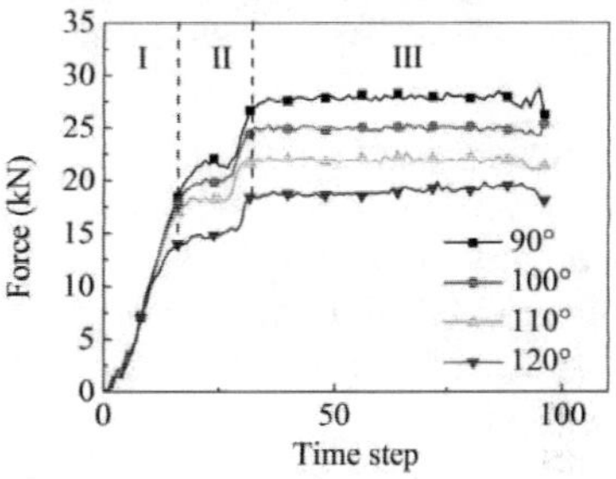


**Figure 3.** Relationship curve of plunger force and time step for ECAP dies with different internal angle.

**Table 1.** FEM results and calculated results of deformation resistance for dies with different internal angle.

| Internal angle Φ | FEM results | Calculated results | Error |
|---|---|---|---|
| 90° | 27894N | 30770N | 9.3% |
| 100° | 25005N | 27351N | 8.6% |
| 110° | 21961N | 23932N | 8.2% |
| 120° | 18965N | 20513N | 7.5% |

## 3. Experiment

In order to study the factors that affect the extrusion load during ECAP process, some experimental measurements were performed. To perform the experimental testing, MTS 311 series electro-hydraulic servo fatigue testing machine shown in figure 4(a) was employed. By using the split die that can be observed in figure 4(b), a cylindrical TA2 billet with size 12 mm in diameter and from 15 to 65 mm in length was extruded, as shown in figure 4(c). The billets were coated with molybdenum disulfide ($MoS_2$) lubricant before ECAP. All billets were processed at 400 °C with constant velocity (v=2 mm/s).

In order to analyze the influence of material properties on the extrusion load, the compression properties of TA2 billets were tested at 400 °C by using a Gleeble 3500 thermal-mechanical simulation tester. The true stress-strain curve was obtained by compression tests with the strain rate 0.1/s.

(a) (b) (c)

**Figure 4.** (a) ECAP pressure equipment; (b) ECAP dies used; (c) billets and billets after processing.

## 4. Results and discussion

In the ECAP process, it is more useful to use the extrusion load instead of the required force. By substituting equation (3) and equation (5) to equation (1), the extrusion load P can be obtained by equation (7):

$$P=\frac{4\mu l_I\sigma_{sI}}{\sqrt{3}D}+\frac{2\mu\sigma_{sII}\psi}{\sqrt{3}}+\frac{4\mu l_{III}\sigma_{sIII}}{\sqrt{3}D}+\frac{\sigma_{sII}\psi}{\sqrt{3}} \tag{7}$$

The formula shows that the extrusion load is related with length-diameter ratio of the billet, friction coefficient, geometric parameters of die and material properties.

As shown in figure 5, the true stress rises with the increase of true strain during the compression process because of work hardening which also occurs during ECAP process. From the analysis of Iwahashi et al. [9] and the fourth strength theory, the relation between the equivalent strain and the true strain is given by equation (8):

$$\varepsilon_e=\gamma/\sqrt{3}=-\frac{2}{\sqrt{3}}\ln(1-\delta)=\frac{2}{\sqrt{3}}\varepsilon \tag{8}$$

where $\varepsilon_e$ denotes the equivalent strain, $\delta$ denotes compression ratio and $\varepsilon$ denotes the true strain. According to the geometry of the model, the true strain corresponding to the maximum equivalent strain is 0.523.

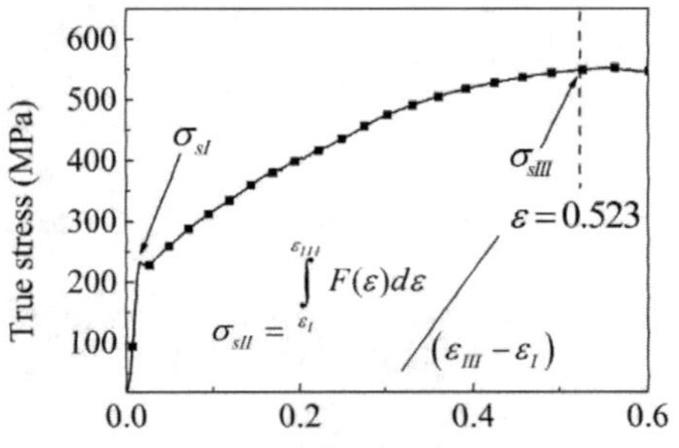


**Figure 5.** The true stress-strain compression curve of TA2 specimen at 400 °C.

**Table 2.** Flow stress of TA2 in different zones.

| $\sigma_{sI}$ (MPa) | $\sigma_{sII}$ (MPa) | $\sigma_{sIII}$ (MPa) |
|---|---|---|
| 233 | 431 | 547 |

It was found that the diameter inflated from 12 mm to 12.24 mm by measuring the tail of the billets. The results show that small plastic deformation appeared in the zone I during the ECAP process. So the yield strength could be used to denoted the flow stress in the first zone, as shown in figure 5. The billet accumulated a large amount of plastic strain by simple shear in the second zone. According to Kim [13], the strain and the external angle are linear increased in the zone II. The flow stress of the zone II is given by equation (9):

$$\sigma_{sII} = \int_{\varepsilon_I}^{\varepsilon_{III}} F(\varepsilon)d\varepsilon \bullet (\varepsilon_{III} - \varepsilon_I)^{-1} \quad (9)$$

where $\varepsilon_I$ denotes the true strain in zone I and $\varepsilon_{III}$ denotes the true strain in zone III. In the third zone, the strain of the billet remained constant, therefore the flow stress of the zone III is the true stress corresponding to the true strain ε=0.523, as shown in figure 5. In order to study the behavior of the extrusion load when material properties changes, table 2 shows a summary of the flow stress in different zone by using the compression curve shown in figure 5.

Figure 6 is the exclusion load versus ram displacement curve. The second stage on the load curve during the ECAP process is observed. Referring to figure 6, the extrusion load increases gradually with increasing the ram movement. The relationship between the exclusion load and the ram displacement shows approximately linear with two different slopes. The positions of the inflection point depend on the geometry of dies. In this study, the ram displacement of the inflection is about 12 mm according the experimental results. At the end of Stage I, the frictional resistance and deformation resistance of the zone II are the same, no matter how long the sample is, as shown in figure 1(a). The extrusion load at the inflection point is given by equation (10):

$$P_I = \frac{4\mu\sigma_{sI}}{\sqrt{3}D}(L-12) + \frac{2\mu\sigma_{sII}\psi}{\sqrt{3}} + \frac{\sigma_{sII}\psi}{\sqrt{3}} \quad (10)$$

Where $L$ denotes the length of the billet. The peaking load of stage I had a linear relationship with the length of the billet, and the intercept and the slope of the fitting line were calculated by origin software, as shown in figure 7. Based on known conditions, the friction coefficient $\mu=0.137$ was calculated by the slope of the fitting line.

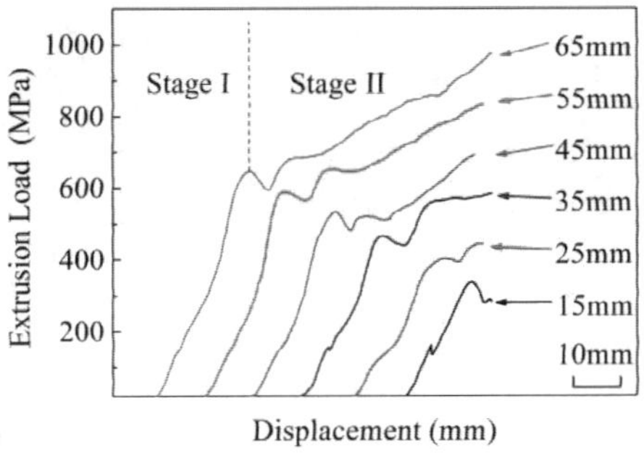


**Figure 6.** Relationship curve of experimental load and ram displacement for different lengths of billet processed at 400 °C.

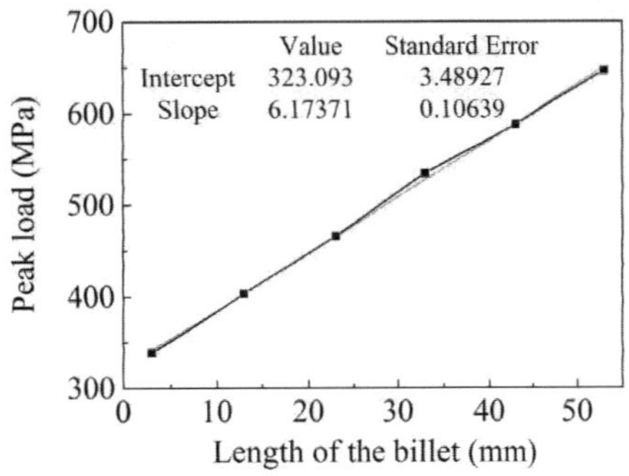


**Figure 7.** The fitting curve for peak load at the end of stage I.

The movement position of the billet at the end of stage II is shown in figure 1(c). Through the comparison and analysis of the peak load at the end of different stages, the slope of the second stage curve $k_{II}$ can be given by equation (11):

$$k_{II} = \frac{P_{II} - P_I}{L-12} = \frac{4\mu(\sigma_{sIII} - \sigma_{sI})}{\sqrt{3}D} \quad (11)$$

The slopes of the fitting curves and the calculated results of the friction coefficient are shown in table 3. The average of the friction coefficient is 0.131±0.003, which is close to the research results of

Perez et al. [14]. By substituting the friction coefficient into equation (10), the frictional resistance of zone II is 68.3 MPa, the deformation resistance of the zone II is 260.6 MPa, the sum of frictional resistance and deformation resistance is 328.9 MPa. It can be stated that the error between calculated results and the intercept of the fitting curve is 1.8%, which means that the formulas are correct and practical. By using all the parameters in equation (7), the peak load of stage II is 1060 MPa, the error between calculated and experimental results is 6%, and it is shown that the analytical formulation predicts the extrusion load accurately.

**Table 3.** The calculated results of friction coefficient in the stage II.

| Length of the sample | Slope of the curve | Friction coefficient |
|---|---|---|
| 65 | 7.78 | 0.129 |
| 55 | 7.61 | 0.126 |
| 45 | 7.96 | 0.131 |

## 5. Conclusion

In this work, it has been found that the extrusion load consists of the deformation resistance and frictional resistance during the ECAP process. The same formula of the deformation resistance has been obtained by the slip line field solution and upper-bound solution, and the calculated results have been verified by FEM simulation. Based on the experimental results, the formulas of extrusion load have been analyzed by considering material properties, geometric parameters of die, length-diameter ratio and friction coefficient. The calculated load has been validated by measuring experimental results. The formulas will provide the support for the die design and the process optimization in the field of ECAP.

## Acknowledgments

The authors acknowledge the support given by the the Fundamental Research Funds for the Central Universities (HEUCMF180205) and the National Natural Science Foundation of China (51701049).

## References

[1] Liang N, Zhao Y, Li Y, Topping T, Zhu Y, Valiev R Z and Lavernia E J 2018 *J. Mater. Sci.* **53** 13173
[2] Valiev R Z and Langdon T G 2006 *Prog. Mater. Sci.* **51** 881
[3] Jia H, Bjørge R, Cao L, Hui S, Marthinsen K and Li Y 2018 *Acta Mater.* **155** 199
[4] Spuskanyuk V, Spuskanyuk A and Varyukhin V 2008 *J. Mater. Process. Technol.* **203** 305
[5] Silva F R F, Medeiros N, Moreira L P, Lins J F C and Gouvêa J P 2012 *Mater. Sci. Eng., A* **546** 180
[6] Sordi V L, De A M F A A, Valio G T, Springer P, Ruber J B and Ferrante M 2014 *Adv. Mater. Res.* **922** 507
[7] Zhang D T, Li Z, Tong Y X, Zheng Y F and Li L 2016 *J. Iron Steel Res. Int.* **23** 1020
[8] Segal V M 1999 *Mater. Sci. Eng.* A **271** 322
[9] Iwahashi Y, Horita Z, Nemoto M, Wang J and Langdon T G 1996 *Scr. Mater.* **35** 143
[10] Luri R and Pérez C J L 2012 *Int. J. Adv. Manuf. Tech.* **58** 969
[11] Medeiros N, Moreira L P 2014 *Comput. Mater. Sci.* **91** 350
[12] Suo T, Li Y L and Zhao F 2009 *Int. J. Mod Phys B* **22** 6088.
[13] Kim H S 2001 *Mater. Sci. Eng.* A **315** 122
[14] Pérez C J L and Luri R 2008 *Mech. Mater.* **40** 617

# AUTHOR INDEX